PROGRESS IN POWDER METALLURGY
VOLUMES 34 & 35

1978 and 1979 National Powder Metallurgy Conference Proceedings

Edited by:
1978 Conference
James Hoffman, Cincinnati Incorporated
1979 Conference
Walter Cebulak, Aluminum Company of America
Erhard Klar, Glidden Metals Div., SCM Corp.

Proceedings of the 1978 National Powder Metallurgy Conference, April 24-26, 1978, Century Plaza Hotel, Los Angeles, Calif., and the 1979 National Powder Metallurgy Conference, June 4-6, 1979, Stouffer's Cincinnati Towers, Cincinnati, Ohio. Sponsored by the Metal Powder Industries Federation and the American Powder Metallurgy Institute.

105 College Road East, Princeton, N.J. 08540

ISBN No. 0-918404-49-5
ISSN No. 0079-6719

PRINTED IN U.S.A. 8C-11/80

CONTENTS:

PART A
PROGRESS IN POWDER METALLURGY, VOLUME 34

PART B
PROGRESS IN POWDER METALLURGY, VOLUME 35

(continued)

Section A

Progress in Powder Metallurgy
Volume 34

Proceedings of the 1978 National Powder Metallurgy Conference — Los Angeles, California

Edited by: James Hoffmann

STATUS REPORT ON ENERGY CONSUMPTION AND DISTRIBUTION IN THE P/M INDUSTRY (1976)(USA)

H. S. Nayar

MPIF Energy Committee*

ABSTRACT

The average energy consumed in the P/M industry in the United States during 1976 to produce one pound of P/M parts was 17,500 BTU. This includes energy to produce powder; heating and air-conditioning the plants and offices; handling, blending and compacting the powder; heating the furnaces, generating atmosphere; and other normal and secondary operations within the fabricator's plants. Natural gas represents 63% of the total energy consumed in the industry. The P/M process is more energy efficient that the competitive processes but not as efficient as it could be. There is considerable room to conserve energy in the use of furnaces, in sintering atmospheres and in heating and air-conditioning offices and plants. Energy management programs are recommended.

INTRODUCTION

The Technical Board of the Metal Powder Industries Federation formed an Energy Committee three years ago to investigate energy matters in the industry. The Committee was to learn how much of what kinds of energy is used to make P/M parts. Further, it was to compare the industry's energy usage to that of competitors and related industries. Finally, it was to recommend ways the industry could become more energy efficient. Such studies, it was felt, would be useful to member companies and to the industry as a whole.

The Committee based its conclusions and suggestions on responses to a questionnaire sent to 117 P/M parts fabricators. Thirty-seven respondents - 32% of the total - provided data on the amount of powder processed, sintering practice, the energy consumed, the types of sintering furnace used, and the methods of heating those furnaces.

---

* Members of the MPIF Energy Committee: Harb Nayar, Chairman (Airco), Ted Crease (Drever), Don Gustafson (Gould), Jim Hoffman (Cincinnati), and Tom Pietrocini (IPM).

The data generally covered January to December 1976. As might be expected, some of the answers showed great variations in the industry; other indicated that patterns of industry-wide usage exist.

SINTERING PRACTICE

In general, the respondents varied greatly in size. About one-third processed less than 1,000,000 lb of powder per year. Another quarter processed more than 5,000,000 lb. The "average" company processed 3,500,000 lb, 92% of which was iron or low alloy steel. The respondents varied considerably also on belt loading, ranging from 5 to 15 lb/ft^2. The average load was 9.3 lb/ft^2 of belt. They varied also in the flow rate of atmospheres. Some ran as low as 30 CFH/inch of belt width, and some as high as 300 CFH. The average was about 100 CFH/inch of belt width.

The companies seemed to agree, however, on the number of hours sintering operations were run. Two-thirds indicated the furnaces were used 6000 to 7000 hr/year to sinter P/M parts. The average, about 6000 hr, works out to five 24-hr days (or 15 shifts) per week. Of the companies using belt furnaces, most used belts between 12 and 18 inches wide. Although one respondent used a 36-inch belt, the average width was 16.5 inches. Most companies produce pieces weighing an average of 1/4 lb (120 g) with an average density of 6.6 g/cc. Detailed analysis of the general sintering practice as revealed by the 32% of the total number of fabricators asked to respond is shown in Figures 1 to 7.

FIGURE 1: POWDER PROCESSED (1976)

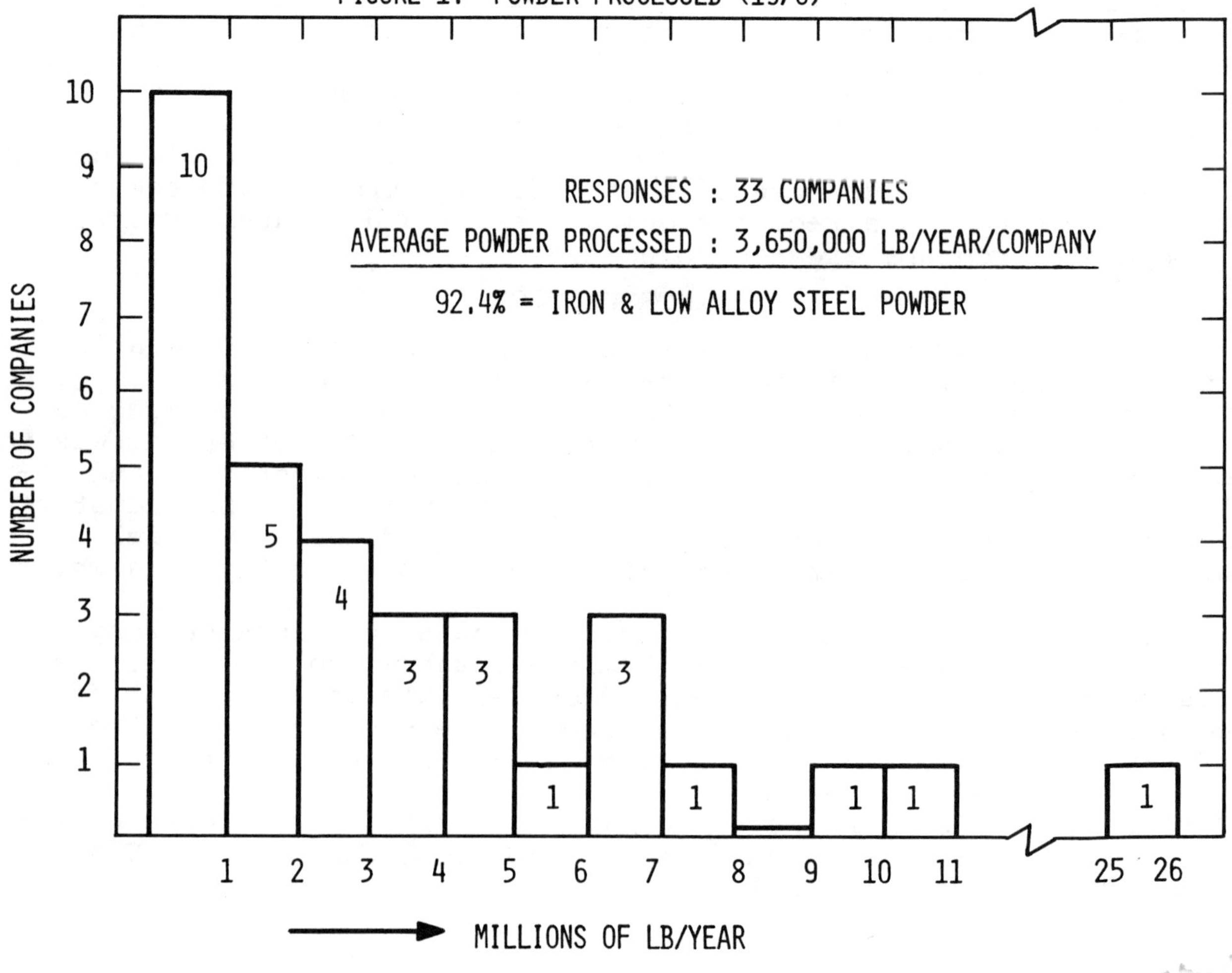

FIGURE 2: SINTERING HOURS/YEAR (1976)

		COMPANIES
RESPONSES	:	27
3000 - 4000 HOURS	:	1
4000 - 5000 HOURS	:	1
5000 - 6000 HOURS	:	5
6000 - 7000 HOURS	:	18
7000 - 8000 HOURS	:	2

"AVERAGE COMPANY" : 5956 HOURS

FIGURE 3: BELT LOADING (1976)

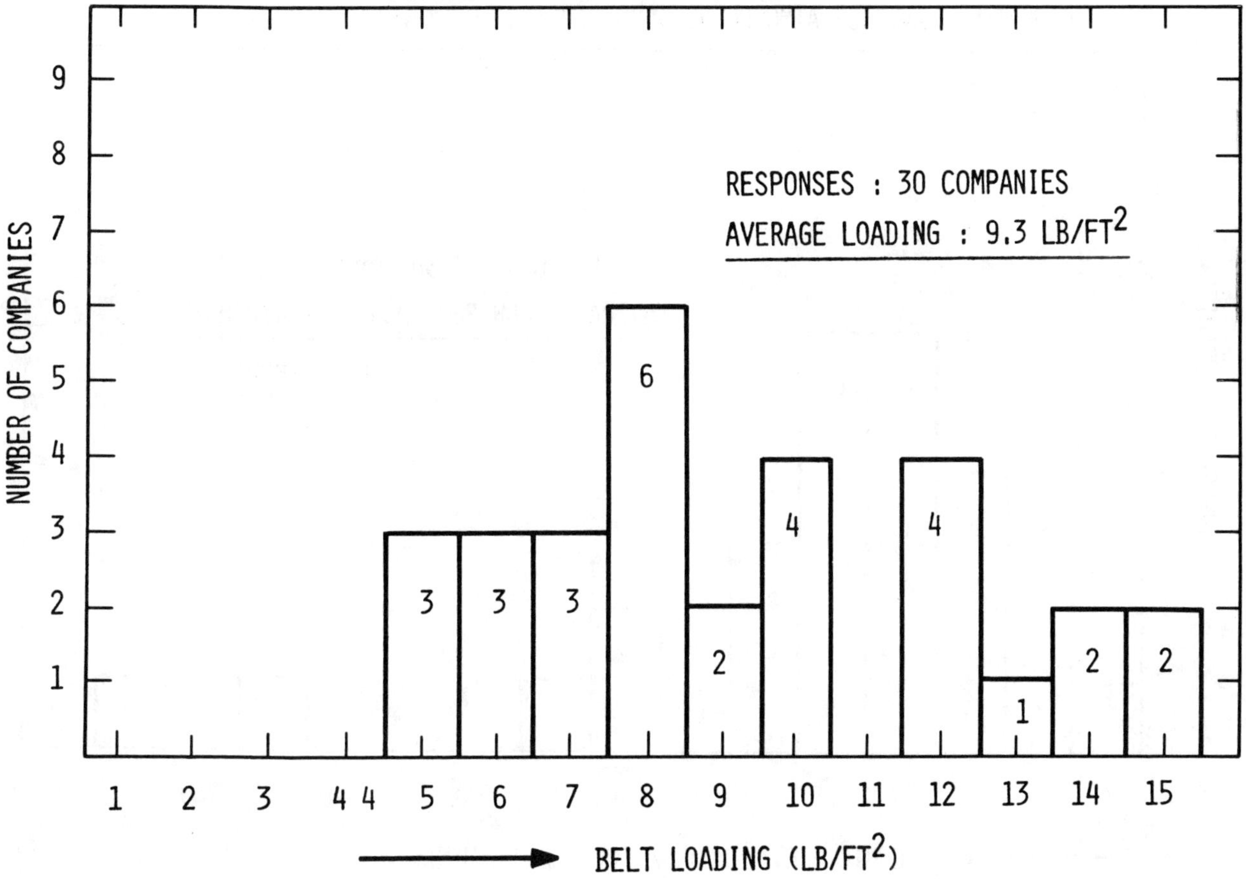

FIGURE 4: AVERAGE BELT WIDTH (1976)

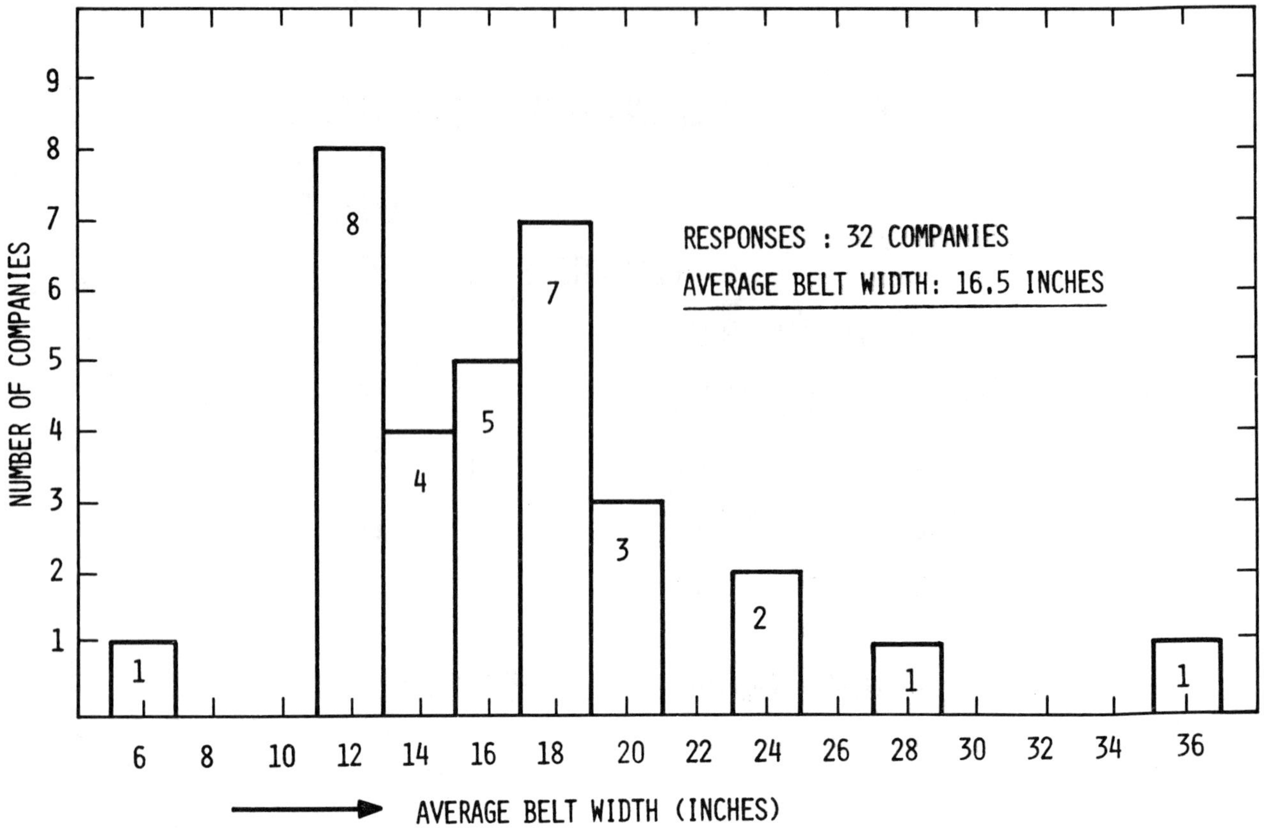

FIGURE 5: AVERAGE ATMOSPHERE FLOW PER INCH BELT WIDTH (1976)

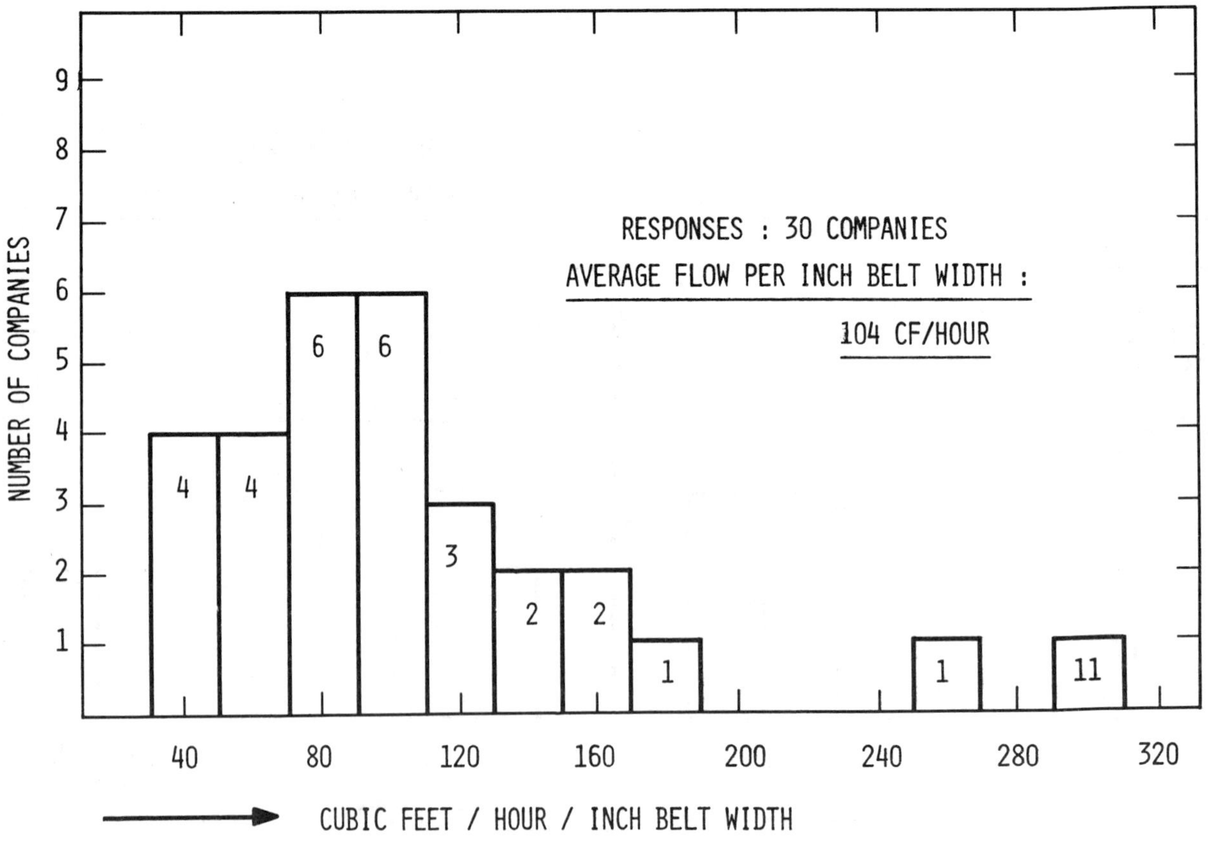

FIGURE 6: AVERAGE P/M PART WEIGHT (1976)

	COMPANIES
RESPONSES :	34
0.01 - 0.1 LB	13
0.1 - 0.3 LB	8
0.3 - 0.5 LB	8
0.5 - 0.8 LB	4
0.8 - 1.0 LB	0
1.0 - 1.6 LB	1

GRAND AVERAGE WEIGHT : 0.264 LB (120 G)

FIGURE 7: AVERAGE P/M PART DENSITY (1976)

	COMPANIES
RESPONSES :	34
5.5 - 6.0 G/CC :	1
6.0 - 6.5 G/CC :	8
6.5 - 7.0 G/CC :	24
7.0 - 7.2 G/CC :	1

GRAND AVERAGE DENSITY : 6.57 G/CC

ENERGY

"Total energy" includes energy used for all operations at the P/M part fabricator's plant: heating and air-conditioning the plant and offices; handling, blending and compacting the powder; heating the furnaces; generating atmosphere for the furnaces and all other secondary operations. It does not include the energy to produce powders. It was arrived at by taking all forms of energy consumed at each plant, converting* these to a common energy unit (i.e. BTUs), and then adding up all forms of energies as expressed in BTUs.

Of great interest was the question on the "total energy" consumed. The respondents indicated that 63% of this energy was natural gas, 21% electricity, and the remainder was propane, oil, ammonia and nitrogen.

The 28 companies responding to the energy question consumed an average of 37.46 billion BTU/company. Pricing this amount as though it were natural gas at $2.65/100 CF results in an average yearly bill of almost $100,000. The energy distribution data is shown in Figures 8 and 9.

FIGURE 8: ENERGY CONSUMED BY P/M FABRICATOR (1976)

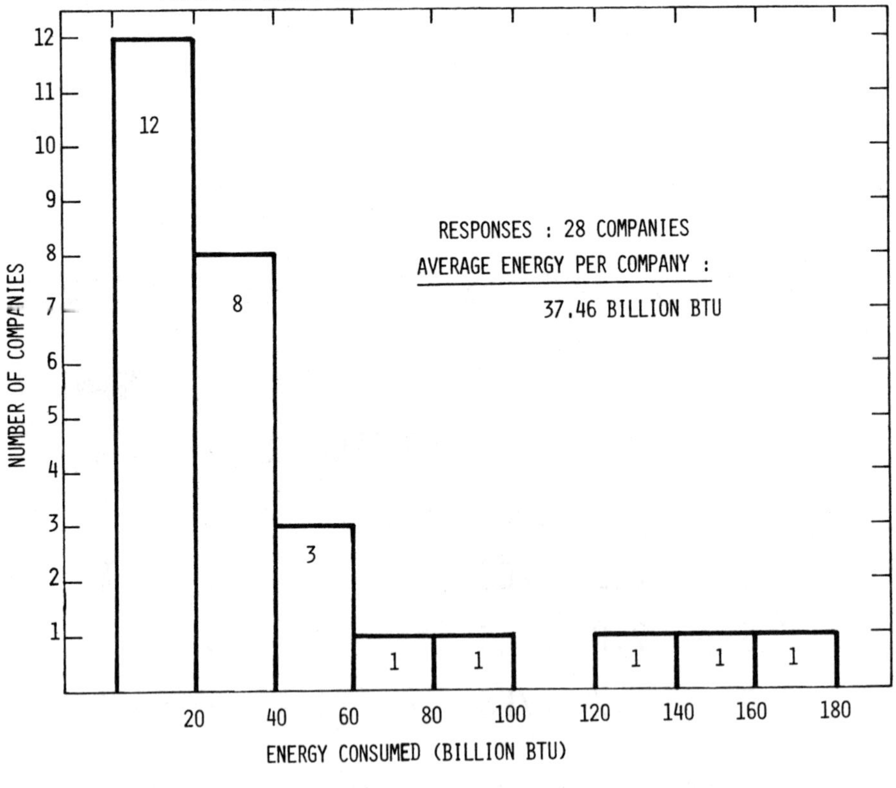

---

* The following scale was used to express all forms of energy in BTUs:

1 kWh of electricity	=	3,413 BTU
1 cubic foot of natural gas	=	1,000 BTU
1 gallon of propane	=	105,000 BTU
1 gallon of oil	=	140,000 BTU
1 lb of ammonia	=	24,000 BTU

FIGURE 9: FORMS OF ENERGY CONSUMED BY P/M FABRICATORS (1976)

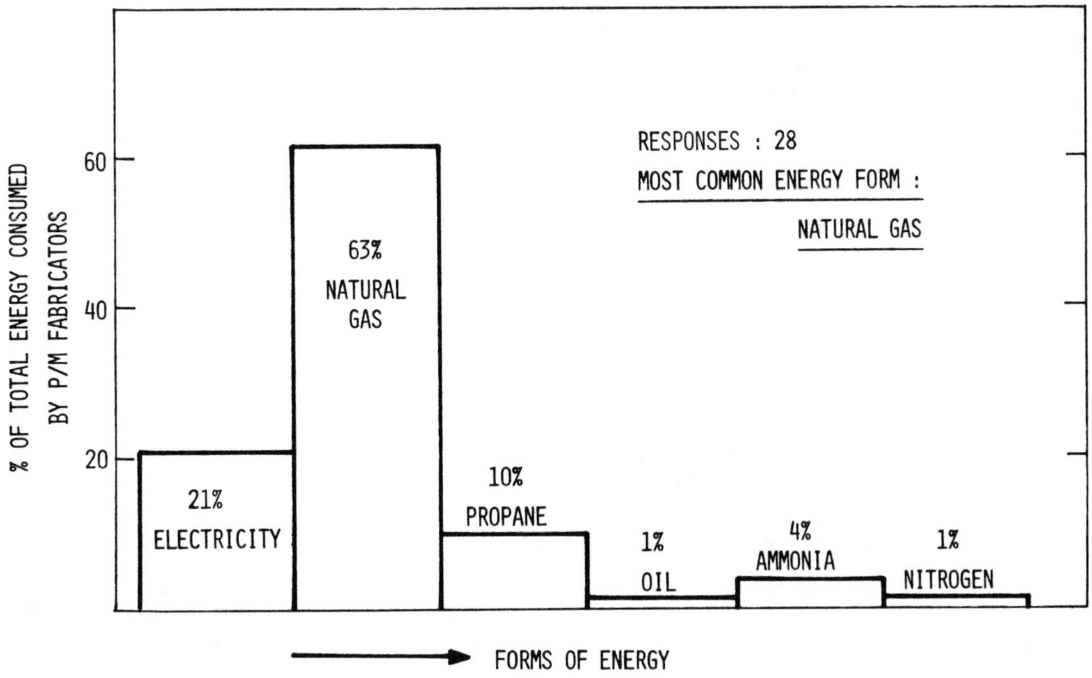

Two-thirds of the respondents heat plants and offices with natural gas; only one-third use oil and electricity as shown in Table 1. In answer to the question how the furnaces were heated, the 30 respondents, with 168 furnaces for an average of 5.6 furnaces per company, split as in Table 2. Belt furnaces with either muffles or brick account for 80% of sintering furnaces. The rest are generally pusher and roller furnaces or the newer vacuum furnaces (see Table 3).

TABLE 1:

PLANT & OFFICE HEATED BY ? (1976)

RESPONSES : 30

NATURAL GAS	:	65%
OIL	:	24%
OTHERS (MOSTLY ELEC.)	:	11%

TABLE 2

SINTERING FURNACES HEATED BY ? (1976)

RESPONSES	:	30
TOTAL NO. OF FURNACES	:	168
NO. OF FURNACES PER COMPANY	:	5.6

NATURAL GAS	:	39%
ELECTRICITY	:	47%
N.G. + ELECTRICITY	:	13%
PROPANE	:	1%

TABLE 3

TYPES OF SINTERING FURNACES (1976)

RESPONSES	:	30
TOTAL NO. OF FURNACES	:	168
NO. OF FURNACES PER COMPANY	:	5.6

BELT - MUFFLE	:	60%
BELT - BRICK	:	19%
PUSHER	:	13%
ROLLER HEARTHS	:	4%
OTHERS	:	4%

ATMOSPHERE

Most of the sintering atmosphere used is endo (83%); the atmosphere requirement for each plant ranges from 250 to 44,800 CFH, averaging 6840 CFH. The most important information tabulated from this questionnaire is that on the average between 11 and 13 CF of atmosphere is used to sinter one pound of P/M product (see Figure 10; not shown in this figure is the atmosphere requirement to sinter one pound of P/M part which ranged between 2.4 and 65.5 CF).

FIGURE 10: ATMOSPHERES USED (1976)

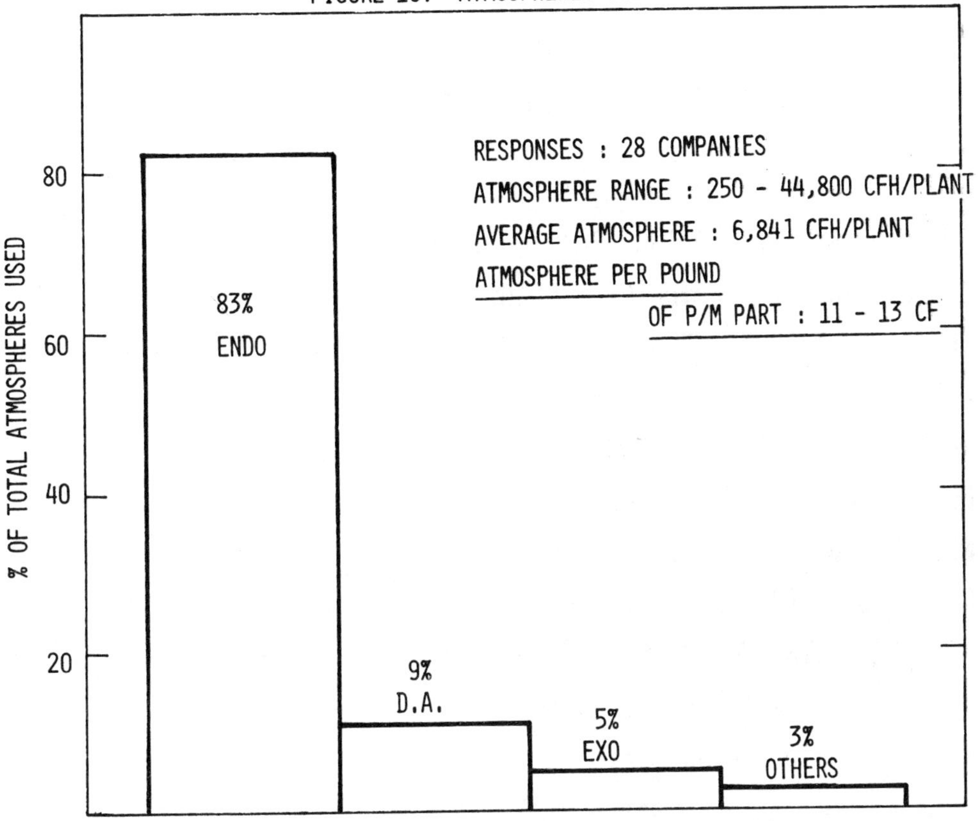

ENERGY CONSERVATION

The Committee was surprised to learn that most P/M fabricators indicated they had energy conservation policies. The overwhelming affirmation (25 to 8), however, may be misleading. Many respondents were in the process of adopting a policy for dealing with the energy crisis but did not expect concrete results for two years or more.

Among the steps being taken, by far the most popular was to divert waste furnace heat to other sections of the plant. Other popular steps were of the housekeeping sort: turning off idling machinery and using thermostats to control temperatures in winter. Changes to furnaces accounted for many of these energy saving steps. These changes ranged from installing flowmeters in order to learn how much energy is being used, to changing over to nitrogen atmospheres or to vacuum furnaces. Other respondents were insulating furnaces, increasing sintering hours per week, increasing belt and furnace loading, and reducing the flame curtains. The degree of the energy conserving measures adopted by the industry are shown by the numbers in parentheses in Table 4.

Among the most interesting responses were the three reporting that the companies had established energy conservation committees. Such committees are signs that the industry is seriously concerned about its use of energy. There should be in each company some group or committee that can look into energy usage, monitor it, recommend changes, or control it in some other way. Indeed, most respondents pointed out that natural gas was the least available and electricity the most available form of energy in their localities (see Table 5).

TABLE 4

ENERGY CONSERVATION & MANAGEMENT (1976)

YES : 25 NO : 8

DIVERT FURNACE & PROCESS HEAT TO OTHER AREAS (7)

TURN OFF INSTEAD OF IDLING MACHINERY (5)

USE THERMOSTATS TO KEEP BUILDING TEMPERATURE LOW IN WINTER (5)

REDUCE & CONTROL ATMOSPHERE FLOWS (4)

CHANGING TO NITROGEN ATMOSPHERE (5)

LOOKING INTO VACUUM FURNACES (1)

INSULATE FURNACES (3)

ENERGY CONSERVATION COMMITTEE (3)

INCREASE SINTERING HOURS PER WEEK (2)

INCREASE BELT & FURNACE LOADING (2)

REDUCE FLAME CURTAINS (2)

TABLE 5

ENERGY SUPPLY (1976)

RESPONSES : 33

	LEAST AVAILABLE	MOST AVAILABLE
NATURAL GAS :	19	9
ELECTRICITY :	1	26
PROPANE :	2	5

"AVERAGE" COMPANY

This then is the profile of an average P/M company:

It produced 3,500,000 lb of powder in 1976, 92% of which was iron or low alloy steel. It sintered about 6000 hr, loading a 16.5-inch belt with about 9.3 lb/ft^2 under an endo atmosphere flowing at 100 CFH/inch of belt width. Typical products weighed 1/4 lb (120 g) with a density of 6.6 g/cc. The company heated its 5.6 furnaces with both electricity and natural gas. The plants were heated mainly with natural gas.

ENERGY PER POUND OF P/M PART

This "average" company used about 12,500 BTU - the spread was between 3000 and 35,000 BTU (see Figure 11) - to process one pound of P/M part. This figure was arrived at by taking the "total energy" consumed by a P/M fabricator (as expressed in BTU units) and dividing the sum by the amount of parts produced. This figure does not include the energy used to produce the powder but it does include everything else as described earlier under the section "Energy". To it, therefore, must be added figures supplied by major powder producers: it takes about 5000 BTU to convert steel scrap into usable powder. The total energy requirement in P/M fabrication, from scrap to finished part, thus averages 17,500 BTU per pound. The range is from 8000 to 40,000 BTU.

Preliminary data from other competitive processes indicates that the powder metallurgy process, despite some shortcomings, is more efficient than most other techniques for producing finished steel products. It should be recalled that the "average" density reported for P/M parts is 6.6 g/cc or about 82% of the theoretical solid density. In terms of number of parts of a given volume, the P/M process produces about 22% more parts for a given weight. This difference in density increases the energy efficiency of the P/M process even further.

Table 6 estimates the distribution of the energy to produce one pound of finished P/M part. In the interests of energy conservation, some reductions in usage should be made in each category. Not all the categories can be reduced or be made more efficient over the short term. However, furnace heating, atmosphere use, and plant and office heating are three areas where important savings can be made. By increasing belt loading to near 15 lb/ft^2, by increasing production hours to 8000 hr/year, by insulating and better maintaining existing furnaces, by carefully designing new furnaces, energy will be used more efficiently for each pound of porduct. (Furnace maintenance can be as simple as plugging leaks and providing curtains to reduce atmosphere flow.) Using atmospheres that consume less energy - nitrogen or vacuum, for instance - can also result in substantial energy savings. Finally, training employees in general energy conservation, and establishing energy committees to monitor, report and control energy consumption within the plant are perhaps the two most important means of assuring energy efficient operations.

FIGURE 11: ENERGY* PER POUND TO PROCESS POWDER INTO P/M PART

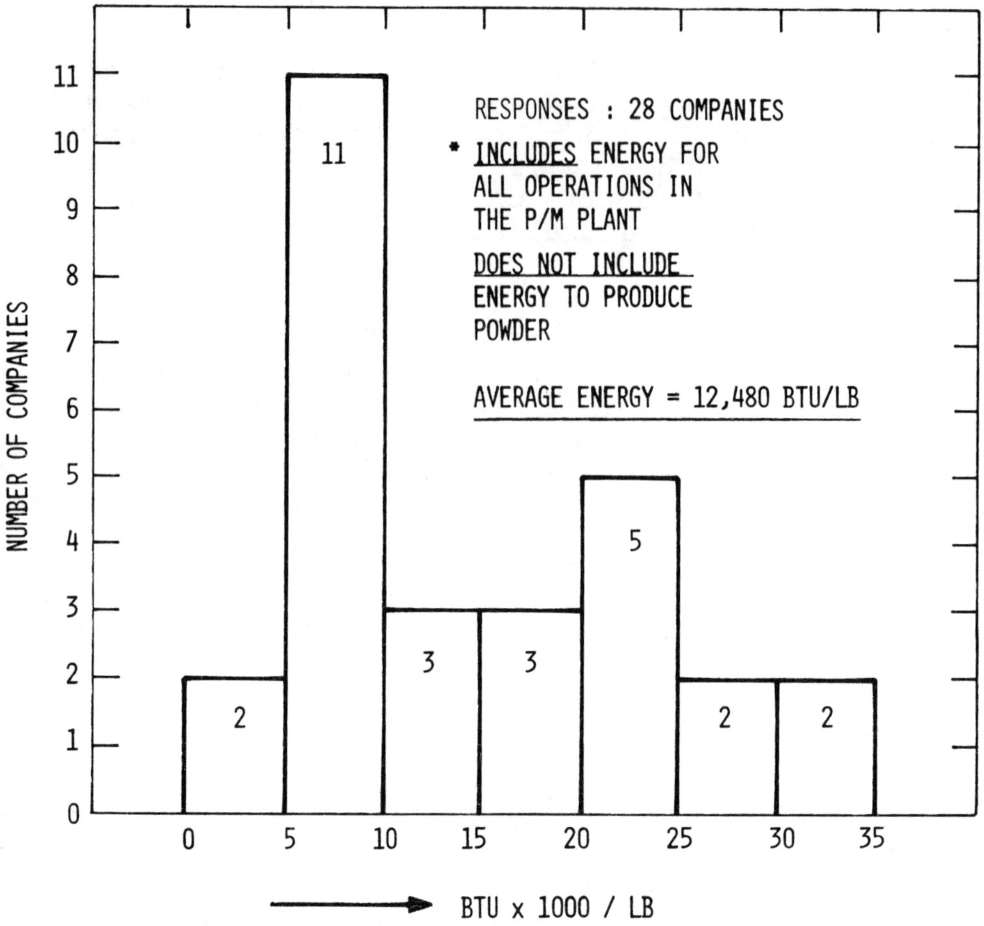

TABLE 6
ENERGY DISTRIBUTION
(ESTIMATION)

POWDER PRODUCTION	:	20 - 30 %
HANDLING, BLENDING, COMPACTING	:	5 - 10 %
HEATING FURNACES	:	15 - 25 %
ATMOSPHERES	:	25 - 35 %
PLANT & OFFICE HEATING	:	15 - 20 %
SECONDARY OPERATIONS	:	5 - 15 %
MISCELLANEOUS	:	0 - 5 %

Although the P/M industry is more energy efficient than its competitors, it is not as efficient as it can be. Energy comparison between the P/M industry and competitive industries is expected to be dealt with in detail by MPIF's Technical Board in the very near future. The P/M industry as a whole can probably grow 12 to 13% per year for the next four years without increasing its use of energy. This assertion assumes care is taken to use less energy per pound in heating of furnaces, in heating and air-conditioning offices and plants, by reducing atmosphere flows and by changing to less energy-consuming sintering atmospheres.

DESIGNING SINTERING FURNACES AND ATMOSPHERES FOR ENERGY SAVINGS

Frank L. Holden
Process Engineer

Sunbeam Equipment Corporation
Meadville, Pennsylvania

Years ago furnaces were designed with the maximum input possible and then adding 10% more if possible. All customers felt that this would make the best possible unit. Of course, there always was a matter of cost involved. Consequently, the refractory walls were rather thin such as 11 inches versus 13 1/2 inches to save costs. Today we are faced with the opposite condition. Of course, price is still paramount. Cost of operation is important but more important is the continuing availability of fuel for uninterrupted operation.

Before we begin a discussion on designing furnaces, we must understand the typical design. These furnaces are classified as pushers, belt conveyors, walking beams and roller hearths. The same design criteria applies to all these furnaces. However, the belt conveyor appears to be the simplest to show the various components and probably the most widely used. We have selected some belt conveyors for our examination. Photo #1 shows the charge end of an electrically heated unit. Here you can see the mesh belt and the free loop for minimal tension and longer belt life. Photo #2 shows the discharge end with the cooled, sintered parts sliding down a chute into another transfer conveyor. These two views showed a completely electric unit.

Photo #3 illustrates a pair of furnaces with a combination of heating methods. The first section is gas-fired while the balance is electric A cross sectional view of the mesh belt sintering furnace is shown in Photo #4. Here we show all the fundamental parts of these furnaces. Not shown, however, is the fact that the entire shell of these furnaces is all welded gas-tight. The main parts of these furnaces are:

a) A purge chamber where parts are charged into the furnace. In this chamber the parts are slightly preheated by the heated atmosphere leaving the furnace. Air is prevented from entering the furnace due to its design with curtains just barely touching the parts.

b) A gas fired radiant tube or electrically heated preheat and burn-off section. Usually this section has a maximum operating temperature of approximately 1500°F. This is the section where the binder is eliminated thus preventing atmosphere contamination of the sintering zone.

c) A high temperature sintering section where the temperature is usually 2100°F maximum. This section is normally heated with Silicon Carbide heating elements.

d) An insulated slow cool transition chamber to drop the temperature several hundred degrees. This prevents shock to the parts as well as preventing excessive maintenance on the first cooling section.

e) A water cooled cooling section to lower the temperature of the sintered parts below oxidation temperature. In some furnaces these sections have fans for circulation to reduce the overall length of the section.

f) A discharge throat similar to the front chamber with curtains to prevent air infiltration back into the furnace.

g) The mesh belt conveyor to carry the parts through the furnace. This is a metallic woven belt and is the limiting factor for for maximum operating temperature. (However, new ceramic designs are now available for higher temperature operation.)

These views have shown a general construction. If we are to consider a design for maximum energy savings we need to take several important points into consideration.

The first of these is the insulation used in the furnace. Since these furnaces are operated over a 24 hour period we do not need to take heat up and cool down factors into consideration. However, heat losses through the shell from lack of insulation are very important. Thus we need the maximum insulation possible consistent with good engineering design. For example, the width of a belt conveyor furnace shell is not as important as the width of a roller hearth shell. Extra costs in alloy rolls can easily price away any insulation savings.

The new fiberwall insulations will sometimes allow us to use thinner walls with minimal heat loss through the shell. A careful analysis of each type of insulation for the specific furnace determines the type to be used.

Next we should look carefully at the manner of heating for the furnace sections. In the sintering section of the furnace the temperature is so high that radiant tubes are practically impossible. Therefore, we usually select Silicon Carbide heating elements and electric heat. Even though operating with a Hydrogen type atmosphere these open elements will have a good long life and a reasonable cost.

In the preheat or burn-off section the maximum operating temperature is low enough for radiant tubes. Thus we can heat these tubes with either gas or electric.

Table #1 illustrates the economy of fuel fired furnaces. In the preheat section which operates at approximately 1500°F we have a stack loss percentage factor of the gross input of approximately 42%. You can see that fuel economy wise it is more advantageous to make this chamber electrically heated. Since we need electric in the sintering section it appears that it would be more reasonable to make the whole furnace electrically heated. From an economic stand-point this might not be true right at this time. However, with the present shortage of hydrocarbon fuels you can see the advantage of eliminating as much as possible the use of hydrocarbons for fuel and substituting electric elements.

Today we are finding many companies wanting to up-date their present equipment and converting to electric heat. At the same time we may be adding two or three inches of blanket insulation thus improving the insulation and maintaining lower heat losses through the shell.

At present time we are converting many gas fired radiant tube or direct fired furnaces to electric heat. When it is necessary to install electric elements in radiant tube we do so.

The photo shows how a bayonet element would appear in a radiant tube. As can be seen the element can be removed from the tube without also removing the tube. Thus replacement can be made without long shut-down delays caused by removing the atmosphere. Since the concentration of Carbon is high in the burn-off section, the element is usually protected by the radiant tube.

Since the temperature of this section is so low, the use of Silicon Carbide elements is not as feasible as the bayonet type element.

Everyone is examining all methods for improving furnace designs especially in the field of energy savings. Actually more savings can be achieved if we plan our furnace useage for maximum efficiency. Following are a few suggestions for energy savings:

1. Shut down heating equipment when not in use for an extended period.

2. Reduce temperature of furnace to the minimum where atmosphere is not required.

3. Use your most efficient furnace at its maximum capacity and less efficient furnaces only when necessary.

4. Rework your less efficient furnaces to more efficient furnaces through reinsulating, and more efficient heating systems.

5. Consider 3 or 4 day round the clock operation rather than one or two shifts a day for 6 or 7 days.

6. Record fuel consumption, gas or electric, on the basis of parts produced. This will give a record of the most efficient equipment and operations.

7. Minimize the operation of equipment required to be maintained in stand-by condition. Planning preventative maintenance shut downs helps cut down on stand-by operation.

8. If gas is used, correct flue gas analyses for maximum efficiency. This could result in a 10% or better saving in fuel.

9. Recuperators on radiant tubes might save as much a 30% in fuel consumption when used to preheat combustion air.

10. Preheat work loads with flue gases to reduce heating time and conserve fuel.

11. The use of percentage timers on electric furnaces can possibly save up to 20% of the input energy.

12. Consider performing several operations with the sintering to save reheating of the work.

It should appear quite clearly that most energy savings can be accomplished by the furnace user. An unloaded belt wastes energy faster than any other condition. Thus it pays extra dividends to stagger break periods, lunch periods, etc. The elimination of hydrocarbons for fuel to allow us material to prepare our sintering atmospheres is being proposed. Over the years the sintering atmospheres commonly used have been:

a) Rich exothermic gases produced by the partial combustion of a hydrocarbon. Usually this has been Methane and more recently Propane. If we partially crack Methane with 6 parts of air to 1 part of gas, we can produce the following atmosphere:

15% Hydrogen
12% Carbon Monoxide
6% Carbon Dioxide
67% Nitrogen (Approximately)

This atmosphere is generally used for Copper Sintering.

b) Endothermic gas produced with a lower percentage of air. Usually this is 2 1/2 air to 1 gas. Due to the high percentage of gas this is an endothermic or heat absorbing process. Normal cracking temperature is approximately 1850 to 2000°F. Since we would like to conserve our hydrocarbons the electrically heated generator is more desirable than gas fired. From this 2 1/2 - 1 ratio we would obtain the following atmosphere:

40% Hydrogen
20% Carbon Monoxide
40% Nitrogen

The residual Carbon Dioxide and Methane will be determined by the cracking efficiency. Usually these percentages are very low.

This atmosphere is used for copper and iron sintering and possibly some Brasses. Because of its Carbon Monoxide and Carbon Dioxide content it can help in the control of the Carbon content of a sintered iron part.

c) Dissociated Ammonia is prepared by cracking Ammonia into its constituents of 25% Nitrogen and 75% Hydrogen. Like the endothermic atmosphere we must also supply heat for the reaction. Temperatures range between 1750 and 1950°F depending upon the catalyst employed.

This atmosphere can be used for practically any sintering except Stainless Steels that may Nitride or special metals.

d) Hydrogen is so expensive in its pure state that it is used in only very special cases.

e) Nitrogen from storage tanks has been and is presently being used especially after blending with the proper additives to give good reducing properties.

f) Vacuum sintering is being done especially with Tool and Stainless Steels.

Chart #1 shows the effect of Hydrogen on heat loss through refractories. As you can see as the Hydrogen increases the heat losses increase. Therefore it is advantageous to do sintering with the minimum amount of Hydrogen in the atmosphere. We can save two ways, namely:

1. Lower cost of the atmosphere.
2. Lower heat loss through the refractory to the furnace sidewalls.

Real conservation measures can sharply reduce our consumption of precious atmosphere. Some of these measures are:

1. Maintain the maximum tightness of the furnace through proper maintenance of seals, etc.

2. Keep furnace door openings to a minimum in height and width or in time on pusher or roller hearth furnace.

3. On belt conveyors or furnaces with constantly open doors use the tightest curtains to prevent air infiltration and excessive gas leakage from the furnace.

4. Control the atmosphere by Water Vapor, Carbon Dioxide or other suitable means to use the minimum quantity of enriching gases.

5. Use the lowest Dew Point gas with the absolute minimum of oxygen pick up from leaky pipes to reduce the volume of gas needed.

6. Blend gases to give the ultimate in reducing properties with the minimum flows.

7. Turn off the gases at the highest temperature possible when shutting the furnace down. Even though small at the time it can add up rapidly over the year.

Furnace manufacturers will work on methods to reduce energy requirements in furnace designs. With the combined efforts of users and manufacturers energy useage and especially hydrocarbon fuel consumption can be reduced to the minimum.

PHOTO #1. Charge End of Sintering Furnace.

PHOTO #2. Discharge End of Sintering Furnace.

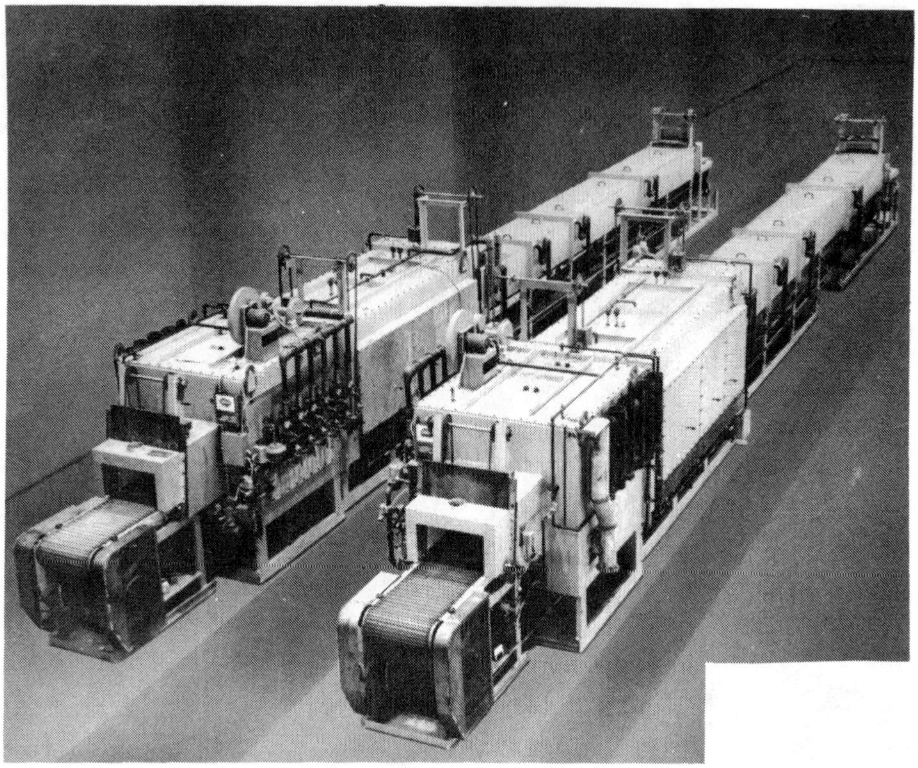

PHOTO #3. Gas Fired And Electrically Heated Sintering Furnaces.

CONVEYOR SINTERING FURNACE

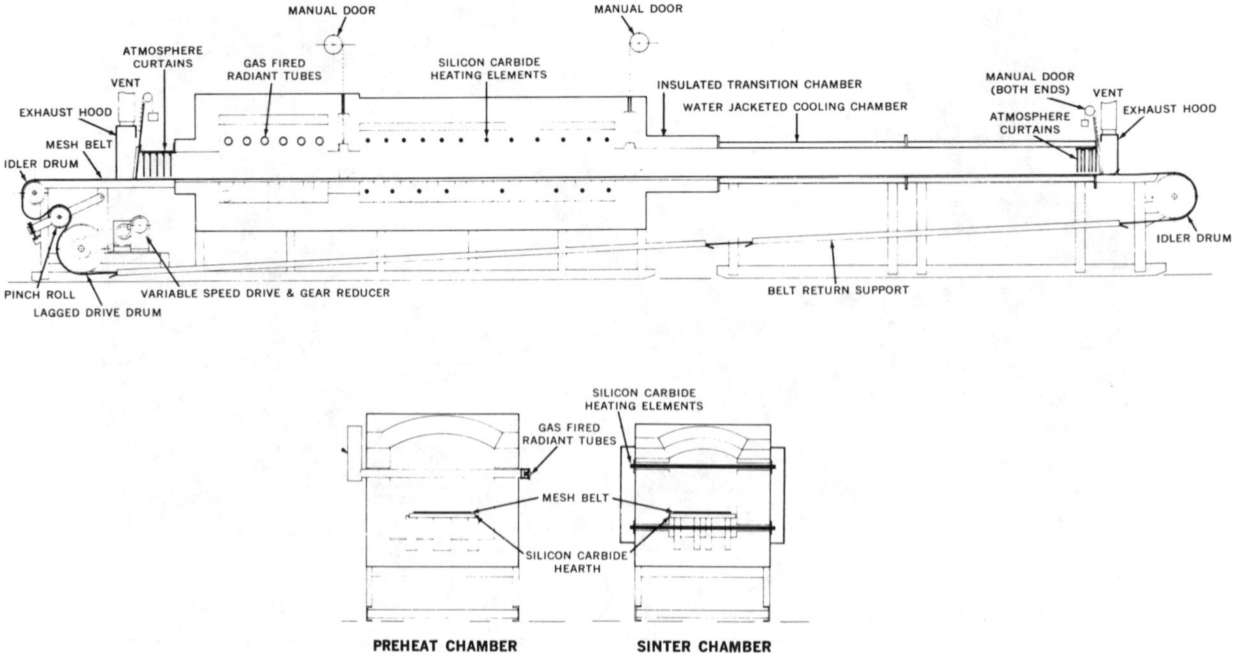

PHOTO #4. Cross Sectional View of A Mesh Belt Sintering Furnace.

ECONOMY OF FUEL FIRED FURNACES

Economical operation of the horizontally and vertically fired radiant tube type furnaces, is limited to operations below 1850^0 F. Above this temperature, stack losses are excessive.

Exhaust Gas Temp. ^{0}F.	Stack Loss % Gross Input	Available Heat % Gross Input
400	16	84
600	21	79
800	26	74
1000	30	70
1200	35	65
1400	40	60
1600	45	55
1800	50	50
2000	56	44
2400	67	33
2800	80	20

TABLE #1.

PHOTO #5. Bayonet Element In A Radiant Tube.

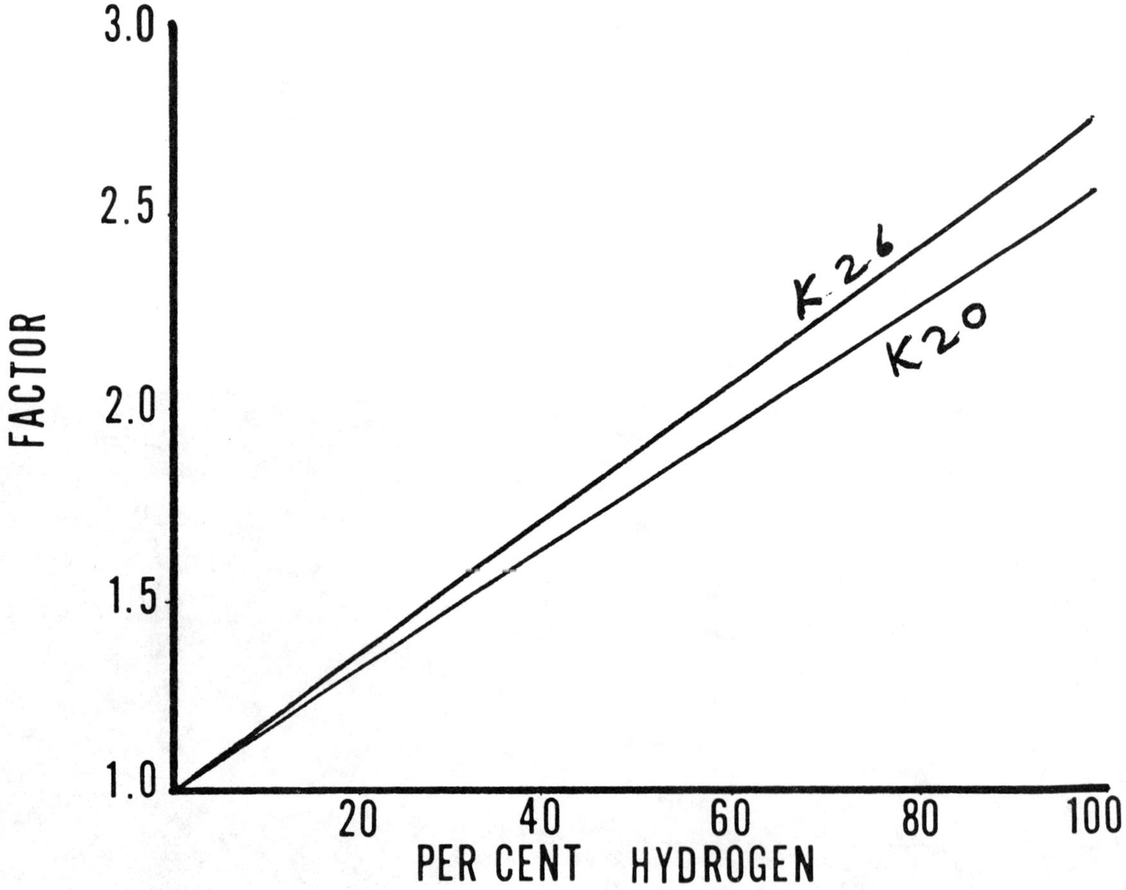

CHART #1. Multiplication Factor For The Increase of Thermal Conductivity Due to the Hydrogen Content.

ENERGY-SAVING APPLICATIONS OF INDUSTRIAL GASES

J. Scott Becker

Air Products and Chemicals, Inc.
Allentown, PA

Summary

The two important areas in which industrial gases can provide energy conservation for the P/M Industry are nitrogen-based atmospheres and oxygen enrichment of combustion air. Nitrogen-based sintering atmospheres replace the more conventional generated atmospheres like endothermic and dissociated ammonia. Nitrogen-based atmospheres are finding increasing popularity because they: (1) are available, (2) conserve energy and natural gas, (3) can provide cost reduction and (4) are a proven technology. Nitrogen-based carburizing atmospheres also replace endothermic atmosphere, providing substantial energy and natural gas conservation. The patented CAP (Carburizing Atmosphere Process) also provides improvement in grain boundary oxide levels relative to endothermic.

Oxygen enrichment of burner combustion air can provide substantial fuel conservation by increasing the efficiency of combustion. The cost of oxygen for enrichment is prohibitive, however, unless on-site oxygen generation is available.

Introduction

There are two main applications for which industrial gases can be utilized to help reduce overall energy consumption in the P/M Industry. The primary application is nitrogen-based sintering atmospheres, and most of this paper will deal with this application. A second application, oxygen enrichment of combustion systems for oil or gas fired furnaces, will also be discussed.

Nitrogen Based Atmospheres

To begin the discussion of nitrogen-based atmospheres one must start with the question, "What purposes do the sintering atmospheres serve?" The answer to this question is twofold. First, the atmosphere must act to exclude air from the furnace and to reduce any oxides present in the furnace. Second, the atmosphere must act to control carbon level in the parts being sintered by maintaining a desired carbon potential in the furnace. Thus the sintering atmosphere serves two distinct purposes; one, an oxide reduction function, and the other, a carbon control function.

The most common generated atmosphere system utilized for sintering is endothermic. A typical "endo" generator is shown in figure #1. In this system natural gas is combusted in a catalyst-filled retort in the presence of an insufficient amount of air (generally 2.4 parts air to 1.0 part natural gas). At this air-to-gas ratio, the gas will not burn unless the container in which the gases are to be burned is heated. Thus the name endothermic gas. The output of an endogenerator is approximately 39% N_2, 39% H_2, 20% CO and 2% impurities.

By contrast, in a nitrogen-based system the basic constituents of the atmosphere are pure gases. These gases are produced away from the sintering plant and stored outside of the plant, usually as liquids. These gases are then blended in the exact proportions necessary for each particular application in a blending panel located near the sintering furnace. Figure #2 shows a schematic of a typical nitrogen-based sintering system. The storage facilities located outside of the plant are normally owned and maintained by the company supplying the industrial gases.

Figures #3 and #4 show the relative compositions of the common generated atmospheres and nitrogen-based atmospheres. The basic difference among the generated atmospheres is the relative percentages of N_2, H_2 and carbon-bearing gases produced. By contrast, the nitrogen-based atmosphere is composed predominantly of pure, inert, nonflammable nitrogen with only the amount of reactive gases (H_2, CO_2, CH_4) necessary to produce the desired furnace operations. This control of hydrogen and hydrocarbons provides both improved furnace control and substantial energy savings.

Advantages of Nitrogen-Based Atmospheres

There are four basic reasons why nitrogen-based atmospheres are finding acceptance for sintering operations. These are:

1) Availability

2) Energy and Natural Gas Conservation

3) Reduced Operating Cost

4) Proven Technology

The next several paragraphs will deal with each of these areas in greater depth.

Availability

Nitrogen is produced by air separation. Figure #5 shows a typical air separation plant. The only raw materials utilized by this plant are air and electricity. The air is taken into the plant and liquified. The liquified air is then fractionally distilled (much like petroleum) in the large cracking tower located at the center of the plant. This tower is known as the cold box.

As the various chemical constituents of the air are taken off, the gases are reliquified and placed in the large storage tanks located at the plant. From the air separation plant the liquid nitrogen is delivered to the customer by over-the-road tankers.

At the customer's plant the industrial gases are stored in tanks located outside of the plant walls (see figure #6). Most gases are stored as liquids and only vaporized as required. Some gases, such as H_2, are stored as gases in pressurized cylinders. All storage facilities are owned and maintained by the industrial gas supplier.

For certain applications, where an extremely large (15,000 SCF/hr) atmosphere requirement exists, it becomes more economical to actually locate a small air separation plant at the customer's facilities. Figure #7 shows a typical on-site plant. These plants are similar in all respects to the larger production facilities, only on a smaller scale. Liquid backup storage is maintained at these on-site plants in the event of mechanical problems or electrical outage. These plants produce an oxygen-rich by-product stream which may be utilized as discussed later in this paper.

Energy and Natural Gas Conservation

Figure #8 shows the total amount of energy consumed to produce 1000 SCF of each of the generated atmospheres and an equivalent volume of nitrogen-based atmosphere. These numbers include all forms of energy that go into producing the various atmospheres, including gasoline for delivery vehicles, raw materials production, and utilities. It can be seen that nitrogen-based atmospheres provide a substantial reduction in total energy consumed relative to both dissociated ammonia and endothermic gas.

In today's environment, even more important than total energy consumption is the type of energy consumed. Figure #9 shows the all too familiar statistics on remaining fossil fuel supplies. Although different experts have varying opinions, the message is clear; as a planet we are in fact running out of hydrocarbon fuels, but we do have substantial reserves of coal. The escalating cost of all forms of energy is clearly shown in figure #10. It is obvious that hydrocarbon fuels are experiencing far more rapid price escalation. The message of this paragraph is clear: hydrocarbon fuels are both scarce and rapidly escalating in price and should be conserved whenever possible.

The relative amounts of natural gas consumed in producing 1000 SCF of the various generated atmospheres and the same volume of nitrogen-based atmospheres are compared in figures #11 and #12. Again these numbers include all natural gas consumed in producing and delivering the various atmospheres. As can readily be seen, the reduction in natural gas consumption made possible by switching to nitrogen-based sintering is very substantial, on the order of 90-95%.

As mentioned previously, the only energy resource consumed in producing nitrogen is electricity. This electricity can come from all the major fuel sources, including coal and nuclear, and is not dependent upon foreign hydrocarbon fuels. Thus, nitrogen-based atmospheres not only reduce total energy consumption, they also all but eliminate dependence on natural gas.

Cost Reduction

The relative cost comparison between generated and nitrogen-based atmospheres depends upon many factors and is different for each P/M fabricator. Generally the cost of nitrogen-based atmospheres is very competitive with generated atmospheres, especially when the cost-savings factors associated with nitrogen-based sintering and listed in figures #13 and #14 are taken into account. For many operations, total processing cost can be reduced by converting to nitrogen-based atmospheres.

Proven Technology

Nitrogen-based atmospheres are currently being utilized in production situations for presintering, sintering, infiltrating and powder reduction and annealing. Materials being processed include all conventional P/M materials. Figures #15 and 16 list the typical atmosphere compositions required for the various processes and materials listed. These compositions represent the considerable in-plant experience of Air Products' engineers and are based upon actual in operation, production situations. As can be seen for most metals (i.e., Al, Fe, Cu, brass, bronze) nitrogen levels run 95% to 98%, balance H_2, CH_4, etc., while for the more exotic metals hydrogen levels run from 5% to 30%, balance nitrogen.

The exact compositions listed above are produced and very accurately controlled by means of a blending panel. Figure #17 shows a typical panel designed and manufactured by Air Products. The inherent simplicity of the gas storage and blending system which is totally pneumatic (operating of its own internal storage pressure) leads to several substantial operating advantages including those listed in figure #18.

Although nitrogen-based sintering is already a proven technology, developmental work continues to further refine the system and its control parameters. Work is especially concentrated in the area of high density parts (6.9 9/cc and higher) where some operating problems have been encountered.

Oxygen Enrichment of Combustion Air

Presently, this paper will briefly diverge from discussing nitrogen-based atmospheres in order to address another energy-saving alternative made possible by industrial gas utilization. This topic is oxygen enrichment of oil or gas fired furnace combustion air.

The air around us is composed of approximately 20% oxygen, 80% nitrogen. When fuels are burned in this atmosphere, substantial amounts of energy are utilized to heat the nitrogen component of the air and combustion is relatively inefficient. By increasing the relative proportion of oxygen in the combustion air, the efficiency of the burning process is substantially improved, and energy and fuel are saved. Typically, air is enriched to 24% to 28% oxygen by adding oxygen gas to the air. This blend is then utilized as the combustion air.

Figure #19 plots the efficiency of energy utilization versus flue gas temperature for various percentages of oxygen in the combustion air. As can be seen, increased oxygen levels do substantially improve the efficiency of energy utilization. Figure #20 converts these increased efficiencies into percentages of fuel savings. As can be seen, these fuel savings run between 2% and 25% for oxygen levels of 22% to 30%.

From an economic standpoint, however, oxygen enrichment is often not warranted. Figures #21 and #22 show fuel cost vs. oxygen cost break-even points. The message of these figures is that in this application for oxygen enrichment to be economically attractive, oxygen cost must be on the order of $.06 to .08 per 100 SCF. Liquid oxygen is not available at this price. However, this is where the oxygen-rich by-product gas from an on-site plant can come into the picture. This oxygen, since it is normally vented to the air, is available at relatively low cost, well below $.06/100 SCF. Utilization of this by-product oxygen can thus lead to further energy and cost savings where an on-site nitrogen plant has been installed for nitrogen-based atmosphere (or other) purposes.

The balance of this paper will return to the discussion of nitrogen-based atmospheres.

Nitrogen-Based Carburizing

This section of the paper will deal with nitrogen-based carburizing. The system described is the patented Carburizing Atmosphere Process (CAP). The reasons for its discussion here are two fold. First, many P/M fabricators perform carburizing as a secondary operation. Second, the technology developed here is finding substantial application in sintering, especially in sintering high carbon and high density parts.

The CAP system is a blend of three gases which serve the functions listed below:

CONCEPTUAL BASIS FOR CAP BLENDS

	Component	Function
1.	Nitrogen	Inert Carrier
2.	Hydrocarbon (often natural gas)	Carbon and Hydrogen Source
3.	Oxidant (often CO_2)	Generates - Carbon Monoxide Controls - Carbon Potential Prevents - Sooting

Figure #23 shows schematically what the CAP system looks like. Again the gases are stored as liquids outside of the plant wall and blended at a panel located near the furnace. This blend can be utilized on both batch and continuous carburizing furnaces and will produce an excellent uniform case depth (see figure #24). This process is currently in production in about a dozen plants.

From an energy and natural gas conservation standpoint, the CAP system is very beneficial. Figure #25 shows the total energy consumption for producing 100 SCF of endo and 100 SCF of CAP based on either liquid or on-site nitrogen supply. As can be seen, the total energy required to produce the CAP blends is substantially less than that for endo. Even more important, as figures #11 and #12 indicate, utilization of the CAP atmosphere reduces natural gas consumption by 70% to 90%.

Finally, figure #26 shows a plot of oxygen concentrations as a function of distance from the surface of a part processed in CAP versus the same part processed in endo. What is apparent is that the CAP processed part shows a very substantial reduction in grain boundary oxides near the surface of the part. Apparently the CAP process reduces grain boundary oxides relative to the core of the part, while endo increases grain boundary oxides. The full implications of this phenomenon, as it affects mechanical properties, are not yet known, but substantial testing has shown the mechanical properties of the CAP-processed parts to be at least as good as those processed in endo.

Conclusions

In brief summary, energy saving industrial gas applications in the P/M Industry are generally confined to two areas. The first and most important application is nitrogen-based atmospheres. Nitrogen sintering atmospheres are rapidly becoming an accepted alternative to generate atmospheres because nitrogen offers several unique advantages. Probably the most significant advantage is freedom from natural gas dependence. Nitrogen carburizing atmospheres are also finding excellent acceptance, due to many of the same reasons. Oxygen enrichment on the other hand is fully proven technically, but limited in application due to economic considerations.

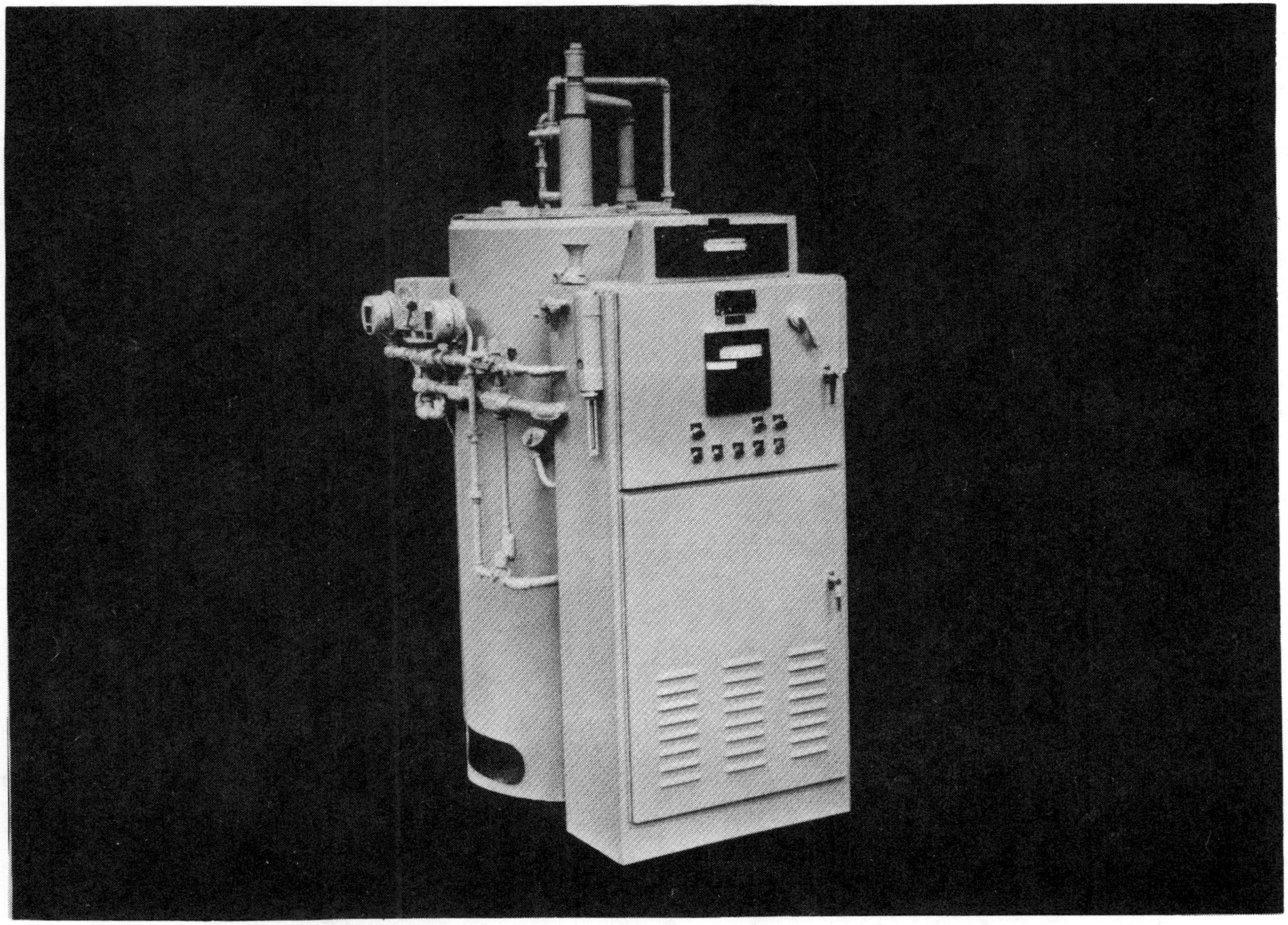

FIGURE #1
ENDOTHERMIC GAS GENERATOR

Figure 2

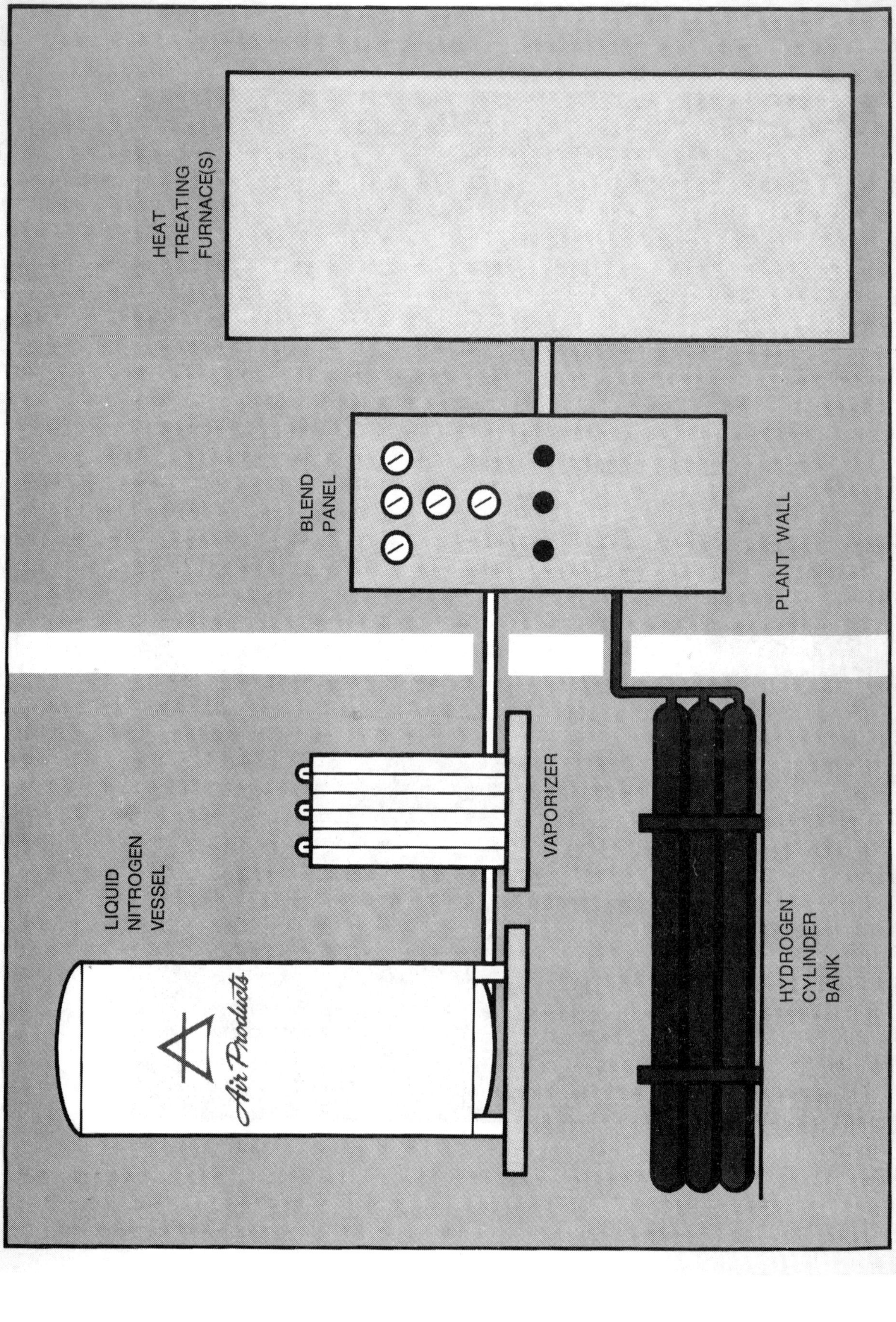

Air Products

General Schematic of an Air Products Protective Atmosphere Installation for Controlled Composition and Flow.

IGDM-26

COMPOSITION
- GENERATED ATMOSPHERES -

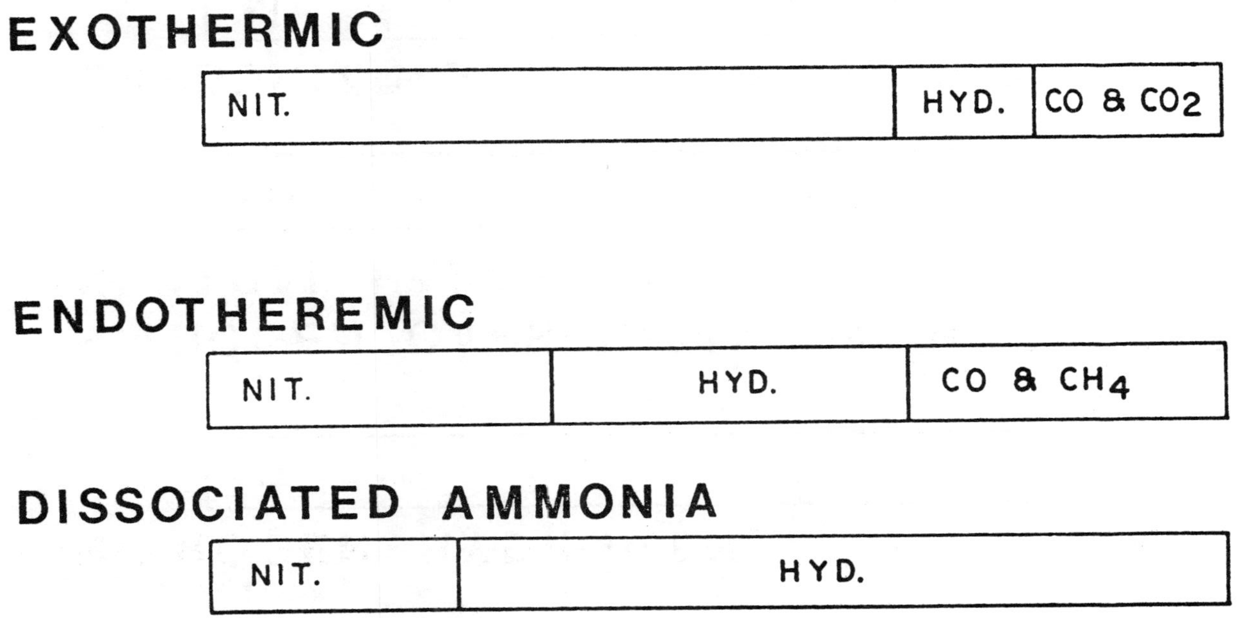

FIGURE #3

COMPOSITION
-NITROGEN BASED ATMOSPHERES-

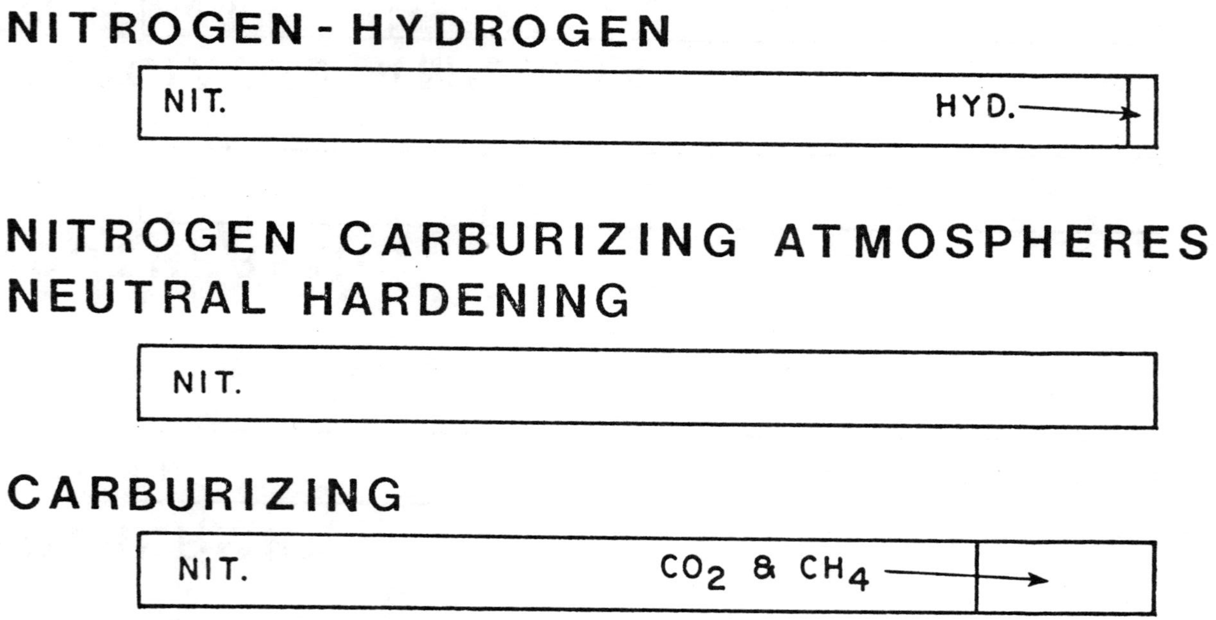

FIGURE #4

FIGURE #5 AIR SEPARATION FACILITY, WHARTON, NEW JERSEY

FIGURE #6
TYPICAL HYDROGEN-NITROGEN PLANT INSTALLATION

FIGURE #7 75,000 SCFH AUTOMATIC NITROGEN GENERATING FACILITY

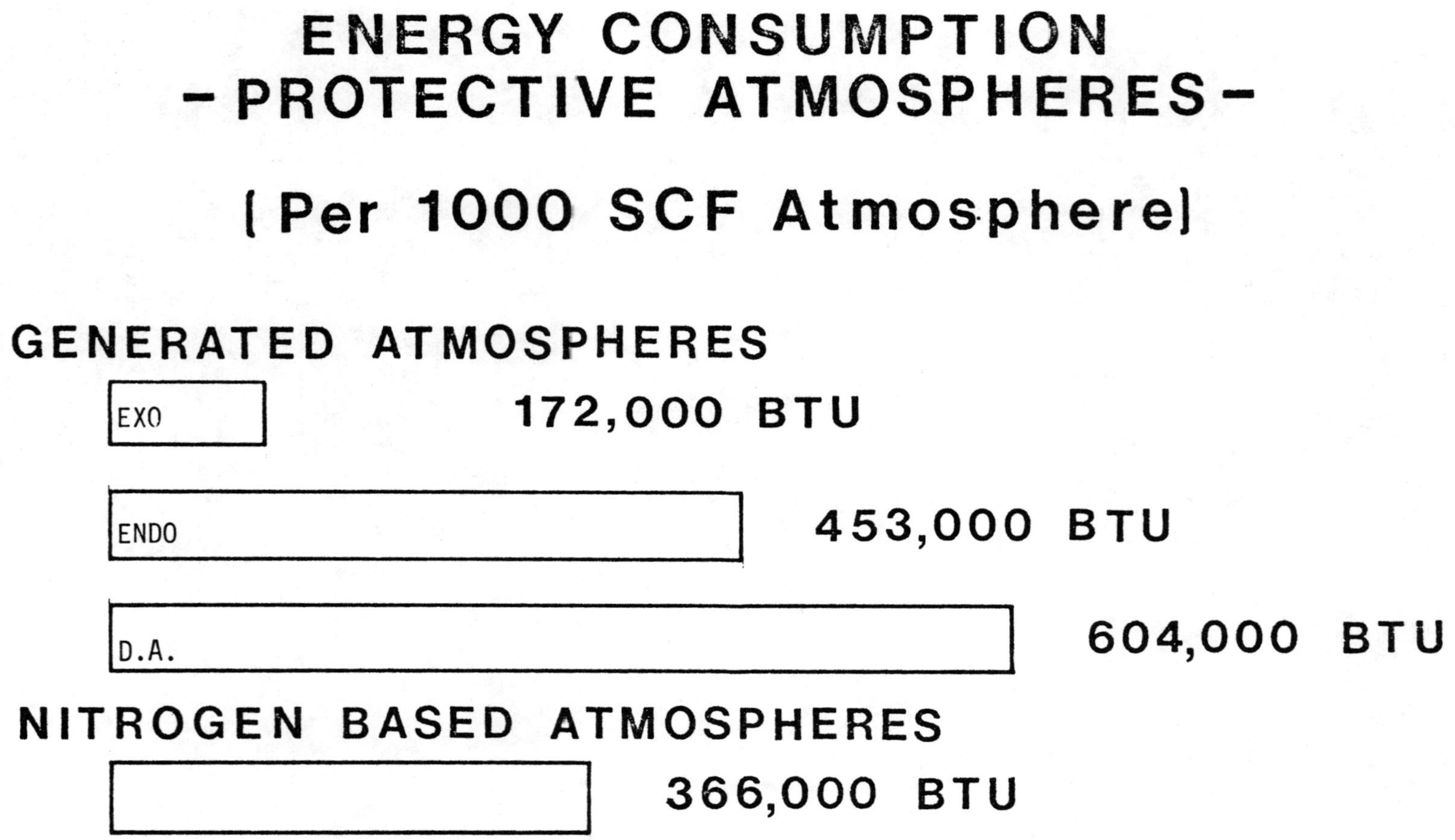
ENERGY CONSUMPTION
-PROTECTIVE ATMOSPHERES-
(Per 1000 SCF Atmosphere)
GENERATED ATMOSPHERES
EXO
172,000 BTU
ENDO
453,000 BTU
D.A.
604,000 BTU
NITROGEN BASED ATMOSPHERES
366,000 BTU

FIGURE #8

U.S. FOSSIL ENERGY RESOURCES*

	TECHNOLOGICALLY RECOVERABLE [10^{15} BTU]	YEARS OF SUPPLY
COAL	36,000	2400
NATURAL GAS	395	13
CRUDE OIL	415	21

FIGURE #9

AVERAGE U.S. PRICES OF CRUDE HYDROCARBONS 1945 - 1974

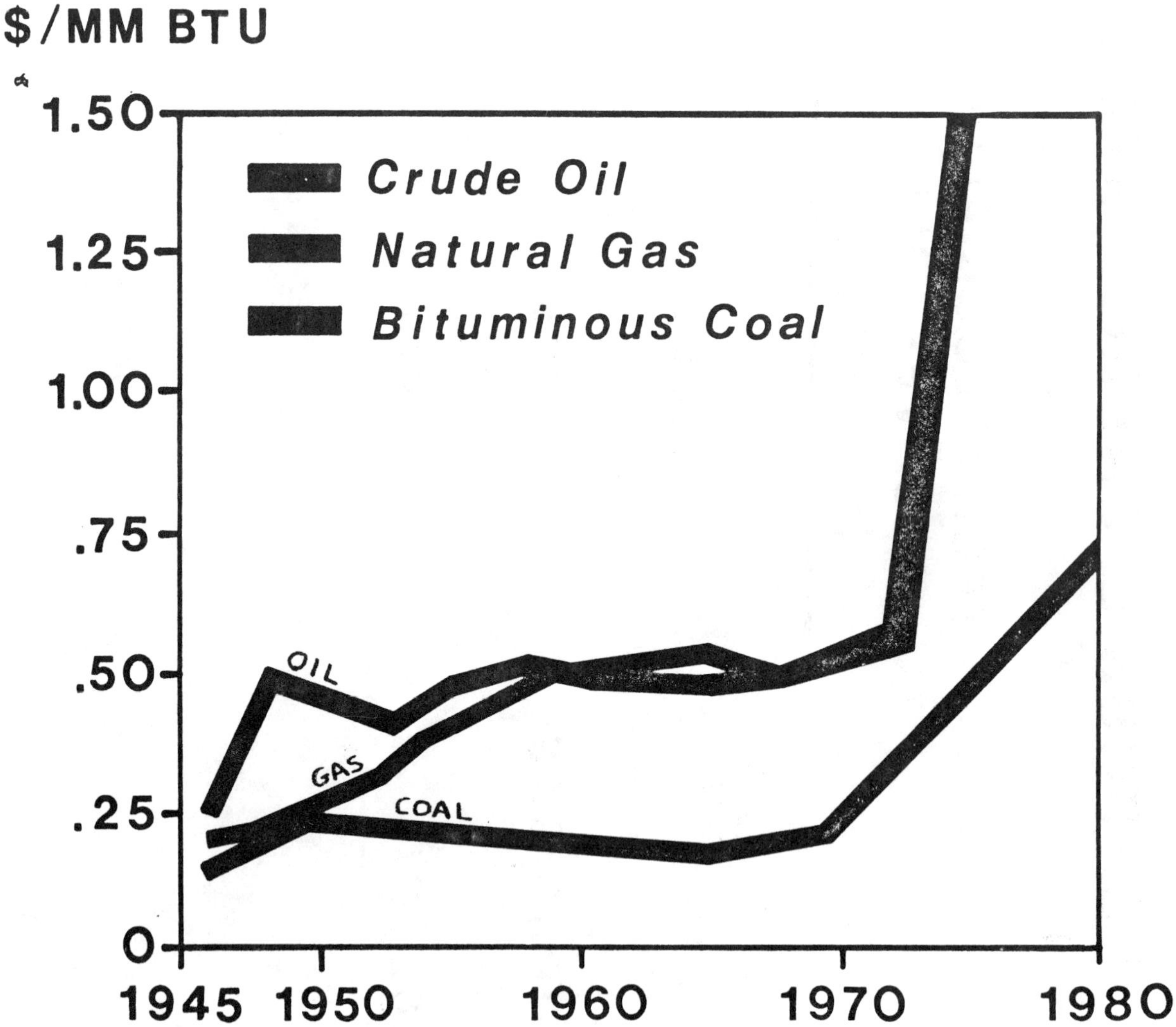

FIGURE #10

NATURAL GAS CONSUMPTION -GENERATED ATMOSPHERES-

(Per 1000 SCF Atmosphere)

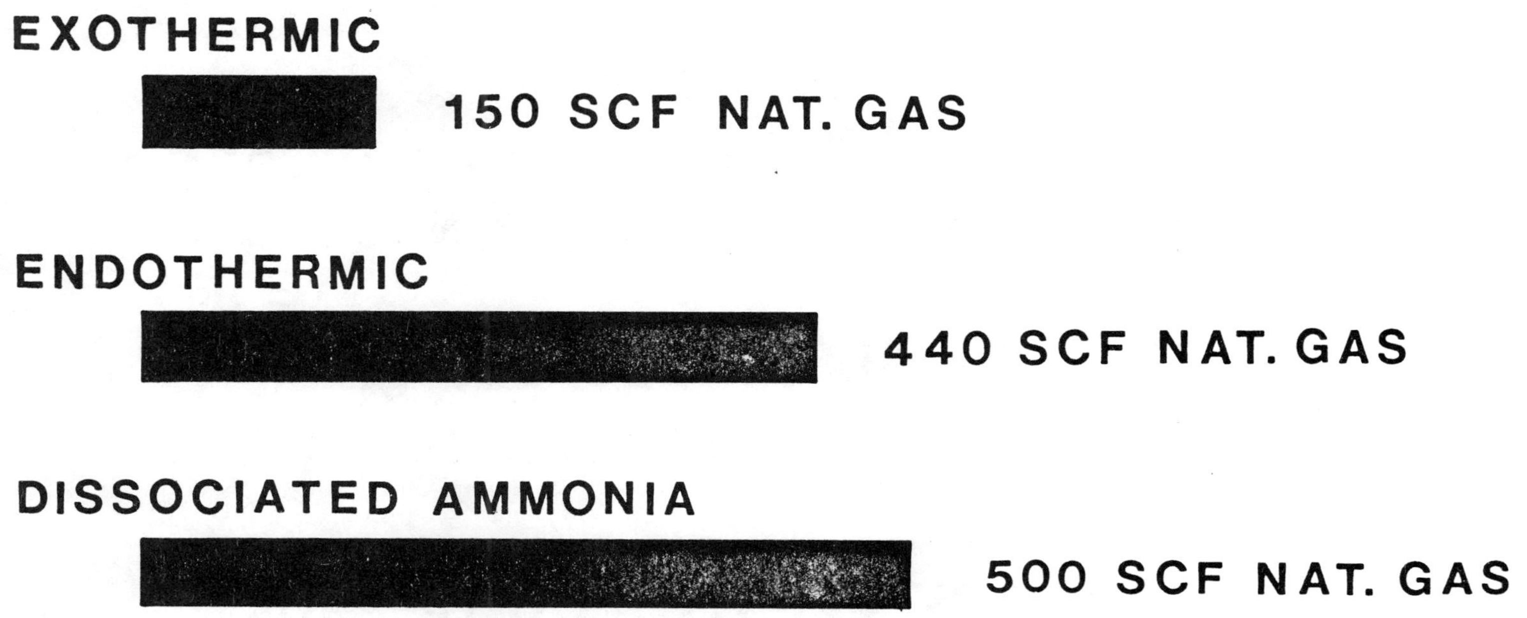

FIGURE #11

NATURAL GAS CONSUMPTION
-NITROGEN BASED ATMOSPHERES-

(Per 1000 SCF Atmosphere)

NITROGEN-HYDROGEN

25 SCF NAT. GAS

NITROGEN CARBURIZING ATMOSPHERES
NEUTRAL HARDENING

30 SCF NAT. GAS

CARBURIZING

140 SCF NAT. GAS

FIGURE #12

ECONOMIC BENEFITS FROM INDUSTRIAL GASES

1. *REDUCED ATMOSPHERE FLOW RATES*
2. *IMPROVED FURNACE BELT LIFE*
3. *MINIMAL MAINTENANCE COSTS*
4. *NO ATMOSPHERE RELATED DOWNTIME*

FIGURE #13

ADDITIONAL ECONOMIC BENEFITS FROM INDUSTRIAL GASES

1. *LOW CAPITAL INVESTMENT*
2. *IMPROVED PRODUCT QUALITY* **and** *YIELD*
3. *INCREASED PRODUCTION*
4. *CLEANING OPERATIONS ELIMINATED*

FIGURE #14

RECOMMENDED H-N BLENDS

	H_2	N_2	CO_2	CH_4
ALUMINUM	—	100 %	—	—
COPPER & BRASS	2 %	98 %	—	—
IRON .5 % C OR LESS	2 %	98 %	—	—
IRON MORE THAN .5 % C	1 %	98 %	—	1 %
IRON - CARBURIZE		85 %	3 %	12 %

FIGURE #15

RECOMMENDED H-N BLENDS
continued

	H_2	N_2	CO_2	CH_4
IRON-INFILTRATE	3%	96%	—	1%
400 STAINLESS	10%	90%	—	—
300 STAINLESS	30%	70%	—	—
PRESINTER TUNGSTEN CARBIDE	10%	90%	—	—
NICKEL ALLOYS	10%	90%	—	—

FIGURE #16

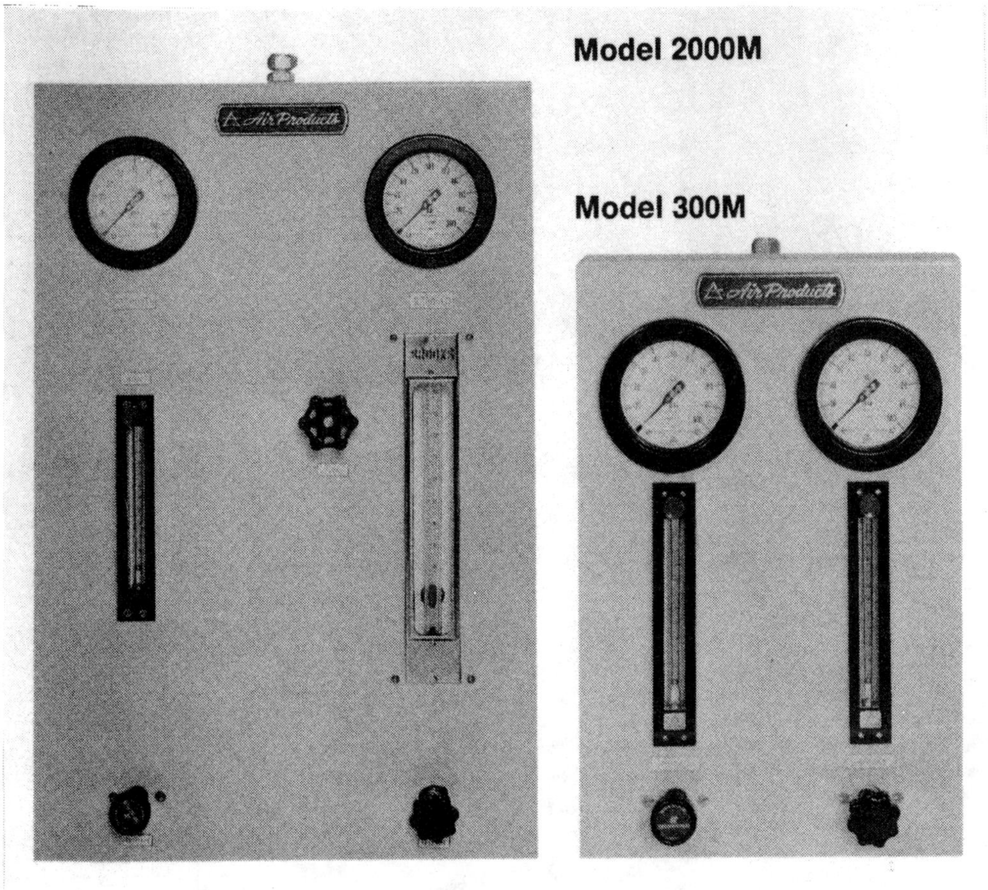

FIGURE #17
TYPICAL H-N BLEND PANELS

OPERATING ADVANTAGES OF INDUSTRIAL GASES

1. *CONSISTENT HIGH ATMOSPHERE QUALITY*
2. *RELIABILITY*
3. *MINIMAL MAINTENANCE*
4. *FLEXIBILITY*
5. *SAFETY*

FIGURE #18

PERCENT AVAILABLE HEAT vs. FLUE GAS TEMP.

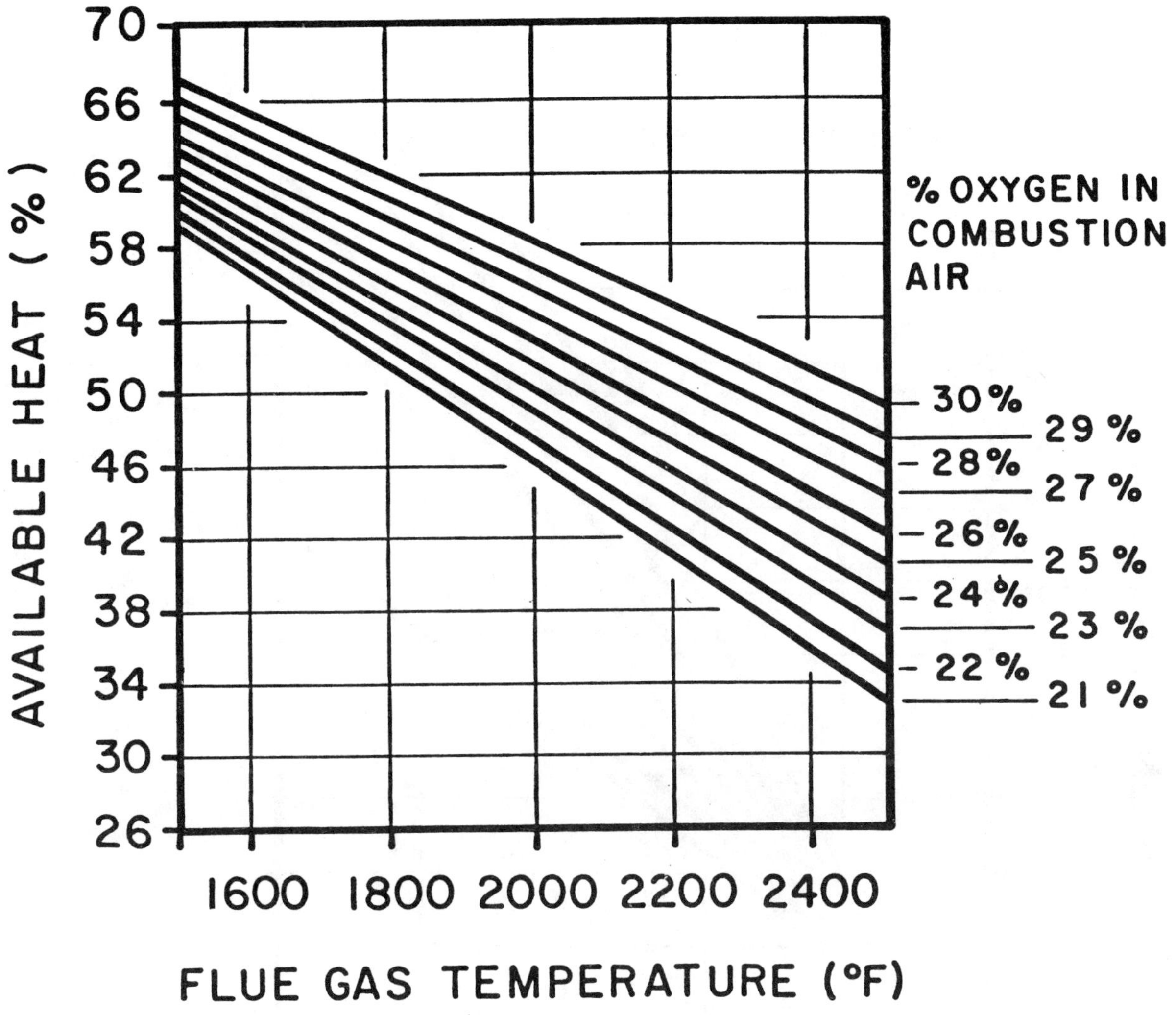

FIGURE #19

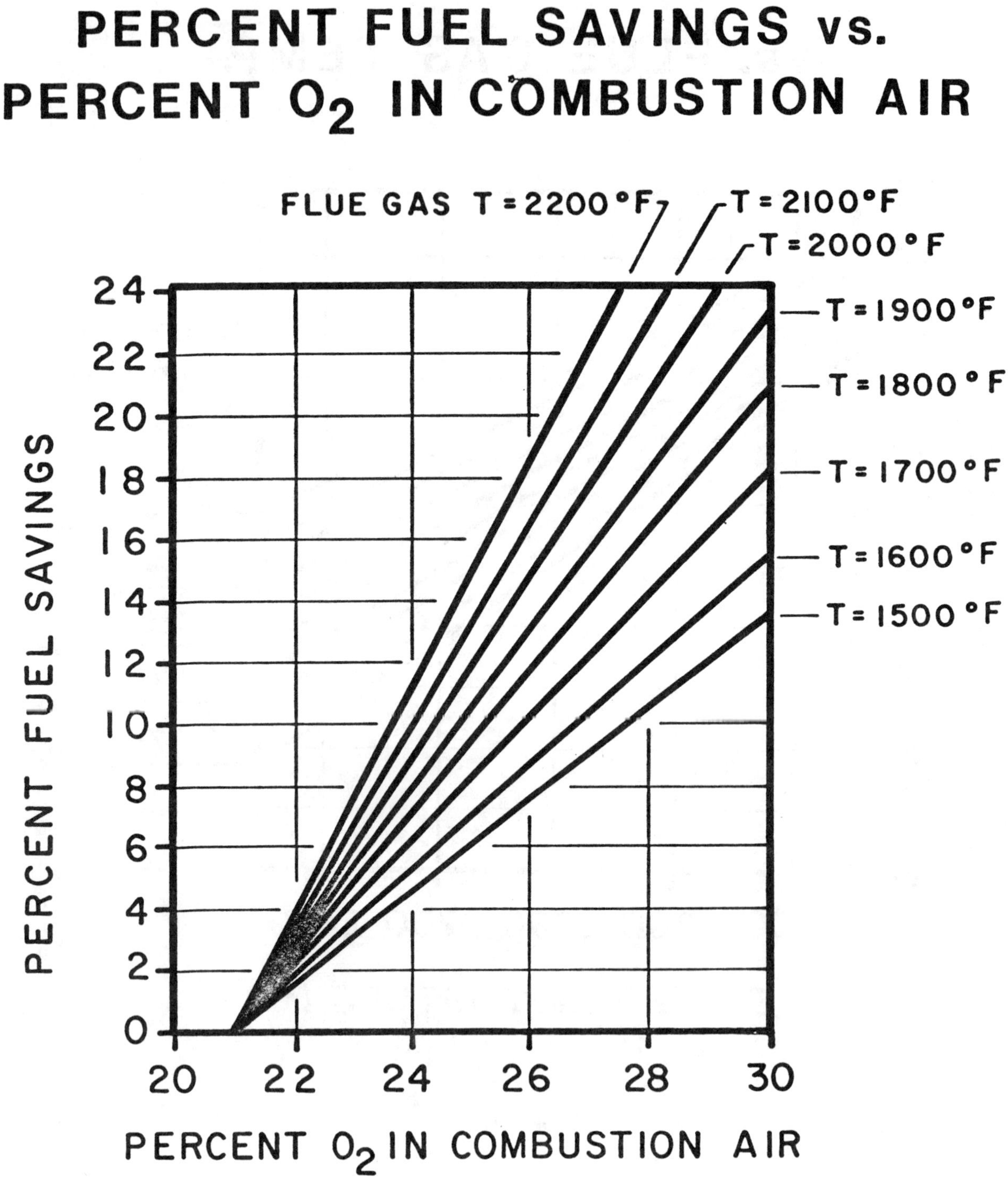
PERCENT FUEL SAVINGS vs.
PERCENT O2 IN COMBUSTION AIR
FLUE GAS T = 2200°F
T = 2100°F
T = 2000°F
T = 1900°F
T = 1800°F
T = 1700°F
T = 1600°F
T = 1500°F
24
22
20
18
16
14
12
10
8
6
4
2
0
20
22
24
26
28
30
PERCENT FUEL SAVINGS
PERCENT O2 IN COMBUSTION AIR

FIGURE #20

FUEL COST vs. OXYGEN COST INDIFFERENCE CURVES

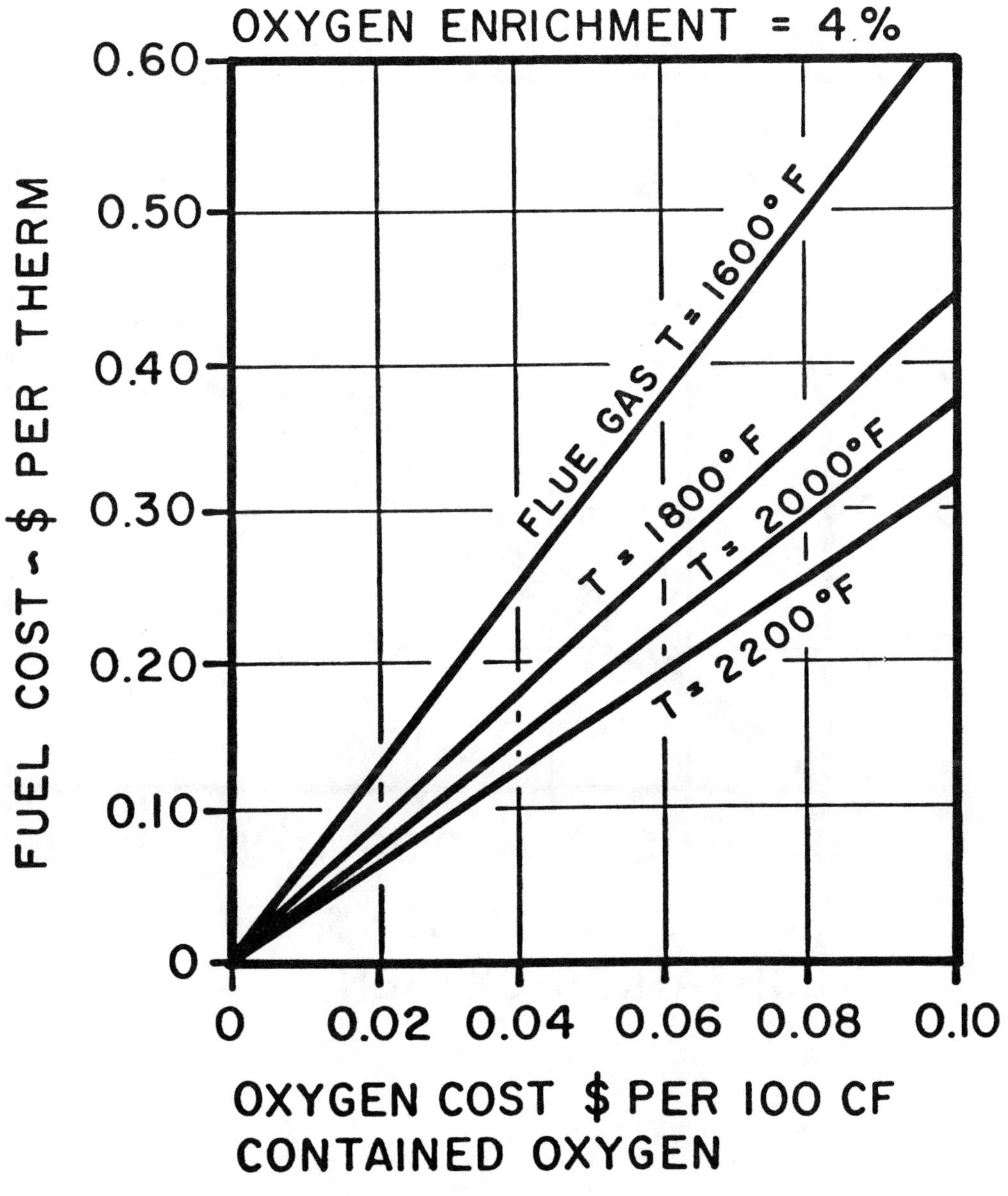

FIGURE #21

FUEL COST vs. OXYGEN COST INDIFFERENCE CURVES

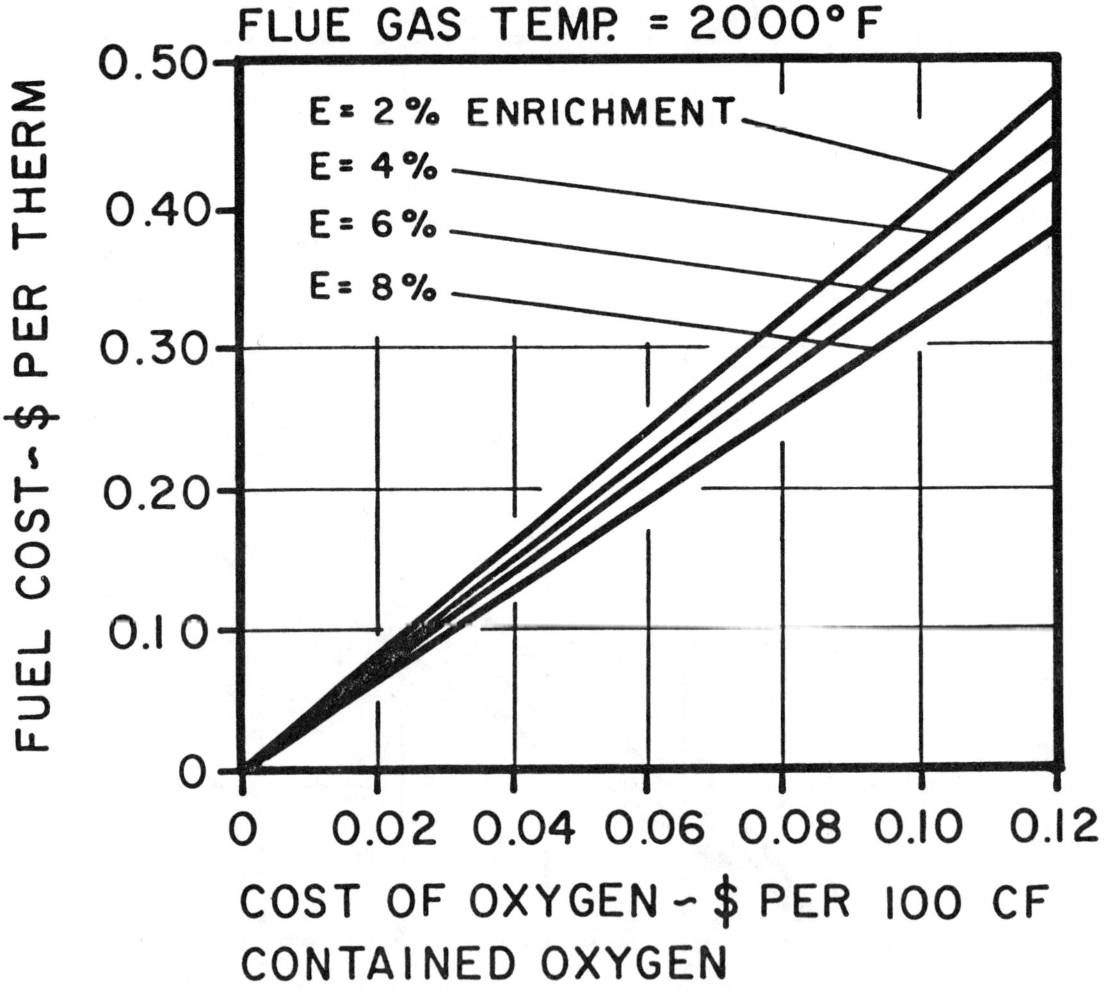

FIGURE #22

CAP INSTALLATION
Nitrogen – Natural Gas – Carbon Dioxide

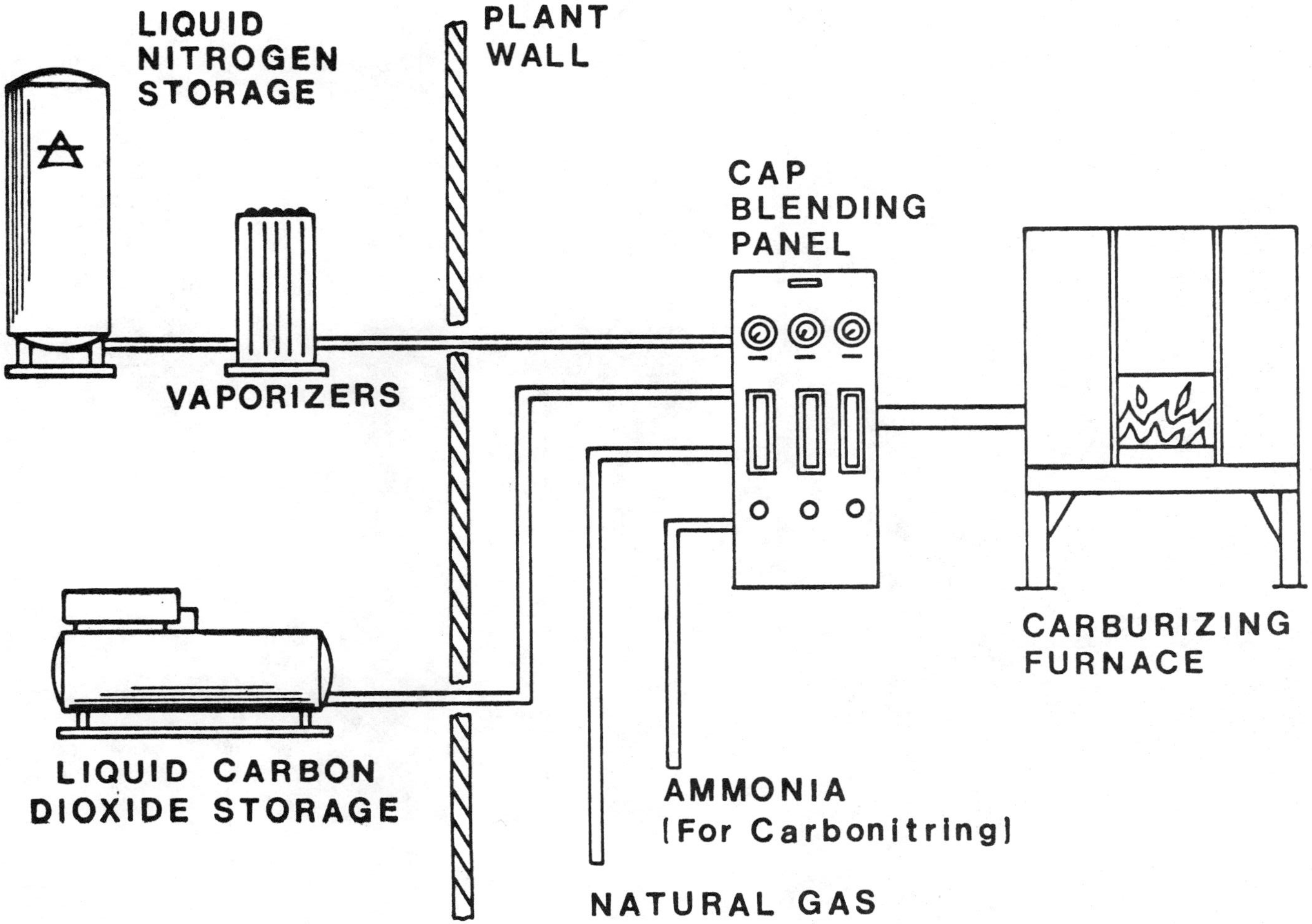

FIGURE #23

FIGURE #24
CARBURIZING ATMOSPHERE PROCESS (CAP) MACRO CASE DEPTH

TOTAL ENERGY CONSUMPTION

Carburizing Atmospheres

Basis: 100 SCFH Endo Flow Rate

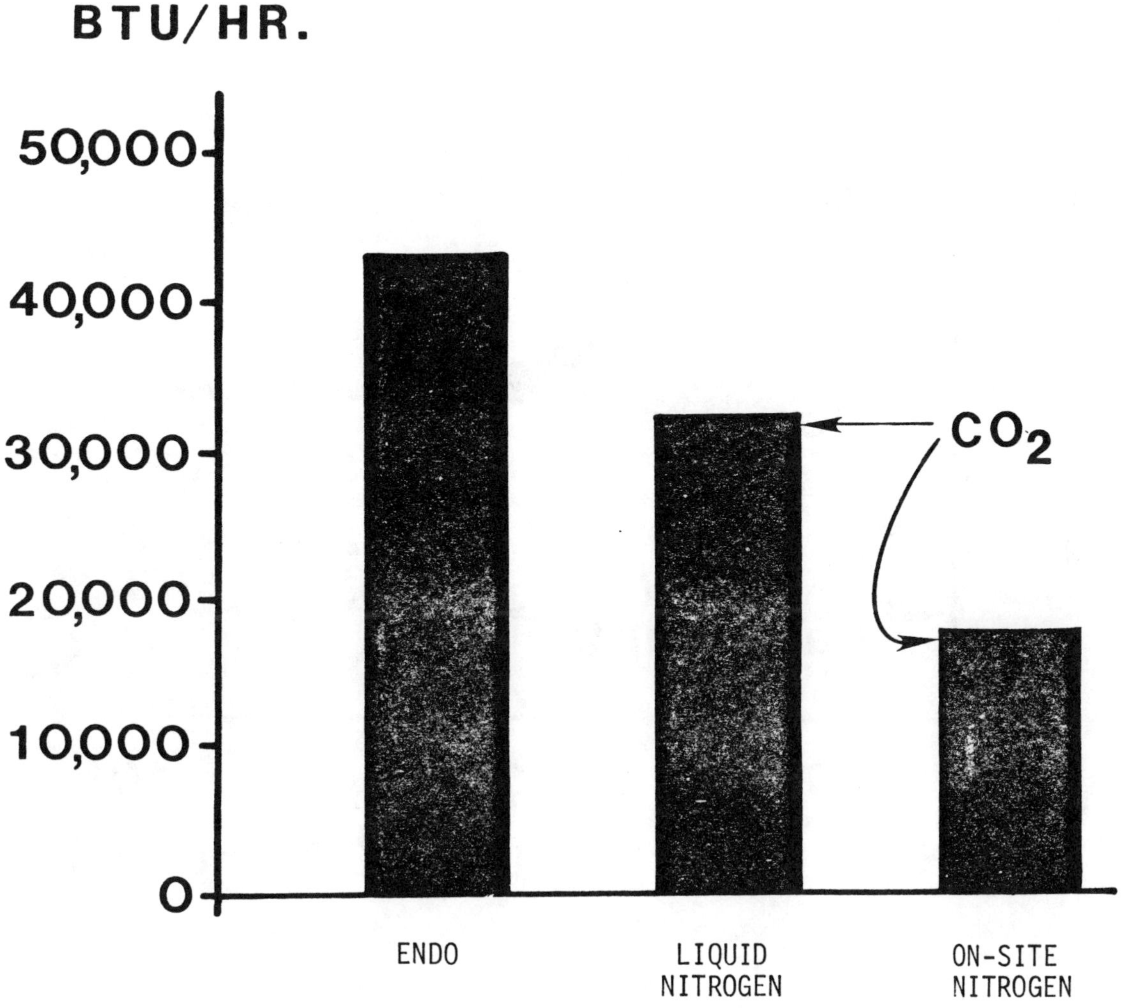

FIGURE #25

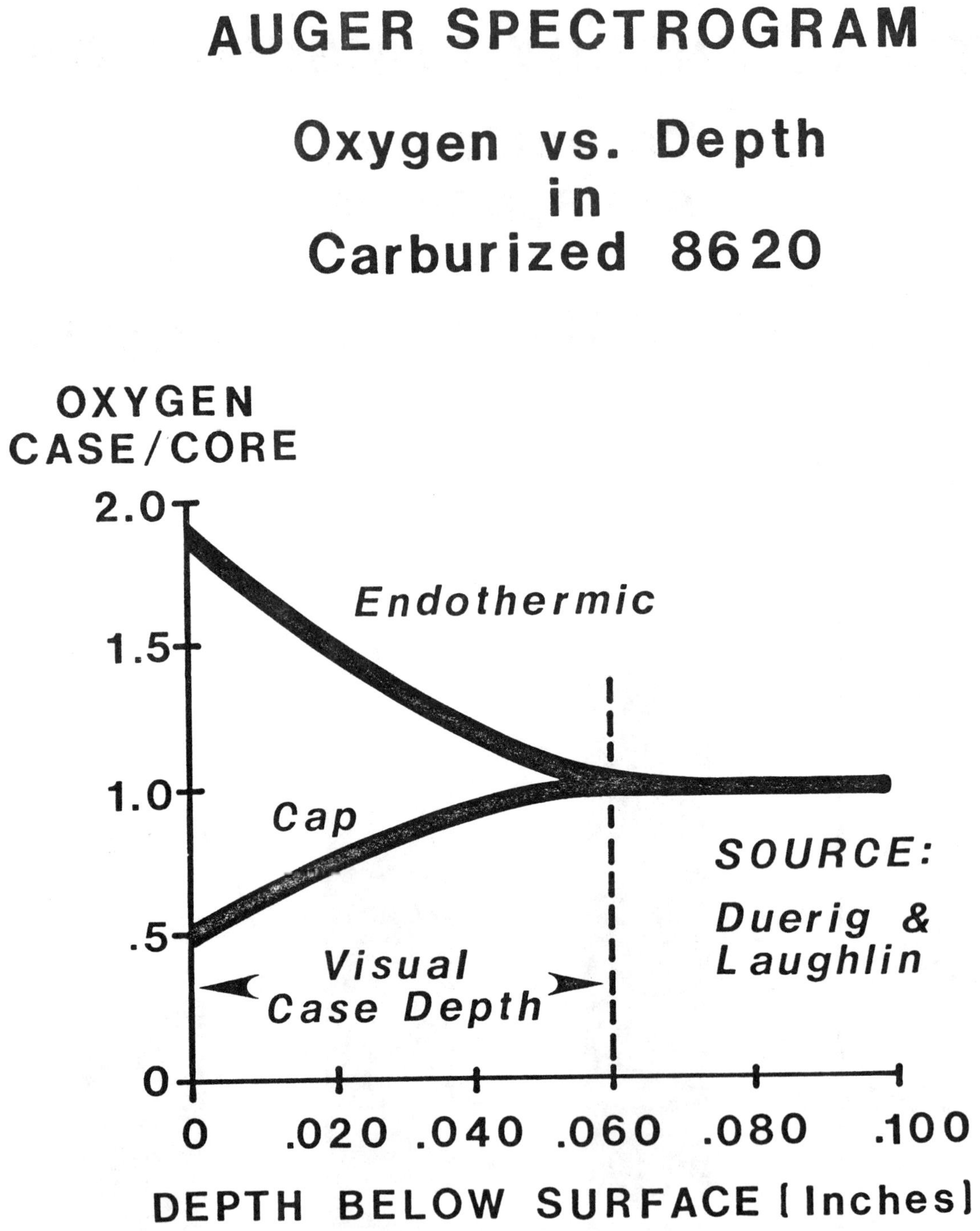
AUGER SPECTROGRAM
Oxygen vs. Depth
in
Carburized 8620
OXYGEN
CASE/CORE
2.0
1.5
1.0
.5
0
Endothermic
Cap
Visual
Case Depth
SOURCE:
Duerig &
Laughlin
0
.020
.040
.060
.080
.100
DEPTH BELOW SURFACE [Inches]

FIGURE #26

THE USE AND APPLICATION OF MINIATURE P/M PARTS

W. T. SULLIVAN
ASCO SINTERING CO.

For many years, the P/M industry has avoided production of parts with small holes and thin walls. Most potential P/M parts with holes or walls of less than .050 diameter or thickness were turned down, primarily because of the extremely fragile tooling needed to make such parts, but also because of difficulty in handling the green parts prior to sintering, as well as the problems inherent in inspection of very small parts.

In the last several years, Asco Sintering Co. has successfully produced a variety of parts that represent a significant advance in the production of small structural P/M parts. A wide range of materials has been used, including high density copper, SS-316-P and SS-316-R stainless steel, CZP-0228-T leaded brass, CT-0010-R bronze, CZ-1818-U nickel silver, FC-0208-R copper alloy steel, FN-0205-S nickel alloy steel and FX-2008-T copper infiltrated steel. The type of parts produced have ranged from very simple shapes to quite intricate and detailed parts. Production quantities have run from as few as 1,000 parts per order to as high as several million per order.

Following are photographs of a cross section of miniature P/M parts, a description of the material from which each is made, as well as some of the unique features, dimensions and tolerances of each part.

Figure 1 - This part, which is used to orient the location of printed circuit boards, is produced from sintered SS-316-P stainless steel. It is a good example of the detail which can be provided by the P/M process. Of particular interest is the wall thickness of this part:

1. The wall thickness from the tip of the O.D. spline to the minor diameter of the I.D. spline is .040.
2. The wall thickness between the O.D. of the outside spline and the root of the I.D. spline is .024.
3. The wall thickness between the root of the O.D. spline and the root of the I.D. spline is .015.

Figure 2 - This part works inside the part illustrated in Figure 1, and is also made from SS-316-P stainless steel. Features of this part include:

1. The wall thickness from the tip of the O.D. spline to the I.D. is .0425, with a wall thickness to length ratio of approximately 7:1. The .030 flange thickness is not included in calculating this wall thickness ratio.
2. The wall thickness from the root of the spline to the I.D. is .0235, with a wall thickness to length ratio of slightly over 12:1, again not including the .030 flange length.
3. This part must index within the mating part through any of the 16 spline positions within .002, non-cumulative.

Figure 3 - This high volume part was produced from CZP-0228-T brass, and was used on an electric, hand controlled toy airplane. The I.D. is .064, with a wall thickness of .0375. Good density was a must, in that this part was staked onto the main propeller drive cable of the toy.

Figure 4 - This part, also produced from CZP-0228-T brass, was used on the opposite end of the drive cable of the toy airplane noted in Figure 3. This part, too, was severely staked to lock the cable in position. The I.D. of this part is .039, and the wall thickness opposite the I.D. is .0195, with a thickness to length ratio of 11:1.

Figure 5 - This high volume part, as well as the part in Figure 6 to follow, are both made from SS-316-R stainless steel and are used in potentiometer turns counting dials. Of interest in this part are:

1. The .034 I.D.
2. The .024 wall between the I.D. and the root of the cog tooth.
3. Each of the 6 cog teeth are .035 wide; 3 are .100 long, the other 3 are .150 long. These lengths must be held within a ±.002 tolerance as sintered.

Figure 6 - When the part in Figure 5 was produced successfully in high volume, a smaller version of the unit in which that part functions became very marketable. Because of this, this part, a modified gear rather than a cog, was tooled. The significant dimensions on this part include:

1. The .034 I.D.
2. The .022 wall between the I.D. and the root of the tooth.
3. The basic tooth width is .028, with each tooth tapering to a .0055 radius at the tip.
4. Three of the teeth are .133 long; the other three are .089 long, with a ±.002 (non-cumulative) tolerance on both lengths.

Figure 7 - These parts are produced from pure copper at 7.8 g/cc minimum density. They are produced in extremely high volume, and are used in the printing mechanism in a computer terminal.

Dimensional considerations include:

1. The .0125/.0135 I.D.
2. The .0115 wall between the I.D. and the periphery of the part.
3. Each of the three lobes is .0225 wide.
4. Cost reduction in converting to the P/M process was 98.3%.
5. The parts weigh 17 grams (.59959 ounces) per 1,000 parts. Three and one-half pounds of powder will produce just over 100,000 parts.

Figure 8 - The part in this illustration as well as in Figure 9 were produced from CZN-1818-T nickel silver. They were used in miniature pressure transducers for determining blood pressure at various points within the veins and arteries prior to and during cardiovascular surgery. Of unique interest are:

1. The major diameter of the part is .054.
2. Parallel grooves, .0305 wide and .010 deep, run for .085 of the .125 overall part length.
3. A slot runs between these grooves, at the small end of the part, with the slot dimension being .020 wide and .010 deep.
4. Filet and corner radii were .002 maximum, and the part was required to be symmetrical within .001.

Figure 9 - This "H" shaped part measures .040 by .0575 by .075 long. Two grooves, .0095 wide and .012 deep run the .077 length of the part on the curved .040 width and two grooves, .035 wide and .0055 deep, run the length of the part on the .0575 dimension. This leaves a web in the center of the part .029 thick. At each end of the part, a slot runs across the .029 web, with each slot being .0095 wide and .015 deep.

Filet and corner radii are .002 maximum, with the part being symmetrical within .001.

Powder metallurgy is growing rapidly in both volume of production and in technology. The type of parts illustrated here will do little to increase the tonnage of metal powder consumed each year in the P/M industry. They do show graphically the advances being made by those seeking to broaden the limits of that which can be done to save time, money and material through practical application of the latest in P/M technology.

Figure 1

Figure 2

Figure 3

Figure 4

Figures 5 and 6

Figure 7

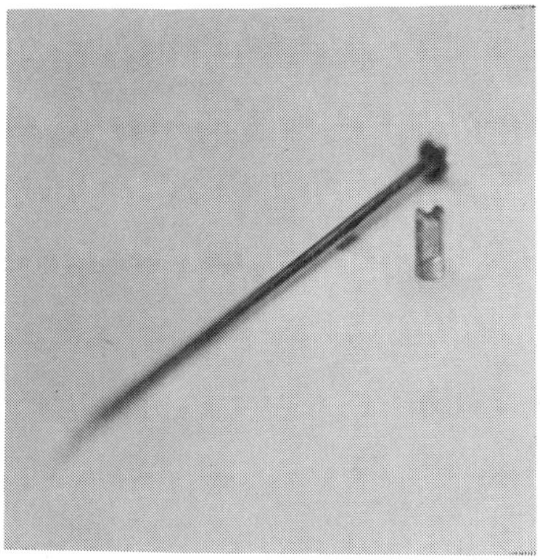

Figure 8

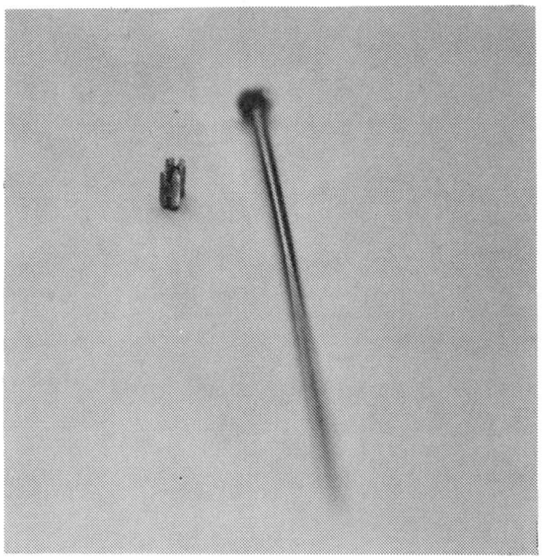

Figure 9

All parts balanced on human finger.

HOT ISOSTATIC PRESSING OF MARAGING STEELS*

J. E. Smugeresky
Sandia Laboratories, Livermore, CA 94550

R. M. German
J. M. Ney Company, Bloomfield, CT 06002

ABSTRACT

Pre-alloyed powders of 250-grade maraging steel have been consolidated by hot isostatic pressing (HIP). Significant ductility values have been obtained with rotating-electrode-process (REP) and with inert-gas-atomized (IGA) powders--both in the as-pressed condition. This result contrasts with those of other researchers, who used post-HIP plastic deformation to achieve ductility. In the present work, ductilities equal to those of wrought material have been obtained by reducing the titanium segregation at prior powder-particle boundaries. This reduction can be achieved either by post-HIP heat-treating material consolidated from REP powders or solely by hot isostatic pressing IGA powders. In both cases the production of very fine titanium-rich precipitates at prior powder-particle boundaries is reduced. Also, the resultant fracture mode is transparticle rather than interparticle, an outcome suggesting that complete bonding across particle boundaries has occurred.

*Work supported by the U.S. Department of Energy under Contract DE-AC04-76DP00789.

Introduction

Hot isostatic pressing (HIP) of metallic powders has recently been used to improve the chemical homogeneity of complex alloys, to consolidate pre-alloyed powders to full theoretical density, and to fabricate complex shapes. The approach, unlike techniques previously described at this meeting, produces fully dense parts and properties equivalent to, or better than, conventionally processed material. The process quite simply entails the simultaneous application of heat and pressure to an encapsulated metallic powder. The pressure is applied through a fluid medium, usually argon, and thus provides isostatic compaction. In contrast with uniaxial hot pressing, pressures at least an order of magnitude higher can be used. Since present hot isostatic press units can routinely develop pressures of 105 MPa (15 ksi), full-density parts can be obtained even in very large, complex shapes.

In the work reported in this paper, we have used hot isostatic pressing to consolidate maraging steels. These steels are low-carbon, iron-based alloys nominally containing Ni, Co, and Mo, together with smaller amounts of titanium and aluminum. Although their chemistry is complex, these steels offer an attractive combination of high strength (1585 MPa or 230 ksi) and high toughness (95 MPa $\sqrt{m}$ or 87 ksi $\sqrt{in.}$) and, therefore, are good candidates for structural applications. However, problems such as casting segregation, lack of a fine grain size, and high material cost have precluded the extensive use of maraging steels. The present study was conducted because powder-metallurgical processing offers the potential for solving these problems.

Previous attempts to powder-process maraging steels have included conventional pressing and sintering of blended elemental powders,[1-7] post-sinter forging,[7] hot extrusion of pre-alloyed,[8-13] and hot isostatic pressing.[8,11,12] Only in cases where extensive post-consolidation plastic deformation was used did the compacted powders have significant ductility.[7,8,11,12] Since this extra processing step means higher costs, we have looked at both modified heat treatments and different powder-making processes as alternative means to obtain a ductile maraging steel in the as-pressed condition. The success of this approach makes it feasible to use the HIP technique to fabricate maraging steels into net-shape complex parts that have significant ductility.

Experimental Procedures and Results

The two processes used to make pre-alloyed 250-grade maraging steel powders were the rotating-electrode process (REP) of Nuclear Metals, Inc., and the inert gas atomization (IGA) process of Udimet Powder Division, Special Metals Corporation. The material was a 250-grade maraging steel with nominal chemistry of 18% Ni, 8% Co, 5% Mo, 0.4% Ti, and < 0.03% C. Actual compositions are listed in Table I. In each case the powders were screened to size and consolidated by hot isostatic pressing at 1100 and 1200°C. The REP powder particles were spherical in shape and were screened to -50+200 mesh (the 300-to-74 μm size range).[14] The IGA powder particles were also spherical, contained agglomerated fines, and were screened to -60 mesh (less than 250 μm size range).[15] The mean particle size by weight was 190 μm for the REP powder and 96 μm for the IGA powder; their shapes are shown in Figures 1, 2. In each case the oxygen content was less than 70 wt ppm.

TABLE I

CHEMISTRY OF STARTING POWDERS COMPARED WITH NOMINAL 250-GRADE MARAGING STEEL

Element	REP Powder	IGA Powder	Nominal 250 Grade
Ni	18.35	18.2	18
Co	8.22	8.4	8
Mo	4.85	4.5	5
Ti	0.43	0.34	0.4
C	0.006	0.023	<0.03
Al	0.06	0.039	0.1
Si	0.03	0.01	<0.10
Mn	0.01	0.10	<0.10
P	0.003	0.007	<0.01
O	0.006	0.0072	--
N	0.004	0.0017	--
Fe	Bal	Bal	Bal

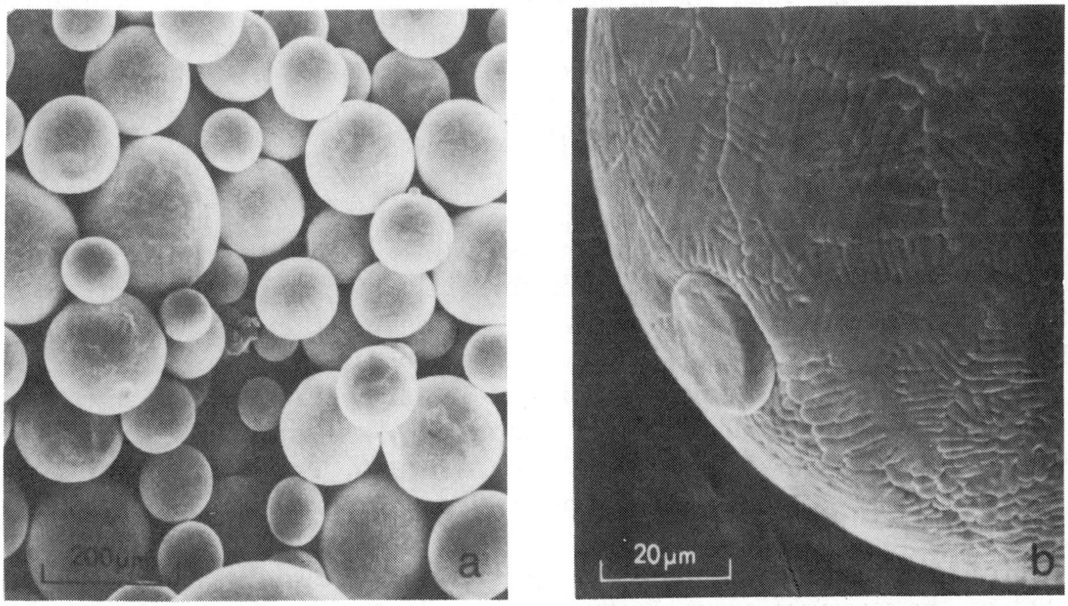

Figure 1. Scanning electron micrographs of the 250-grade maraging steel rotating electrode powder showing (a) the spherical shape and (b) the surface dendritic pattern.

Figure 2. Scanning electron micrograph of the 250-grade maraging inert-gas-atomized powder showing the generally spherical powder with smaller satellite agglomerated spheres.

The powders were vibrationally loaded in outgassed 304 stainless steel tubes, and the loaded tubes were further outgassed and then sealed. Both powders were hot isostatically pressed, the REP at 210 MPa (30 ksi) and the IGA at 105 MPa (15 ksi), and the temperature and pressure were maintained for three hours. Since full densification was achieved, the effects of different pressures is assumed to be negligible. Post-pressing heat treatments in all cases consisted of the conventional two-step solutionizing[16] at 925°C for one hour, followed by 815°C for one hour. The subsequent aging treatment consisted of the conventional 480°C for three hours or modified treatments that pre-aged the samples for 10 minutes at temperatures up to 560°C.

The mechanical properties after the conventional treatment are given in Table II. Despite reasonable strength values for the REP powder product, no usable ductibility was achieved. However, a substantial portion of the wrought-product ductility was achieved for the IGA powder products, with the 1200°C HIP condition giving a slightly better ductility. Using a 525°C pre-age for 10 minutes plus the conventional age of 480°C for 3 hours resulted in a significant increase in the ductility for the REP powder product. As shown in Table III, 47 percent reduction in area (RA) was observed for the REP powder product.

The fracture mode for conventionally aged REP powder samples, as compared with that of the wrought equivalent, is shown in Figure 3. In the powder product, failure is along prior powder-particle boundaries. Closer examination of the HIP fracture surface (Figure 4) shows that each of the planar surfaces contains small, finely spaced ductile dimples. Associated with each dimple is a second-phase precipitate particle in the 0.5 to 1.0 μm range.

TABLE II

MECHANICAL PROPERTIES OF HIP MARAGING 250 STEEL AFTER CONVENTIONAL SOLUTIONIZING AND AGING*

	UTS (MPa)	YTS (MPa)	RA (Pct)
REP**	1730	1700	0
IGA (1100)	1635	1615	42
IGA (1200)	1720	1710	52
WROUGHT **	1750	1710	59

*Solutionize at 925°C for 1 hr and at 815°C for 1 hr; aging is at 480°C for 3 hrs.

**Average of 2, others are single specimen data.

TABLE III

MECHANICAL PROPERTIES OF HIP MARAGING 250 STEEL USING MODIFIED AGING TREATMENT*

	YTS (MPa)	RA (Pct)
REP	1655	47
IGA (1100)	1700	38
IGA (1200)	1665	55
WROUGHT**	1705	62

*Solutionize at 925°C for 1 hr and at 815°C for 1 hr; preage at 525°C for 10 minutes and age at 480°C for 3 hrs.

**Average of 2, others are single specimen data.

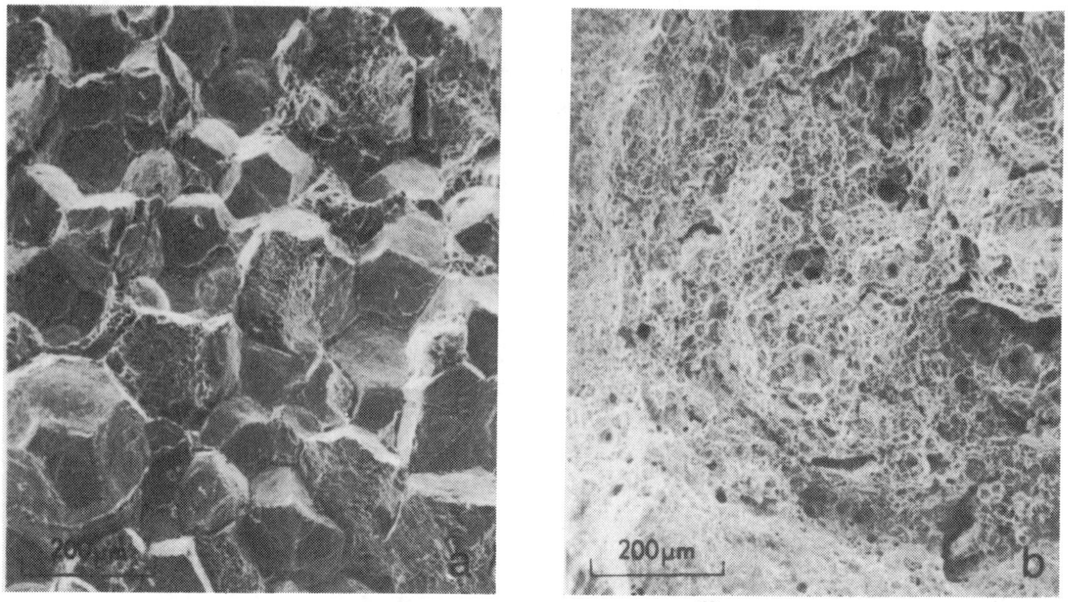

Figure 3. Fracture surfaces of (a) rotating-electrode-process powdered product after hot isostatic pressing, compared with (b) wrought product after conventional aging at 480°C for three hours.

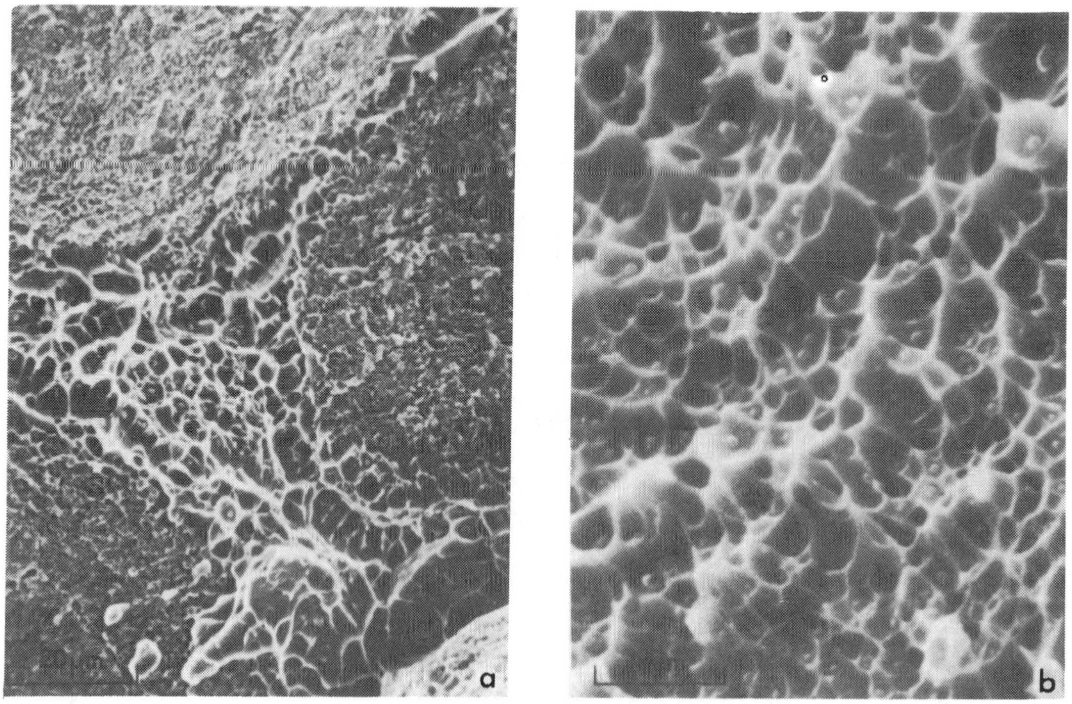

Figure 4. Closer examination of the powder-particle junctions in Figure 3a show dimpled fractures both at powder-particle junctions (a) and on the faceted regions of the fracture surface (b). Note the association of second-phase precipitates at the bottom of the larger dimples.

An optical micrograph of the prior powder-particle boundary (Figure 5) shows that the boundary is decorated with submicron-sized precipitates. The computer-enhanced electron microprobe scan for titanium across a typical prior powder boundary triple point (Figure 6) shows evidence of titanium enrichment at these boundaries. No nitrogen, carbon, or oxygen segregation was detected for the fine prior powder boundary precipitates. The large, blocky region is a typical titanium carbide inclusion that was occasionally observed. These inclusions are common in wrought material.

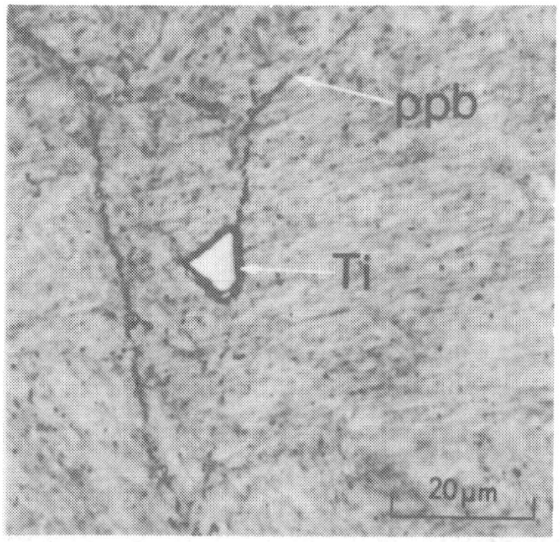

Figure 5. Typical prior powder boundary (PPB) decoration with 0.2 μm precipitates for HIP maraging steel after conventional aging. These precipitates are believed to be those associated with the dimples in Figure 4. The Ti-rich inclusion contains carbon and is typical of inclusions found in wrought material.

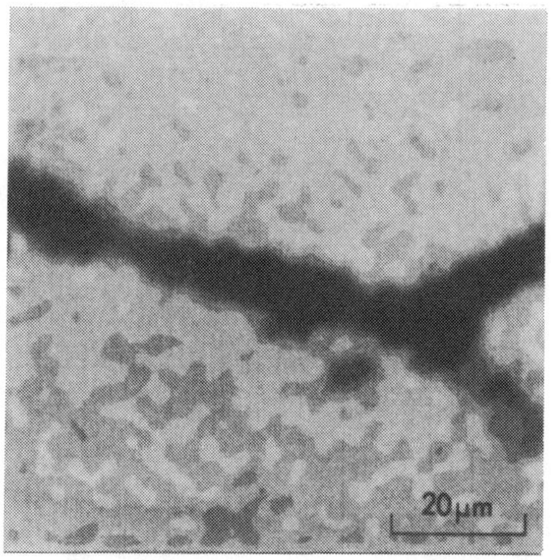

Figure 6. Computer-enhanced microprobe scan for titanium over an 80 x 80 μm area of a powder-processed sample after conventional aging. The dark areas correspond to the highest titanium concentrations.

Auger analysis of the fracture surfaces shows a correlation between the titanium concentration and the measured ductility. The data are shown in Figure 7 for the REP samples and in Figure 8 for the IGA samples. The differences in the intensity ratio scale are related to differences in absolute titanium content and to the experimental conditions. The data suggest that increased titanium concentration on the fracture surface results in decreased ductility.

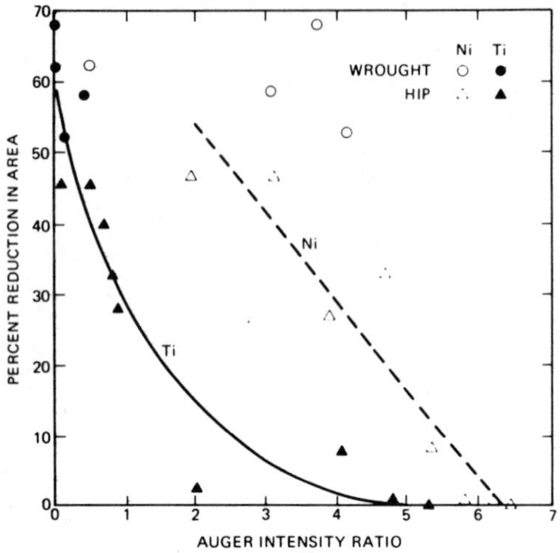

Figure 7. Correlation between the ductility and the titanium and nickel concentration on the fracture surfaces of hot isostatically pressed rotating-electrode-process powders. The relative concentration is normalized by dividing the titanium and nickel intensities by the iron intensity. The data suggest that both nickel and titanium are associated with the low-ductility fracture surface of HIP material.

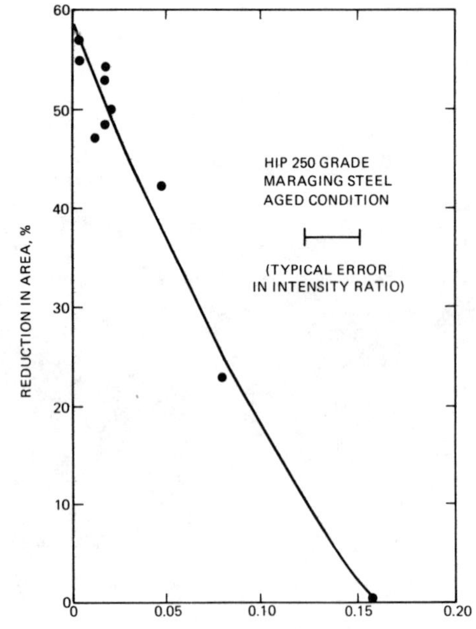

Figure 8. Correlation of ductility with titanium concentration on the fracture surfaces of hot isostatically pressed inert-gas-atomized powders. No nickel concentration variation could be detected for these samples.

Discussion

In previous attempts to powder-process maraging steels to full density, the most significant shortcoming has been a lack of ductility except when thermomechanical deformation is incorporated either during or after consolidation. This lack of ductility appears to occur because the metal powders do not form a good metallurgical bond. Evidence for this is suggested by the interparticle separation[12] fracture mode. Such fracture behavior has been attributed to a surface film on the powder; however, the nature of this film has been subject to controversy. Although it may be an oxide, the only experimental evidence reported in the literature is the high oxygen content of the HIP material as measured by chemical means.[12] Based on the electron microprobe data of Figure 6 and the Auger data of Figures 7 and 8, we have concluded that it is titanium segregation, not oxygen, at the prior powder boundaries that is responsible. Furthermore, it has been demonstrated[10-13] that a hot deformation process such as extrusion as a consolidation procedure produces powder-processed material with ductilities equal to those of wrought material. On the basis of these two facts, we conclude that the segregation of titanium at the prior powder boundaries was reduced by homogenization during extrustion and led to significant ductility in the powder-processed material.

There are three ways to reduce the segregation: (1) reduce the titanium concentration in the alloy, (2) reduce the size of the powder particles so that the surface concentration of titanium is reduced, and (3) subject the samples to thermal treatments that reduce titanium supersaturation at the powder boundary. Comparison of the data for these parameters is shown in Table IV. As the overall titanium concentration was reduced, the ductility in the conventionally aged condition was improved. This improvement can also be attributed to the reduction in the median particle size (96 μm) of the IGA powder. Finally, the modified aging treatments introducing a pre-age step at a temperature above the conventional aging temperature, resulted in a significant recovery of ductility, even when the mean particle size was large and the titanium content was high.

TABLE IV

COMPARISON OF FACTORS THOUGHT TO AFFECT THE DUCTILITY OF POWDER-PROCESSED MARAGING STEELS

Maraging Grade	Powder Process	Mean Particle Size (μm)	Ti Content (wt%)	Oxygen Content (wt%)	RA (Pct) Conventional Age	RA (Pct) Pre-Age
300	REP[12]	400-600	0.66	0.004	0	--
250	REP[14]	190	0.43	0.006	0	47
250	IGA[15]	96	0.34	0.007	52	55
250	WROUGHT[14]	--	0.43	--	62	67
250	WROUGHT[15]	--	0.34	--	57	58

A change in the extent of prior powder boundary decoration resulting from the pre-age treatment is shown in the optical micrographs in Figure 9. Conventional aging produces a nearly continuous film of precipitates at the prior powder boundaries, while pre-aging produces a different second-phase morphology. This difference is reflected in the change in fracture mode (Figure 10) to one consistent with changes in boundary morphology, from an almost continuous film of very fine precipitates to a more discrete distribution of larger precipitates as previously discussed by Nes and Thomas.[17] In the case of IGA products, the fracture mode is most similar to wrought product, showing the largest amount of transgranular, transparticle, ductile dimple fracture. This fracture mode, however, will not be evident if the HIP temperature is too low (Figure 11). Evidence for some prior powder-boundary decoration, even for the highly ductile condition, is shown in Figure 12 for a specially prepared fracture surface.[18,19] Comparing the underlying electropolished microstructure to the fracture surface details shows that the ductile dimple fracture propagates through the powder particle rather than at the prior powder boundary.

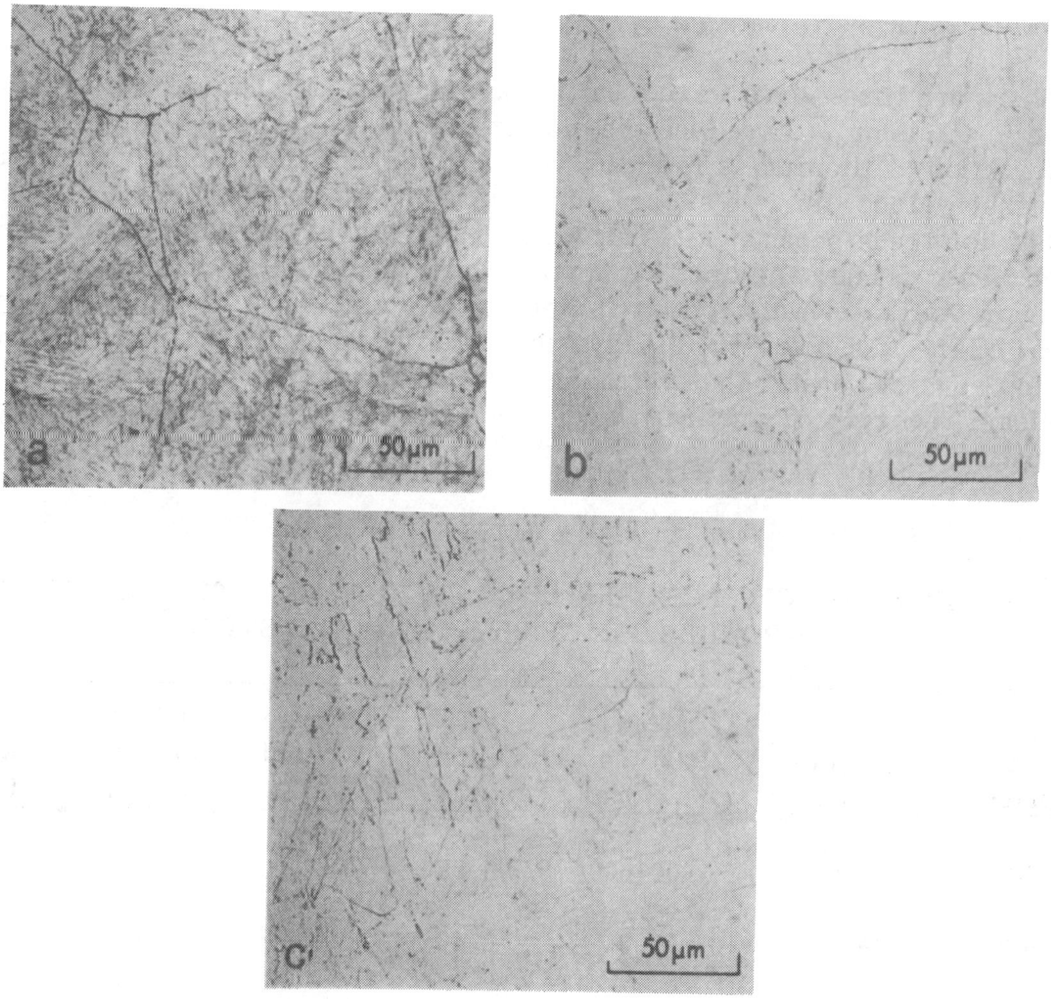

Figure 9. Effect of aging treatment on the prior powder boundary precipitates in hot isostatically pressed REP powders. (a) Conventional age (0% RA), (b) 525°C pre-age plus conventional age (47% RA), (c) 560°C pre-age plus conventional age (47% RA).

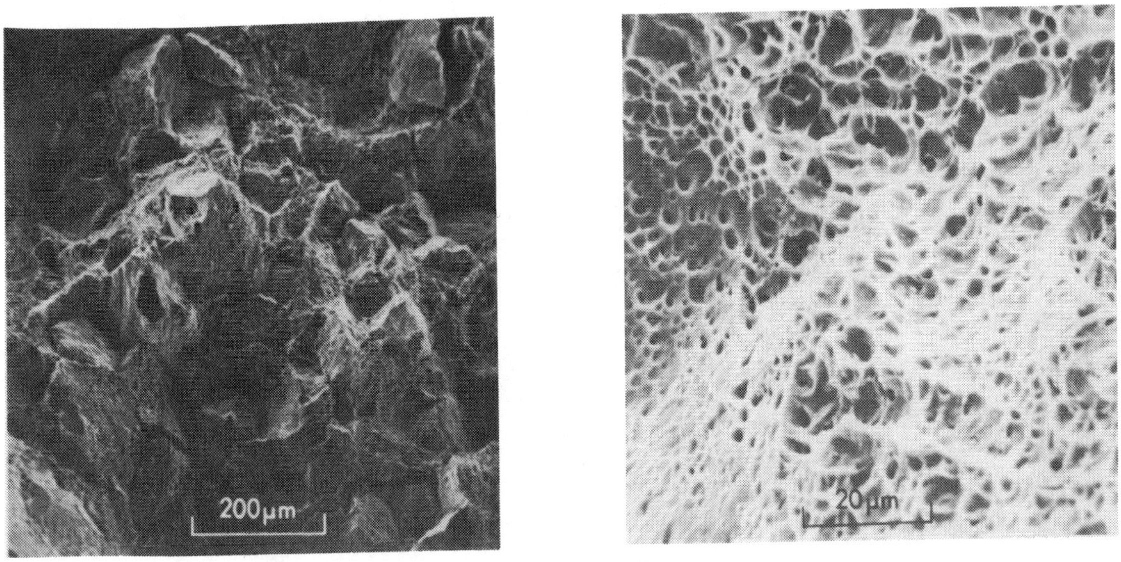

Figure 10. The 525°C pre-age (47% RA) sample shows mixed-mode fracture, now having a large portion of transparticle ductile dimple fracture behavior. Compare with Figure 3a.

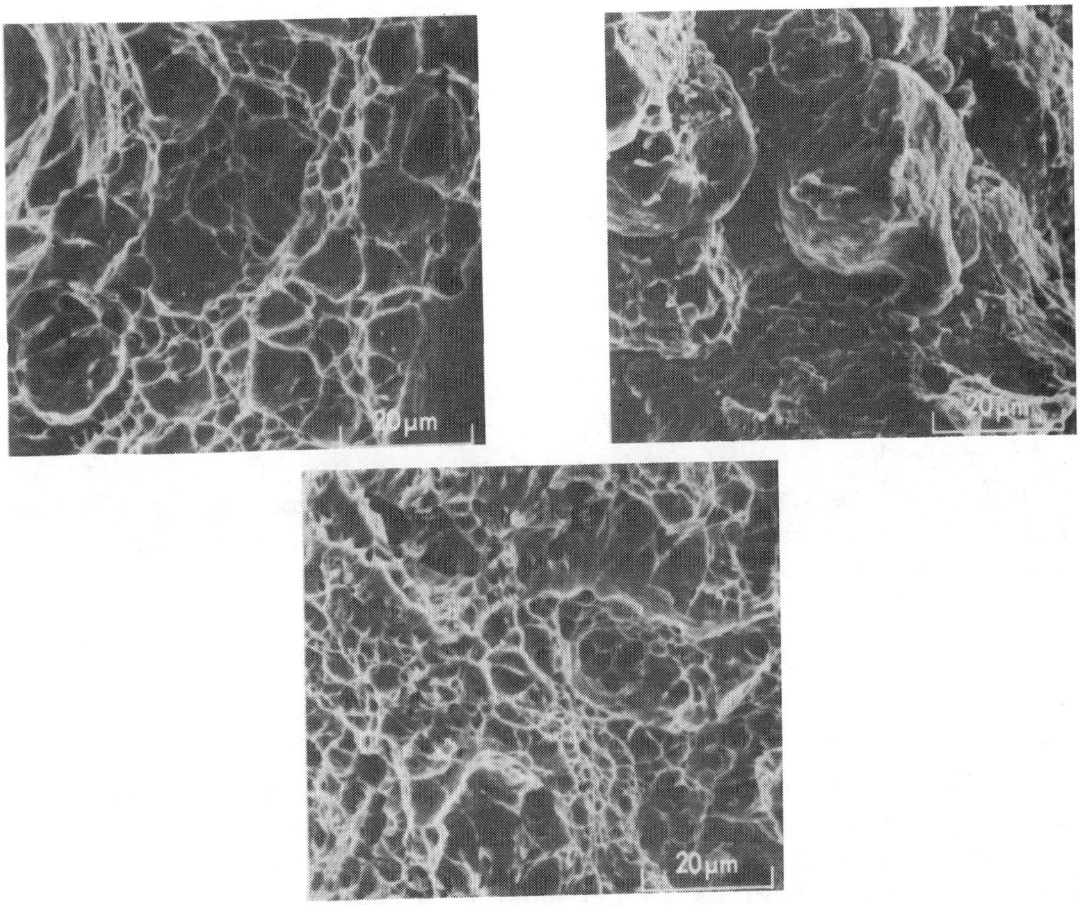

Figure 11. Fracture surfaces for conventionally aged materials (a) wrought, (b) inert-gas-atomized powder hot isostatically pressed at 1000°C, and (c) inert-gas-atomized powder hot isostatically pressed at 1250°C, revealing transgranular dimpled rupture in (a) and (c), and interparticle separation in (b).

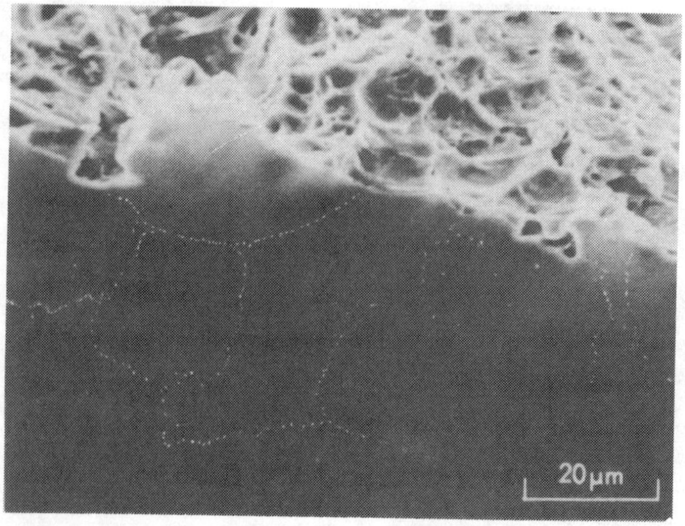

Figure 12. Electropolished fracture surface showing prior powder boundary decoration in inert-gas-atomized powder hot isostatically pressed at 1150°C. Note that the fracture path is transgranular rather than interparticle.

In our work, the reduced titanium content and the reduced powder particle size both provide lower titanium concentration at the prior powder boundaries. However, it is not yet clear whether reduced size or the use of a different powder-making process is the dominant factor in decreasing the titanium segregation. It is evident, however, that using the inert-gas-atomization process together with a smaller mean particle size can result in significant ductilities in HIP-consolidated maraging steels. In fact, nearly 91 percent of wrought-state ductility is achieved in this case. Although pre-age modifications to conventional aging procedures slightly improve the ductility of IGA-processed materials, they substantially improve the ductility of rotating-electrode-processed powder products. This result suggests that the homogenization associated with the pre-age is greater for REP than for IGA products; however, the improved surface condition could also be due to a faster solidification rate for IGA. Experiments aimed at examining these effects separately are in progress.

Summary

The advantages of powder metallurgy techniques to achieve near-net complex parts are now available for 250-grade maraging steels. The unique combination of high strength (1585 MPa) and high ductility (55 percent RA) may be combined with the potential economic advantage of hot isostatic pressing to reduce machining costs in complex parts. For the first time, significant ductility has been produced in HIP material without requiring a post-HIP plastic deformation process step. This result has been achieved in two ways: using post-HIP modified aging treatments to achieve ductilities that are approximately 75 percent those of wrought material, and using inert gas atomization to achieve a finer mean particle size to achieve ductilities approximately 91 percent those of wrought material. In both cases, the titanium segregation at prior powder boundaries has been reduced. In the first case, a thermal homogenization treatment was used, and in the second case, powder particle size and overall titanium concentration were reduced to decrease the segregation. Powder produced by inert gas atomization and mean particle sizes less than 100 μm and titanium concentrations less than 0.4 percent can be hot isostatically pressed at temperatures above 1100°C and pressures of 105 MPa to fabricate fully dense powder parts that have the strength and nearly the ductility of wrought parts. The economic advantage of producing large parts to near-net shape by hot isostatic pressing with minimal final machining can now be explored for specific geometries.

References

1. J. J. Fischer, Int'l. J. Powder Met. (1966) vol. 2, #4, pp. 37-42.

2. T. Kinoshita, Y. Tokunga, and H. Kobayashi, J. Japan Soc. Powder and Powder Met. 13 (1966) pp. 228-235.

3. T. Kinoshita, Y. Tokunga, and H. Kobayashi, ibid., pp. 236-42.

4. R. Kinoshita, Y. Tokunga, and I. Taniguchi, ibid., (1967) 14, pp. 213-220.

5. V. A. Tracey and R.S.K. Raman, Powder Met. (1969) 12, pp. 131-156.

6. J. S. Hirschhorn and D. A. Westphal, Modern Developments in Powder Metallurgy, Vol. 5, H. H. Hausner, ed. Plenum Press, N.Y., 1971, pp. 481-490.

7. V. N. Antsiferov and Y. M. Kolbenev, Sov. Powder Met. Metal Ceram., 1972, Vol. II, pp. 287-289.

8. R. M. Pelloux, Powder Metallurgy for High Performance Applications, J. J. Burke and V. Weiss, eds., Syracuse Univ. Press, Syracuse, N.Y., 1972, pp. 351-363.

9. E. Snape and F. J. Veltry, Inter. J. Powder Met. (1972) 8, pp. 193-302, or Powder Met. (1972) 15, pp. 332-345.

10. E. P. Abrahomson II, Processing and Properties of 18 Ni Maraging Steel by Powder Metallurgy, Army Materials & Mechanics Research Center, Watertown, MA, Report No. AMMRC-TR-73-4, February 1973.

11. N. J. Grant, R. M. Pelloux, M. C. Flemings, and A. S. Argon, Structure and Property Control Through Rapid Quenching of Liquid Metals, MIT, Cambridge, MA, Report AD-775-225, June 1973.

12. L. F. Van Swam, R. M. Pelloux, and N. J. Grant, Powder Met. (1974), 17, pp. 33-45.

13. E. P. Abrahamson II, Processing and Properties of 13 Ni (400) Mar-Aging Steel by Powder Metallurgy, AMMRC TR-74-14, June 1974.

14. R. M. German and J. E. Smugeresky, Met Trans A, 9A (1978) pp. 405-412.

15. R. M. German and J. E. Smugeresky, Mat Sci and Eng 36 (1978) pp. 223-230.

16. C. S. Carter, Met. Trans. 2 (1971), pp. 1621-26.

17. E. Nes and G. Thomas, Met. Trans. A, 7A (1976) pp. 967-975.

18. J. C. Chesnutt and R. A. Spurling, Met. Trans. A, 8A (1977) pp. 216-218.

19. R. M. German, J. E. Smugeresky, and C. W. Karfs, Powder Metal. Int'l., 9 (1977) pp. 178-179.

COLD ISOSTATIC PRESSING OF COMPLEX P/M SHAPES
TITANIUM AND OTHER SPECIAL METALS

Stanley Abkowitz
Dynamet Technology Inc.
Burlington, Massachusetts

ABSTRACT

This paper discusses the latest technology of cold isostatic powder metal shape manufacture. The recent progress in shape making capability, along with enhancement of mechanical properties, is discussed. Concentration of data is on titanium alloys, but the technology is demonstrated to apply to a wide range of special metals. The application of cold isostatic pressing, followed by vacuum sintering to produce complex titanium alloy shapes with properties equal to or exceeding those of castings in heavy to thin sections is demonstrated.

The significant economic and technical advantages of combining cold isostatic pressing (CIP) with hot isostatic pressing (HIP) to produce full density materials are shown. The capabilities for production of unusual powder metal shapes such as impellers, and the mechanical properties of shapes achieved when manufactured from such materials as titanium alloys, aluminum alloys, stainless steels, refractory metals, wear resistant materials such as Tribaloy containing steels, and Hastelloy C are briefly discussed.

INTRODUCTION

Cold isostatic pressing (CIP) of metal powders differs from conventional mechanical pressing in several ways that offer specific technical advantages.

(1) Since hydrostatic (water-oil) pressure is applied uniformly on all surfaces of the part to be consolidated, the density achieved is of significantly greater uniformity.

(2) With this uniform pressure the size of the part to be compacted is limited only by the dimensions of the pressure chamber.

(3) Since the mold material is elastomeric (as opposed to punch and die hard tooling) no lubricant need be added to the powder to minimize galling and die wear.

(4) With the elimination of lubricant no pre-sinter for lube removal is necessary and lubricant contamination of the reactive metal powders need not be contended with.

(5) With the use of elastomeric tooling, complex shapes beyond the capabilities of conventional pressing are feasible.

In the past, isostatic pressing was employed to compact ceramic materials at relatively low pressures and to form simple billets of difficult to melt metals such as tungsten and beryllium. More recently the technology has progressed to the manufacture of complex shapes from a variety of metals. The greatest economic advantages of isostatic pressing lie in its application to metals that are relatively expensive and which are rather difficult to machine (since the process requires less input material and reduces the extent of machining required). Certainly one of the metals that fulfills these characteristics is titanium and its alloys.

AS-SINTERED PROPERTIES AND FORGING PREFORMS

Intensive developmental efforts of recent years [1][2] have shown that titanium and its alloys could be cold compacted and vacuum sintered to achieve satisfactory density and useful mechanical properties. This was accomplished using inexpensive elemental powders in combination with master alloy powders as opposed to the more expensive pre-alloyed powders which in addition to their high cost are not readily cold compactable.

The tensile properties achieved with commercially pure titanium and TI-6Al-4V alloy are reviewed in Table I, below:

TYPICAL PROPERTIES

C.P. Titanium (.12% oxygen)			Ti-6Al -4V (.12% oxygen)		
	Sintered (94% dense)	Forged Preform (100% dense)		Sintered (94% dense)	Forged Preform (100% dense)
UTS	62,000 psi	66,000 psi	UTS	120,000 psi	133,500 psi
YS	49,000 psi	53,000 psi	YS	107,000 psi	122,000 psi
Elong.	15%	23%	Elong.	5.0%	11.5%
R.A.	23%	30%	R.A.	8.0%	24.7%

TABLE I

The as-sintered properties represent a 94% of theoretical density product produced by isostatic pressing at 60,000 psi followed by vacuum sintering at 2250°F for 3 hours. The further improvement in properties achieved by forging from a 94% dense preform is also apparent from this data. The as-sintered tensile properties compare well with those properties achieved by conventional casting and offer an economic process for like applications. Where higher density and further improvement in mechanical properties is required, forging the pre-shaped P/M preform may be employed. Other metal working techniques which have been proven to fully densify the as-sintered preform are rolling and extrusion. Typical forging preforms and finished forged parts are shown in Fig. 1 and Fig. 2

By designing the elastomeric tooling in conjunction with a mandrel, cored shapes can be produced with configurations pressed and sintered close to size. The large titanium valve ball of Fig. 3 and the titanium alloy liner of Fig. 4 were produced by isostatic pressing around a mandrel which was removed from the "green compact" prior to vacuum sintering. The rotor cannister shown in Fig. 5 demonstrates the producibility of very thin walls (.035"). These parts are subsequently finish machined on the outside to final dimensions. In the above applications the components possess the desirable mechanical properties and corrosion resistance in the as-sintered condition. It is interesting to point out that in addition to the considerable material and machining time saved by the preformed shape, further savings are offered in the finish machining operations which must still be performed. This is due to the free machining aspects of the P/M material. This ease of machinability is indicated in Fig. 6. At 100 hole "tool life" the cutting speed for annealed Ti-6Al-4V improves from 25 feet per minute for the wrought product to 60 feet per minute for the P/M material[3].

SUBSEQUENT HOT PRESSING - "CHIP"

It has been pointed out that for certain applications where full density is required, the hot working operations of forging, extrusion, or rolling can be performed on P/M preforms. More recently Dynamet has demonstrated that cold isostatic pressing and sintering can be followed by hot isostatic pressing (HIP) to produce fully dense shapes. The microstructures of Figs. 7A and 7B demonstrate the closing of porosity for the Ti-6Al-4V transformed beta structure achieved by "HIPing" the as sintered P/M shape.

The chemical analysis and mechanical properties achieved after HIP are shown in Table II below and compare favorably with conventional 100% dense wrought product.

TABLE II - MECHANICAL PROPERTIES AND CHEMICAL ANALYSIS SINTERED AND HIP PREFORM [CHIP] OF Ti-6Al-4V

TENSILE TEST AT ROOM TEMP:	
Tensile P.S.I.	132,797
Yield .2% offs.	120,024
Elongation in 4D	13.0%
Reduction Area	26.0%

QUANTITATIVE ANALYSIS:	
Iron	.085
Vanadium	4.07
Aluminum	5.97
Carbon	.035
Hydrogen-Core	.0018
Nitrogen	.0109
Oxygen	.1978

This combination of CHIP (CIP plus HIP) is a highly economical technique for achieving the maximum density, particularly for titanium alloys. The prior cold isostatic pressing and sintering permits use of the elemental master alloy powder as opposed to the more expensive pre-alloyed powder and in addition, only the elastomeric tooling need be employed in the consolidation since the hot pressing step can be accomplished on the preform sintered shape without the need for any expensive, non-reusable hard tooling. In other words, the sintered preform can be HIPed as inexpensively as many commercial titanium castings are currently densified.

COMPLEX SHAPES

The technology of both CIP and CHIP processing permits significant advantages in the manufacture of complex shapes such as impellers. Figs. 8 and 9 show the detail that can be achieved in the complex configurations of radial and axial impellers. Cost savings in the manufacture of such shapes can be as high as 70%, even where subsequent machining might be required. Fig. 10 indicates the potential (with recent developments in tooling) for production of multi-holed structures and perforations in both thin and thick sections. The titanium alloy disc shown has over 1000 holes and is 3/16 inch thick. Larger sections with square, triangular, or any shaped hole could readily have been incorporated into the tooling.

OTHER MATERIALS

The technology above has been applied to materials other than titanium[4,5]. Stainless steels, iron base alloys, aluminum alloys, copper, nickel, Hastelloy C and the refractory metals among others have all shown technical promise and economic advantage with these processes. Increasing acceptance and significant utilization of this technology is anticipated in the immediate future.

REFERENCES

(1) S. Abkowitz, J. M. Sierjiej, and R. D. Reagan "Titanium P/M Preforms, Parts and Composites". Proceedings of 1970 International Powder Metallurgy Conference, Vol. 4, Pages 501-511.

(2) S. Abkowitz "Cost Savings In the Application of P/M Titanium and P/M Aluminum Alloys". 1974 National Powder Metallurgy Conference Proceedings, Vol. 30, Pages 85-101.

(3) Metcut Report to Dynamet Technology "Drilling Pressed and Sintered Titanium 6Al-4V". Metcut Report No. 1023-12894-1, February 1969.

(4) "Developments in Titanium and Aluminum P/M Technology", Precision Metal Magazine, September 1974, Pages 44-48.

(5) "Cold Pressing Opens the Door to Complex Shapes", Iron Age Magazine, May 9, 1977, Pages 39-41.

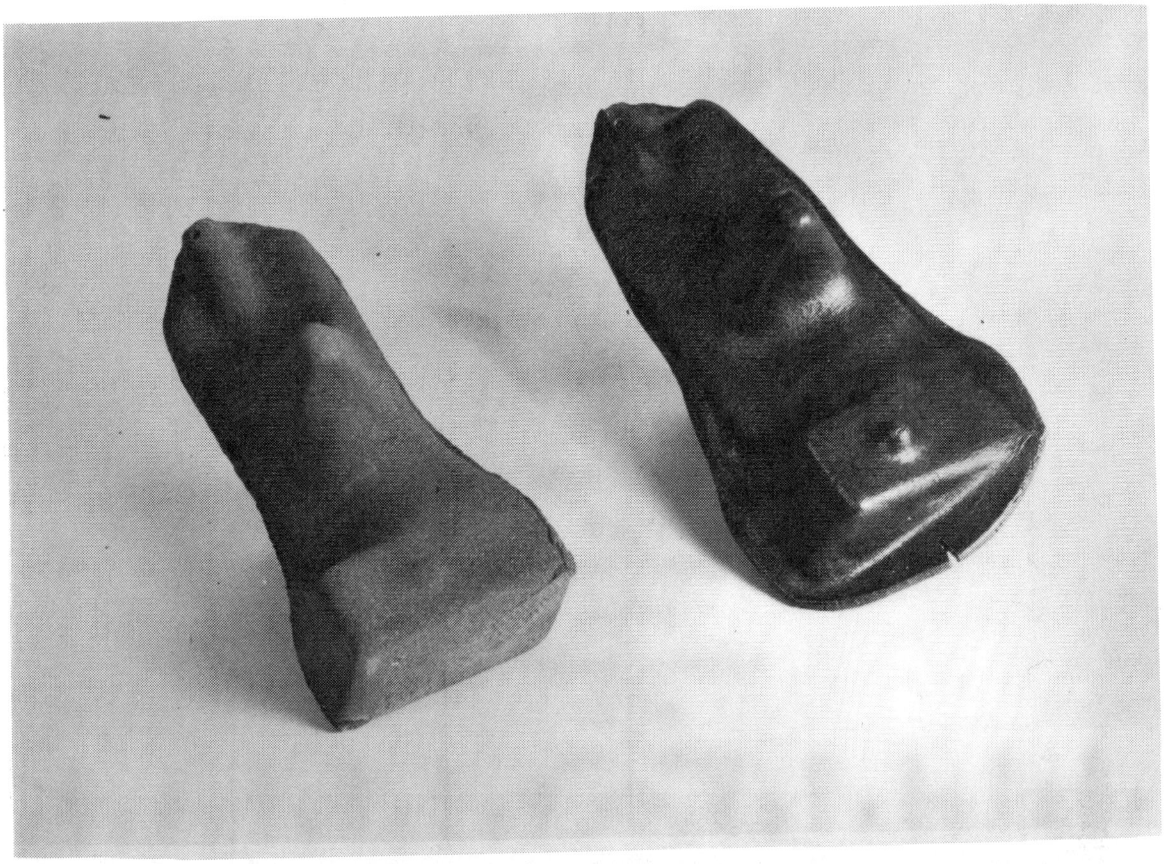

FIG. 1 - TITANIUM ALLOY FORGING PREFORM AND ROUGH FORGED PART

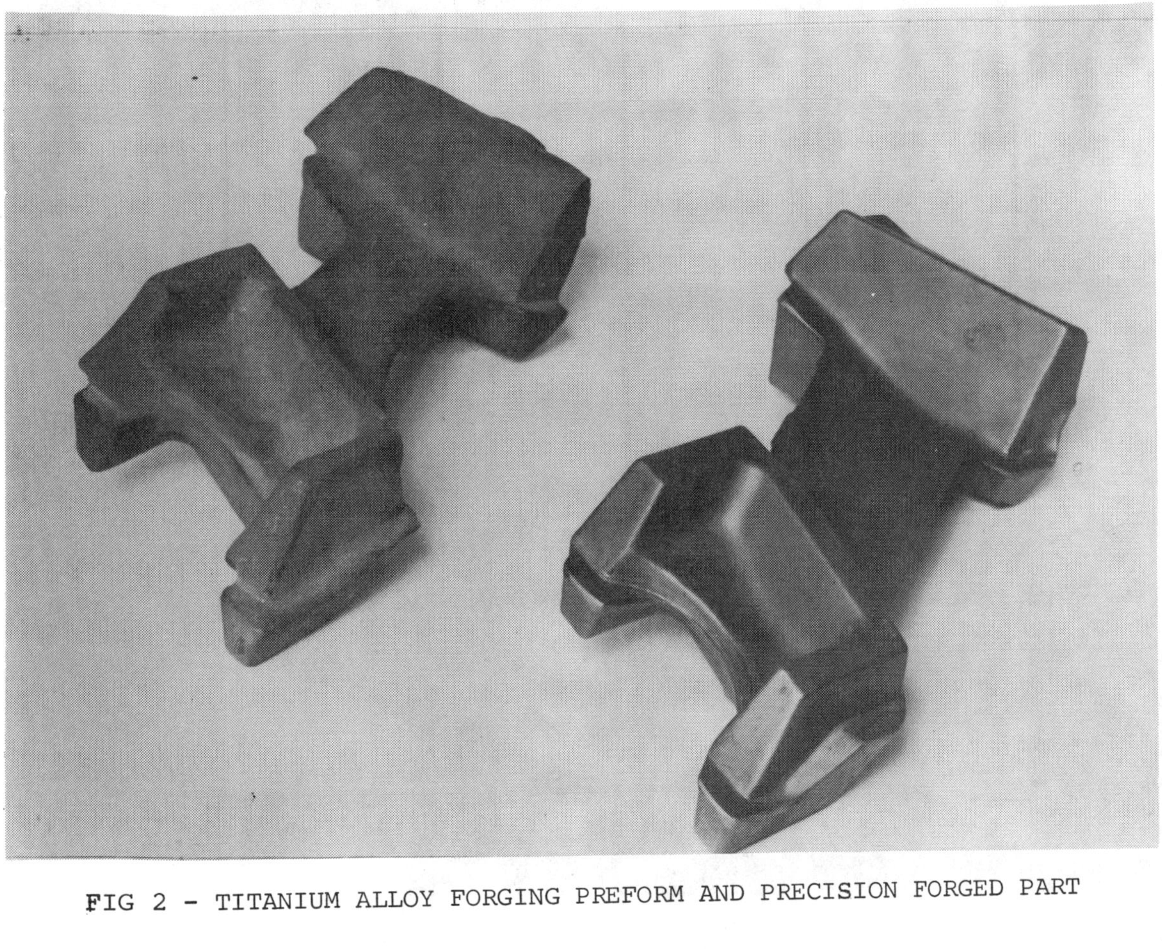

FIG 2 - TITANIUM ALLOY FORGING PREFORM AND PRECISION FORGED PART

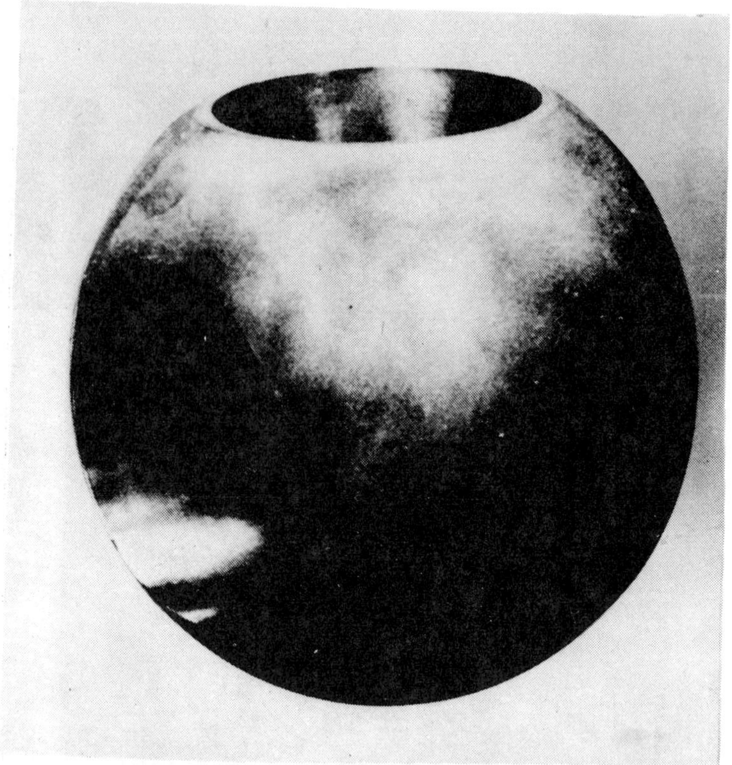

FIG. 3

TITANIUM VALVE BALL (9" Dia.) AS PRESSED AND SINTERED

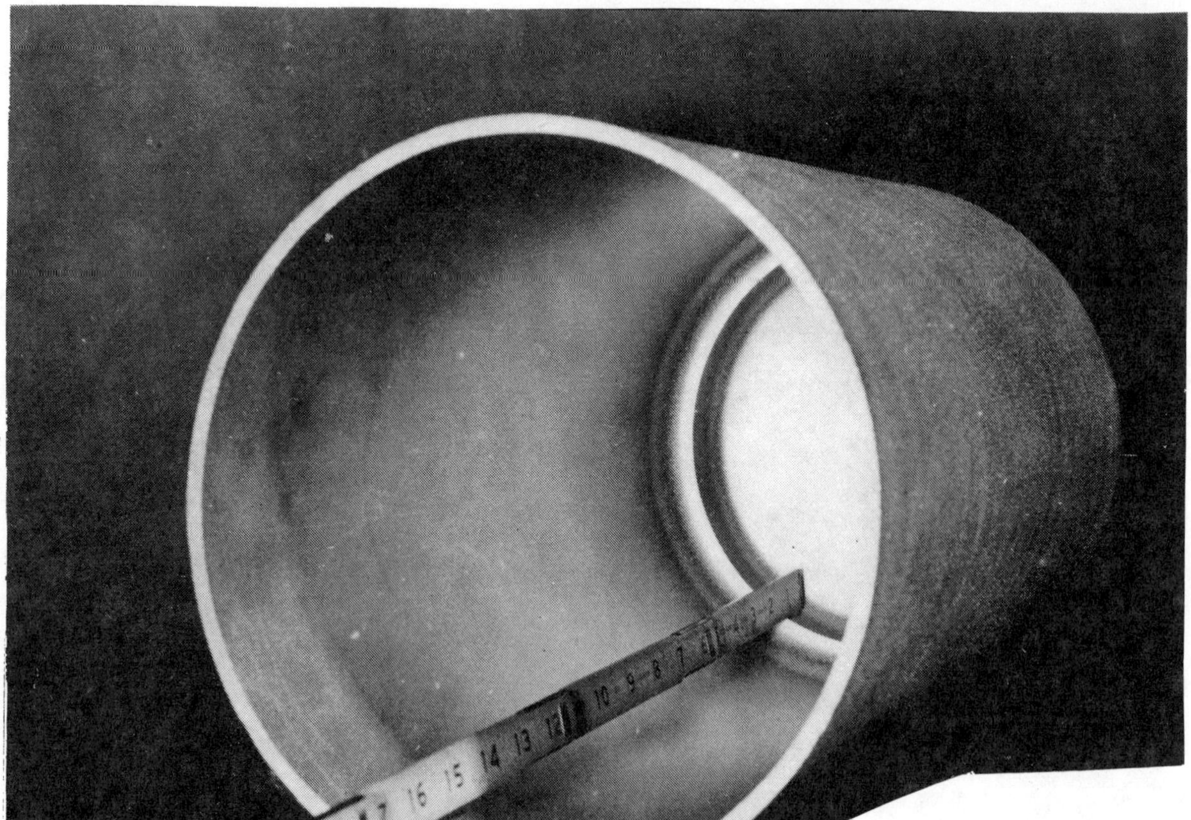

FIG. 4

TITANIUM LINER (8" Dia.) AS PRESSED AND SINTERED

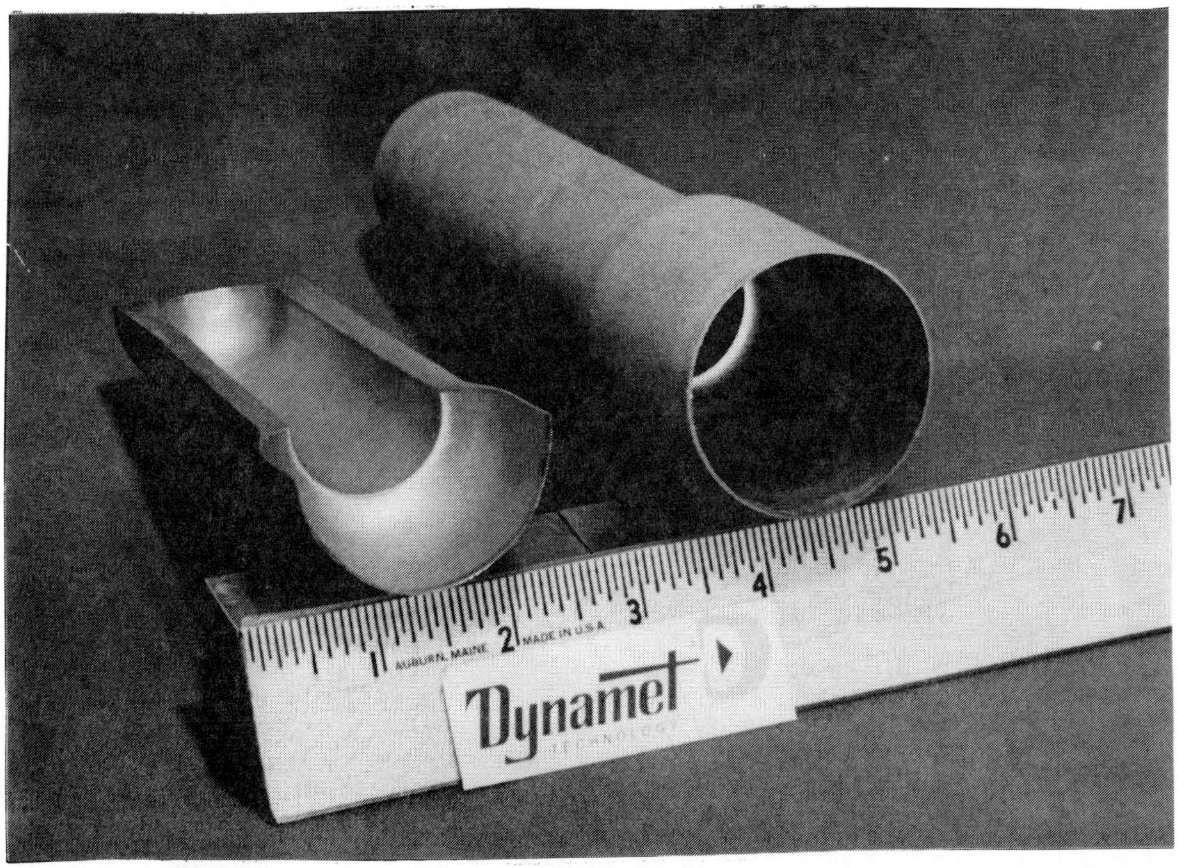

FIG. 5

Ti-6Al-4V ROTOR CANNISTER AS PRESSED AND SINTERED

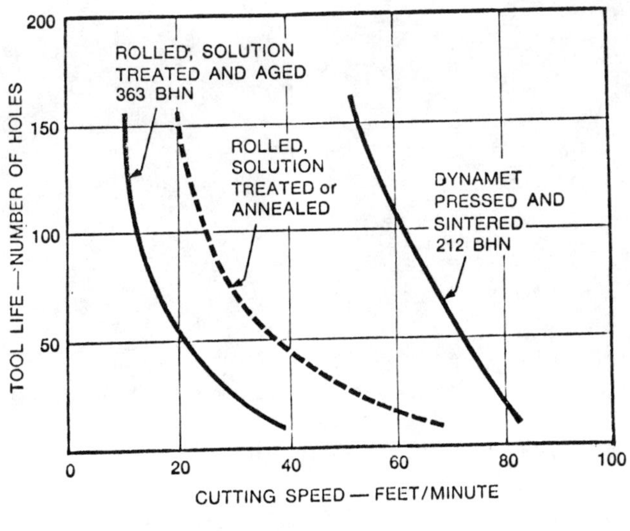

FIG. 6

MACHINABILITY STUDY ON P/M
TITANIUM 6Al-4V ALLOY

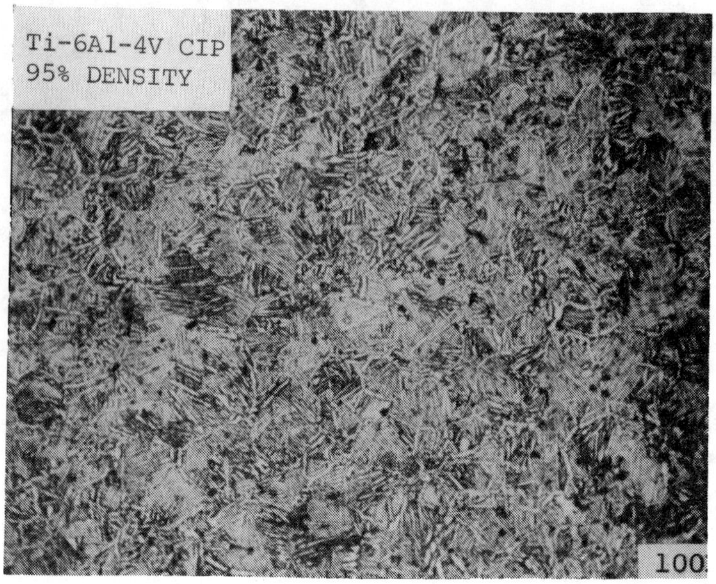

FIG. 7A

MICROSTRUCTURE Ti-6Al-4V AS PRESSED AND SINTERED

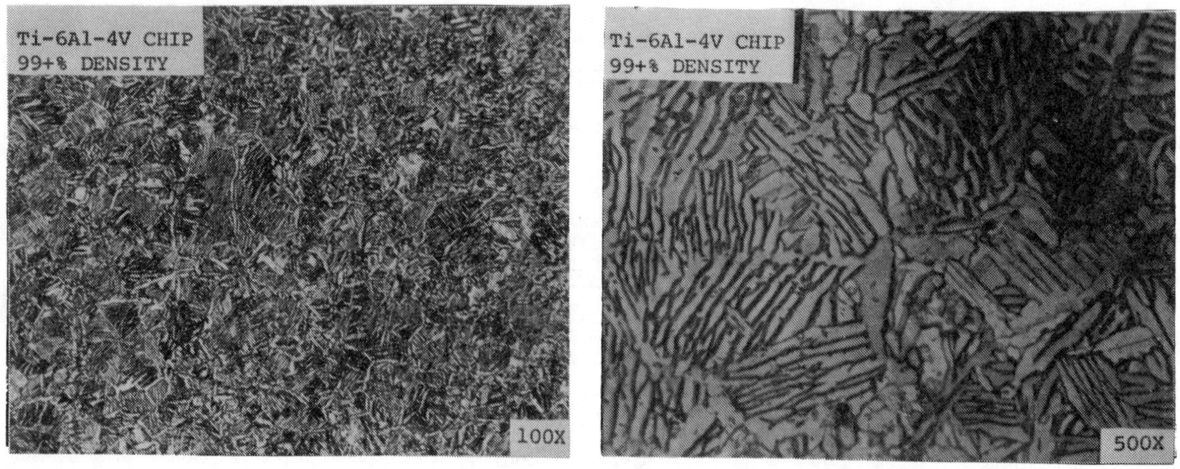

FIG. 7B

MICROSTRUCTURE Ti-6Al-4V COLD ISOSTATIC PRESSED AND SINTERED FOLLOWED BY HIP

FIG. 8

Ti-6Al-4V RADIAL ROTOR

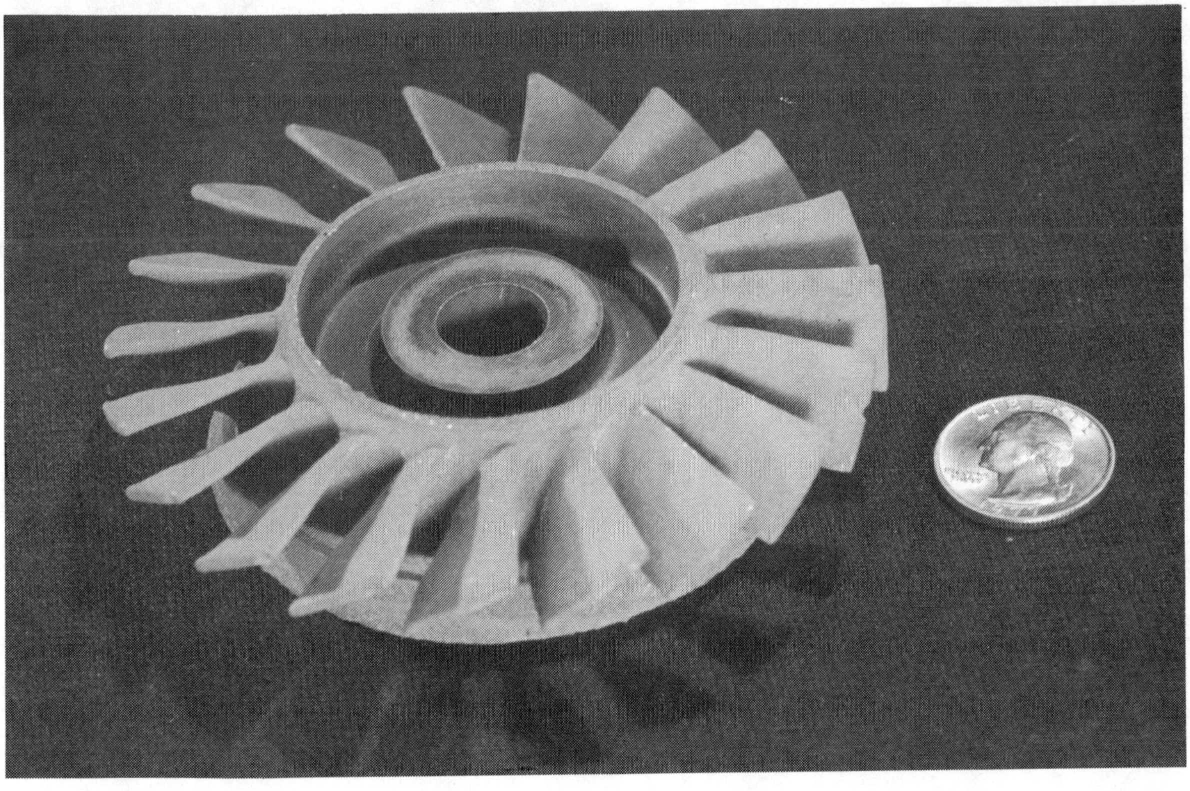

FIG. 9

Ti-6Al-4V AXIAL ROTOR

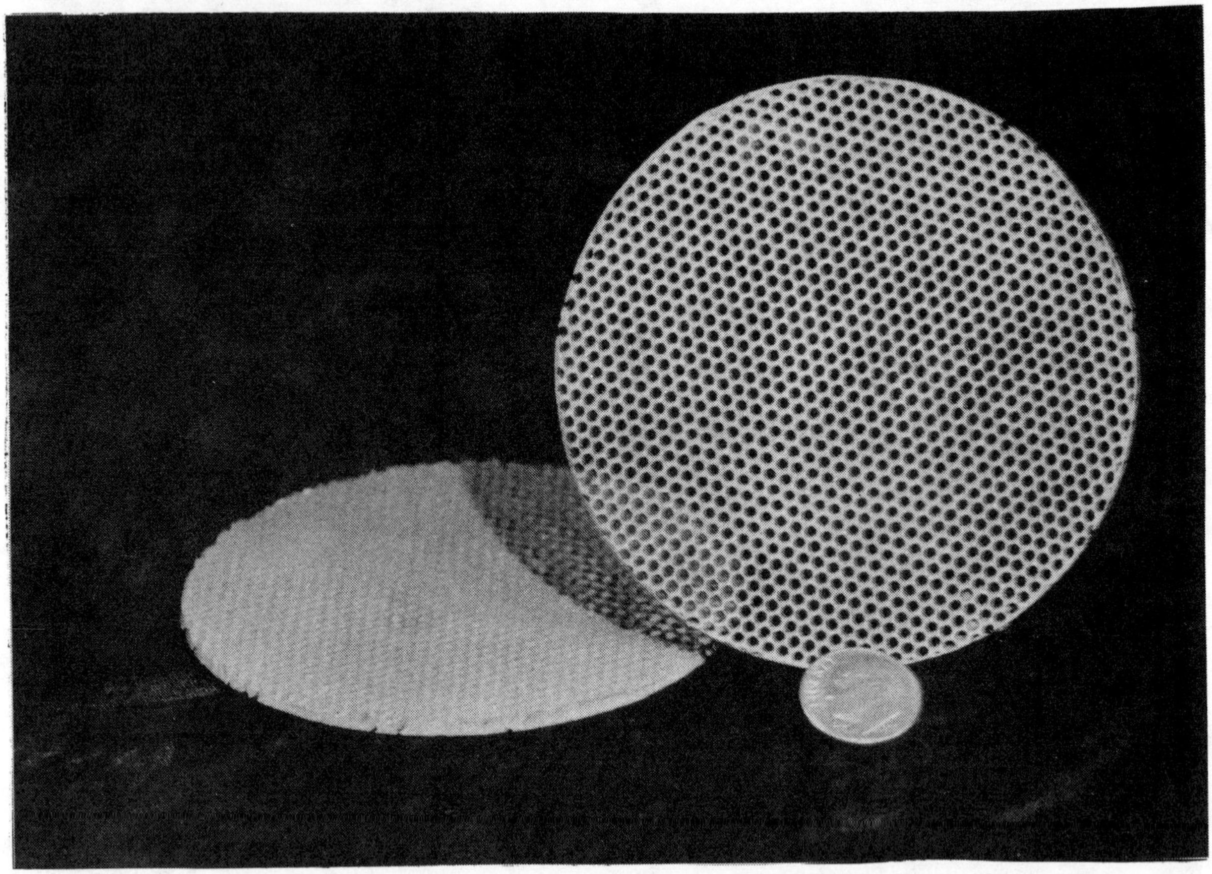

FIG. 10

Ti-6Al-4V PERFORATED DISC 3-1/2" DIAMETER
BY 3/16" THICK WITH OVER 1000 HOLES

BIOGRAPHY

Mr. Stanley Abkowitz is President of Dynamet Technology Inc. of Burlington, Massachusetts. He received his degree in Chemical Engineering from Massachusetts Institute of Technology in 1948. His career has covered pioneering work in the field of metallurgy where he has 14 patents covering several commercial titanium and nickel base alloys and metallurgical processes as a result of his work at Reactive Metals Inc., Special Metals Inc., and Nuclear Metals Inc. He is co-author of the book "Titanium in Industry".

Mr. Abkowitz's recent efforts in powder metallurgy at Dynamet have resulted in the enhancement of isostatic pressing technology to a new level of technical versatility.

Mr. Abkowitz is a member of ASM, APMI (American Powder Metal Insistute) and the Metallurgical Society of AIME, where he was past Chairman of the Titanium Committee. He is also a director of the Boston Chapter of the ASM

FORMING P/M PARTS BY CONSOLIDATION AT ATMOSPHERIC PRESSURE

Vince N. Di Giambattista

Coast Metals Inc., Little Ferry, NJ

In July, 1973, at the "Fourth International Powder Metallurgy Conference" in Toronto, Canada, a paper entitled "Pressureless Process For Manufacture Of P/M Preforms", was presented. That paper presented for the first time a newly patented P/M Process for "Consolidating @ Atmosphere Pressure", metal powders such as, superalloys, titanium, beryllium and tool steels to a theoretical density of 99 plus percent.

Since that time, a considerable amount of development effort relative to my U.S. and Canadian patents has been put into the Process. The net result being, that the process is currently being used to produce large diameter disk rings up to 24" in diameter, which are being isothermally pressed by one of the largest forging companies in the U.S. Another product being manufactured by consolidating at atmospheric pressure is low carbon Astroloy sheet.

Literally, thousands of parts have been manufactured from various alloy metal powders, such as Rene 95, IN 100, Astroloy, X-40, IN 718, Waspalloy, Hi-Ni-Cr-Co Alloys, including a variety of tool steels and high speed steels.

The parts can be essentially any shape, size, and weight, including internal coring, and consistently exhibiting a theoretical density of 99% and greater.

In order to achieve 100% of theoretical density and develop maximum physical and mechanical properties, parts produced by "Consolidating @ Atmospheric Pressure" have been successfully hot rolled, forged, extruded, isothermally pressed, and "hot isostatically pressed". In all instances, the parts conform to aircraft engine manufacturers specification requirements both at room temperature and elevated temperature, including low cycle fatigue specifications.

The Process is predicated on the removal of all surface oxides, and absorbed gas molecules which may penetrate a given particle, 3 to 4 angstroms in depth, with subsequent activation of the particle surfaces (by heat) in a vacuum, and in the presence of a boron electron donor, that an ionic and/or interatomic exchange occurs between the powder particles that are in intimate contact. It is this reaction which metallurgically bonds the metal powder particles together.

This reaction is depicted by the following formulas.

The hydrated form of boric oxide, more commonly known as "boric acid", has a structure in the Ortho Form, which places the boron atom in the center of an equilateral triangle with Oxygen atoms at each apex. As the temperature is increased @ atmospheric pressure a water of hydration is lost, and the compound goes to the meta form (HBO_2)

$$H_3BO_3 \xrightarrow{100^oC} H_2O(g) + HBO_2$$

The meta form undergoes a transition to the tetraboric acid (H_2B_4) with the loss of a water of hydration.

$$4HBO_2 \xrightarrow{169^oC} H_2B_4O_7 + H_2O(g)$$

The tetra form goes to boric oxide at elevated temperatures with the loss of the last water of hydration.

$$H_2B_4O_7 \xrightarrow{300^oC} 2B_2O_3 + H_2O(g)$$

Boric acid is solvated in methanol. As the methanol is evaporated a state of saturation is reached in the liquid, and the boric acid will precipitate.

B_2O_3 combines @ elevated temperatures with metal oxides to form meta borates,

$$\text{E.G., Mo } (O_2) + 2B_2O_3 \longrightarrow \text{MO } (BO_2)_4$$

Producing a P/M Part by "CONSOLIDATION AT ATMOSPHERIC PRESSURE"

STEP ONE (1):

A predetermined amount of boric acid is solvated in methanol.

STEP TWO (2):

A predetermined quantity of -100M superalloy powder such as Rene 95, or IN-100 is poured into a blender, to which is added a predetermined amount of boric acid - methanol.

As the methanol is evaporated a state of saturation is reached in the liquid, and the boric acid will precipitate, leaving a very thin film on each powder particle.

STEP THREE (3):

The blended and coated powder is now poured into a borosilicate mold to a predetermined tap density, the mold having the shape of the desired article of the P/M part.

STEP FOUR (4):

The tubulated borosilicate mold, containing the powder is connected to a suitable vacuum system, and heated to approximately 800°F in order to remove entrapped gases. The mold is then sealed in vacua.

STEP FIVE (5):

The sealed mold is then placed into a suitable crucible, and this assembly is then placed into a "Standard Heat Treating Furnace", and heat treated @ a predetermined temperature and cycle below the solidus, depending on the alloy system.

STEP SIX (6):

The assembly is removed from the heat treating furnace at some intermediate temperature and air cooled.

At about 600°F, the borosilicate mold spalls off and the result is a part which is approximately 99% of theoretical density and of course is an identical replica of the shape of the borosilicate container, and about 35% to 40% smaller due to densification.

The resultant part is then sandblasted and is ready to receive additional work such as Hipping or isothermal pressing to the finished article @ 100% density.

"CONSOLIDATION @ ATMOSPHERIC PRESSURE"

TYPICAL PHYSICAL & MECHANICAL PROPERTIES OF RENE-95 PARTS @ R.T. & E.T.

ROOM TEMPERATURE PROPERTIES:

SPECIMEN NO:	UTS	YS	E%	RA%
#1 D	249.4	202.1	12.1	14.3
#2 D	256.3	205.7	11.7	13.8
#3 D	251.7	203.2	12.4	14.7
SPECIFICATION:	230.0	180.0	10.0	12.0

ELEVATED TEMPERATURE PROPERTIES @ $1200^{o}F$

SPECIMEN NO:	UTS	YS	E%	RA%
#1 D	225.4	186.4	10.2	11.7
#2 D	241.1	192.7	9.8	11.9
#3 D	236.5	193.2	11.1	12.1
SPECIFICATION:	207.0	167.0	8.0	10.0

CREEP RUPTURE TESTS @ $1200^{o}F$, STRESS @ 150,000.

SPECIMEN NO:	HRS. DURATION	RA%	E%
#1 D	131.6	8.3	3.8
#2 D	145.2	8.3	3.9
#3 D	144.7	8.1	3.6

"Consolidation @ Atmospheric Pressure", represents a new approach for mass producing near net shape Jet Engine Turbine components, by making fabrication simpler, and effective cost reduction a reality.

The Current Process being used to form metal powders into high strength parts, is Hot Isostatic Pressing, or HIP. Many companies have expressed an interest in HIP, by making a substantial investment in purchasing these presses, such as Kawecki-Berylco Industries, Inc., Kelsey-Hayes Co., Wyman-Gordon Co., Colt Industries, Special Metals Corp., Carpenter Technology, and Pratt & Whitney Aircraft Div. of United Technologies Corp.

In HIP, a metal container which is a facsimile of the shape of the article to be made is filled with loose powder, evacuated of entrapped air, sealed, and consolidated under heat and pressure to 100% of the materials theoretical density.

Simple as it may sound, powder consolidation by HIP is a slow and tedious process. For example, most HIP units can manage only one cycle per day, and the number of parts produced per cycle varies from 1 to 4. Depending on the size, temperature range, and pressure desired a HIP press costs from about $100,000 for a small unit to several millions of dollars for large units. The process also has certain vagaries, for instance, it is not uncommon for a leak to develop in the metal can while pressurizing, in which case the can is no longer usable, and the powder contents is generally oxidized. There have also been catastrophic failures where the powder filled metal can suddenly ruptures, severely damaging the internal area, rendering the HIP press useless for months while effecting repairs. In addition, the can containing the compacted powder article, must be either machined off, or pickled in acid.

The new generation jet engines are going to powder metallurgy, as evidenced by Pratt & Whitney specifying the process for the manufacture of turbine disks for the JT8D Engine and General Electric's decision to use (6) rotating parts made by powder, for the T-700, F-404 and the CFM-56.

The preceding statements are not to be construed as demeaning to HIP technology. The crux of the problem, is with the current powder processes used in conjunction with the HIP press, and it is this that presents problems not only for the HIP press, but for the engineers as well, who want to use a HIP press to its maximum capabilities.

"Consolidation @ Atmospheric Pressure", is therefore a process whereby a number of near net shape superalloy or other alloy P/M parts exhibiting 99% of their respective theoretical density can be isothermally pressed or Hipped. In addition it provides effective cost reduction and increased productivity, by maximum utilization of the HIP press.

For example, twenty (20) parts are produced in one heat treat cycle into near net shape parts. The parts, which can be removed from the furnace at almost any temperature, are colled and sand blasted. The parts which are 99% of theoretical density, are then put into the HIP press, utilizing the full capacity of the unit, heat, pressurize, and remove parts that are now 100% of theoretical density. This procedure precludes leaks and ruptures of metal cans as well as machining or pickling off the can, and reduces maintainence and the amount of argon gas used to pressurize.

"Consolidation @ Atmospheric Pressure" has developed National and International interest regarding consulting and licensing arrangements.

Cyclops Corporation, Pittsburgh, Pa., has acquired a non exclusive license for the patented process.

For licensing information, contact Ryan & Bowser, 425 Porter Bldg., Pittsburgh, Pennsylvania 15219, (412)-281-6580.

Section B

Progress in Powder Metallurgy
Volume 35

Proceedings of the 1979 National Powder Metallurgy Conference — Cincinnati, Ohio

Edited by: Walter Cebulak and
Erhard Klar

PREPARATION OF BINARY AND TERNARY ALLOYS
FROM MIXED OXIDE CERAMIC PREFORMS

M. Robbins, M. L. Green, R. C. Sherwood,
and V. G. Lambrecht, Jr.

Bell Laboratories
Murray Hill, N.J. 07974

Binary and ternary alloys of high density have been prepared by the direct hydrogen reduction of chemically mixed oxide ceramic preforms. This process incorporated a number of advantages with respect to standard powder metallurgical practices. Since all of the preparative steps involve stable metal salts the complete procedure, with the exception of the final reduction reaction, can be carried out in air. Also, chemical mixing of the metal salts leads to the production of alloys with a high degree of homogeneity. The alloy systems which have been investigated here are Fe-Ni and Cu-Ni-Co. It was found that alloy densification behavior is a function of the total internal surface area of the oxide preforms. Variables effecting the internal surface area, which include particle size and morphology, oxide green density and oxide sintering, have been investigated. It was found that any process which increased the total internal surface area of the oxide preform increased the densification rate of the reduced alloy. The effect of variation in sintering techniques on alloy densification behavior is discussed.

Introduction

In a previous paper[1] it had been reported that metal and alloy forms with relatively high densities could be prepared by the direct H_2 reduction of mixed oxide ceramics. As an alternative powder metallurgy technique, the direct reduction of ceramic oxides offers certain advantages in materials handling and possibly in processing. These include:

a) all preparative steps with the exception of the final reduction can be carried out in air;

b) composite oxides are mixed on a chemical scale to improve homogeneity;

c) the necessity to handle and store metal or alloy powders which can undergo surface oxidation is eliminated.

During the reduction of the mixed oxide ceramic to the alloy, considerable shrinkage occurs. This makes the direct reduction method of limited use where metal parts with tight dimensional specifications are utilized. However, for preform production, where the metal or alloy undergoes a secondary shaping process (e.g., rolling or wire drawing) or where there are relaxed dimensional requirements, the direct reduction method holds promise.

In this paper experiments are described which are designed to study the effects of oxide specific surface area, method of oxide preparation and oxide on the densification of the reduced alloy.

Experimental

The alloys studied in this investigation were Cu50-Ni21-Co29 wt % and Fe-50 w/o Ni. The general preparative method is shown in Fig. 1. When using the spray dry technique the solutions were usually dissolved sulfates or ammonium citrates. For the case of the ball mill (ceramic) process, metal carbonates and Fe_2O_3 were used. The sulfates used were reagent grade $CuSO_4 \cdot 5H_2O$, $NiSO_4 \cdot 6H_2O$, $CoSO_4 \cdot 7H_2O$ and $FeSO_4 \cdot 7H_2O$. Ammonium citrate solutions were obtained by reacting elemental Fe powder in ammonium citrate followed by the addition of $CuCO_3$, $NiCO_3$ or $CoCO_3$ which react to evolve CO_2 and form the ammonium citrate in solution. Both the spray dry mixed salts and the mixed carbonates were reacted in air at 1000°C for ∿4 hours. At 1000°C the spray dried salts decomposed to mixed oxides (not necessarily single phase) and the mixed carbonates reacted to form mixed oxides. In this work halowax dissolved in CCl_4 was used as a binder. This was blended with the mixed oxide powder to evaporate the CCl_4, at which time oxide and binder mixture was sieved using a 40 mesh screen.

Cu-Ni-Co

Powders of mixed Cu, Ni and Co oxides plus binders were pressed in the form of bars. The binder was burned off by heating the bars at the rate of 100°C/hr to 600°C. The bars were then sintered in air at various temperatures for four hours in order to increase the oxide grain size (decrease the specific surface area) which is accompanied by an increase in the ceramic oxide density. Figure 2 shows both the increase in oxide density and decrease in surface area as a function of temperature. These mixed oxide bars were then reduced in H_2 at 1150°C for four hours followed by quenching. The quench was achieved by pushing the boat containing the bars into the cool part of the furnace while maintaining the H_2 flow. X-ray diffraction and scanning electron microscopy indicated the presence of only one phase. Alloy density was determined by dimensional measurement and where possible by immersion, the two measurements yielding similar results (within 5%). Densities of alloy bars produced from ceramic oxides sintered at various temperatures are also shown in Figure 2. As shown in Figure 2 the density of the alloy decreases with decreasing oxide specific surface area except for slight maximum at 700-800°C. A similar effect was observed in Cu-Ni-Fe prepared in a manner similar to the Cu-Ni-Co.

Fe-50 w/o Ni

Experiments intended to yield information on the process of densification were undertaken using the alloy Fe-50 w/o Ni. A solution of Ferrous and Nickel sulfates was spray dried, decomposed to the mixed oxide and combined with binder.

All experiments were carried out using this stock of powders. Oxide pellets were heated to 600°C to burn off the binder and two of each pressing pressure were reduced and sintered in hydrogen for 16 hours at the desired temperature. Alloy density as a function of temperature for each of the oxide pressing pressures is shown in Figure 3. It can be readily observed from Figure 3 that the rate of densification decreases with increasing pressing pressure in the temperature region of 700-800°C where most of the densification occurs. Photomicrographs (Figure 4) of polished and etched samples sintered at 800°C show that the rate of grain growth, which is greatest in the sample pressed at 20,000 psi, decreases with increasing oxide pressing pressure as well. Figure 4 also shows that the grain size in the samples heated at 800°C are fairly uniform. This is to be compared to samples heated at 750°C, the structure of which are not at all uniform (Figure 5). The 750°C samples exhibit a hard outside shell (Figure 5a), a relatively dense core (Figure 5c) and a low density, almost powdery region (Figure 5b) between them. Additional oxide samples were reduced in H_2 at 600°C for 15-20 minutes and sealed in evacuated quartz tubes and heated at 700°C for 16 hours. As shown in Figure 5d the samples annealed in sealed quartz possess a uniform structure.

In each experiment a separate sample was used for X-ray analysis. X-ray diffraction analysis always showed that the alloy was formed with no residual oxide. Chemical analysis verified the composition of the alloy.

Discussion

The results of the experiments on the Cu-Ni-Co (Fig. 1) alloy show that the rate of densification varies directly with specific surface area, as calculated from knowledge of the presintered oxide grain size. This relationship between densification and surface area is a function of the surface energy and specific surface area of the newly formed alloy. Smaller oxide grains upon reduction, yield smaller alloy grains. Samples with these smaller metallic grains, having greater total surface energy, will exhibit an enhanced rate of densification.

As can be deduced from Figure 5, the densities shown for the Fe-50 w/o Ni alloys in Figure 3 are average densities as the structure of pellets, reduced and annealed at temperatures <800°C, are far from uniform. We hypothesize that the hard alloy skin shown in Figure 5a is caused by the heat of surface reduction followed by reduction of the central portion in which the heat of reaction increases the local interior temperature. The increased heat at the interior causes increased densification and shrinkage away from the hard surface shell resulting in a "moat effect" between surface and center. When the mixed oxide pellet was reduced at 600°C for 15 minutes the surface "skin" did not have an opportunity to form and subsequent heating at 700°C in vacuum resulted in more uniform densification. At 800°C reduction is so rapid that the reaction is carried out uniformly throughout and the resulting alloys are homogeneous.

Conclusion

It has been shown that the densification rate of alloys formed by direct reduction of ceramic mixed oxides is directly related to the specific surface area of the oxide preform. The detailed mechanism of reduction and densification in the critical temperature region (∿650-800 for Fe-50 w/o Ni) is only partially understood and is the subject of continued research.

Reference

1. M. Robbins, A. Staudinger and S. Shabaka, submitted to J. Mats. Sci.

Figure Captions

Fig. 1 Flow diagram for preparing mixed oxide ceramic parts for reduction experiments.

Fig. 2 Density of Cu--Ni--Co bars as a function of oxide grain boundary specific surface area.

Fig. 3 Density of Fe-50 w/o Ni alloy pellets as a function of reduction temperature of oxides pressed at 20, 40, 60 and 80 KSI.

Fig. 4 Photomicrographs of Fe-50 w/o Ni pellets reduced and annealed at 800°C for 16 hours from mixed oxide pellets pressed at 20, 40 and 60 KSI.

Fig. 5 Photomicrographs of Fe-50 w/o Ni samples reduced in flowing hydrogen at 750°C for 16 hours (a) rim; (b) porous "moat"; (c) center; and (d) reduced in flowing hydrogen at 600°C for 15 min, then sintered in vacuum at 700°C for 16 hours (center).

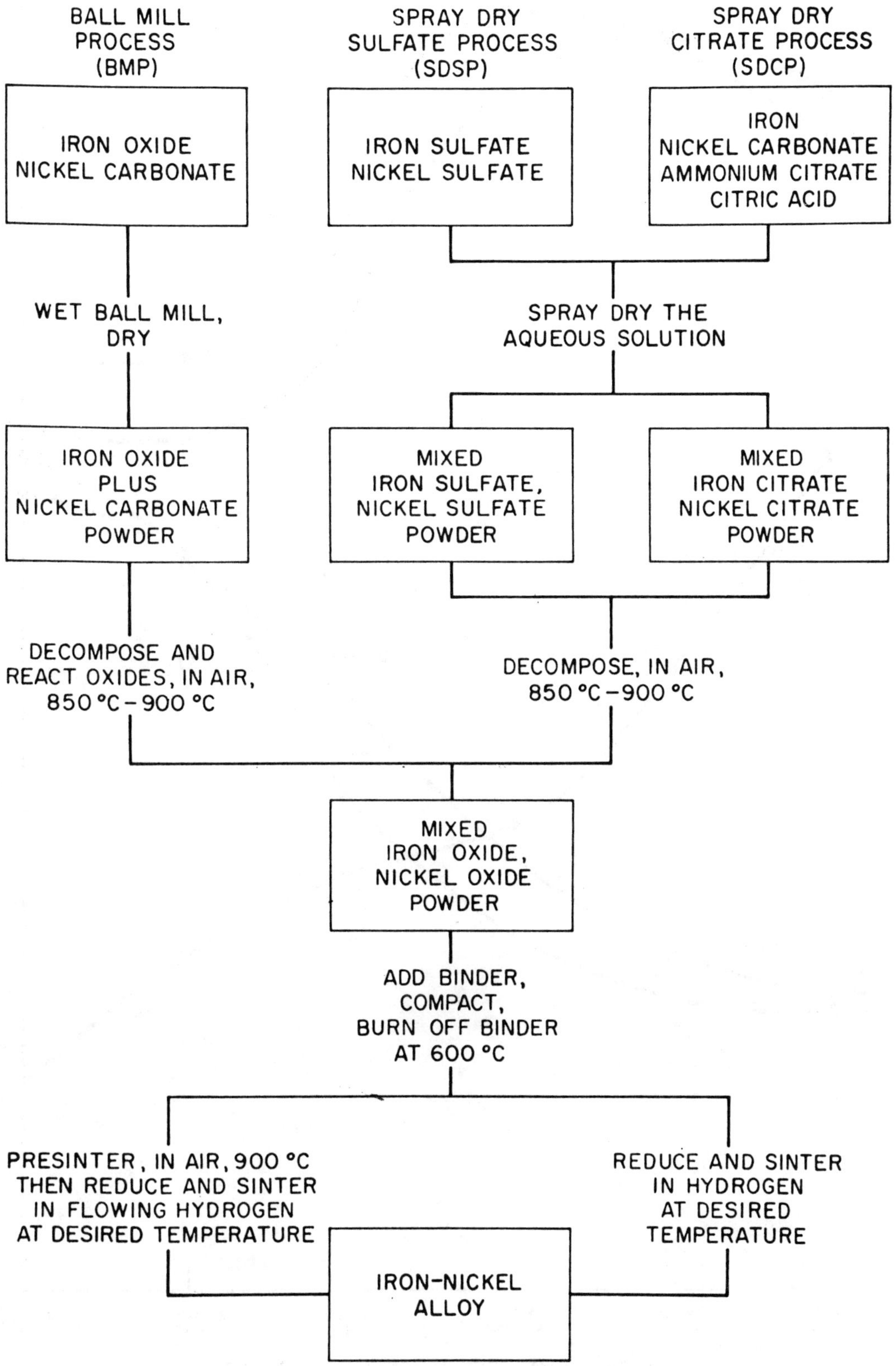
BALL MILL
PROCESS
(BMP)
SPRAY DRY
SULFATE PROCESS
(SDSP)
SPRAY DRY
CITRATE PROCESS
(SDCP)
IRON OXIDE
NICKEL CARBONATE
IRON SULFATE
NICKEL SULFATE
IRON
NICKEL CARBONATE
AMMONIUM CITRATE
CITRIC ACID
WET BALL MILL,
DRY
SPRAY DRY THE
AQUEOUS SOLUTION
IRON OXIDE
PLUS
NICKEL CARBONATE
POWDER
MIXED
IRON SULFATE,
NICKEL SULFATE
POWDER
MIXED
IRON CITRATE
NICKEL CITRATE
POWDER
DECOMPOSE AND
REACT OXIDES, IN AIR,
850 °C - 900 °C
DECOMPOSE, IN AIR,
850 °C - 900 °C
MIXED
IRON OXIDE,
NICKEL OXIDE
POWDER
ADD BINDER,
COMPACT,
BURN OFF BINDER
AT 600 °C
PRESINTER, IN AIR, 900 °C
THEN REDUCE AND SINTER
IN FLOWING HYDROGEN
AT DESIRED TEMPERATURE
REDUCE AND SINTER
IN HYDROGEN
AT DESIRED
TEMPERATURE
IRON-NICKEL
ALLOY

Fig. 1

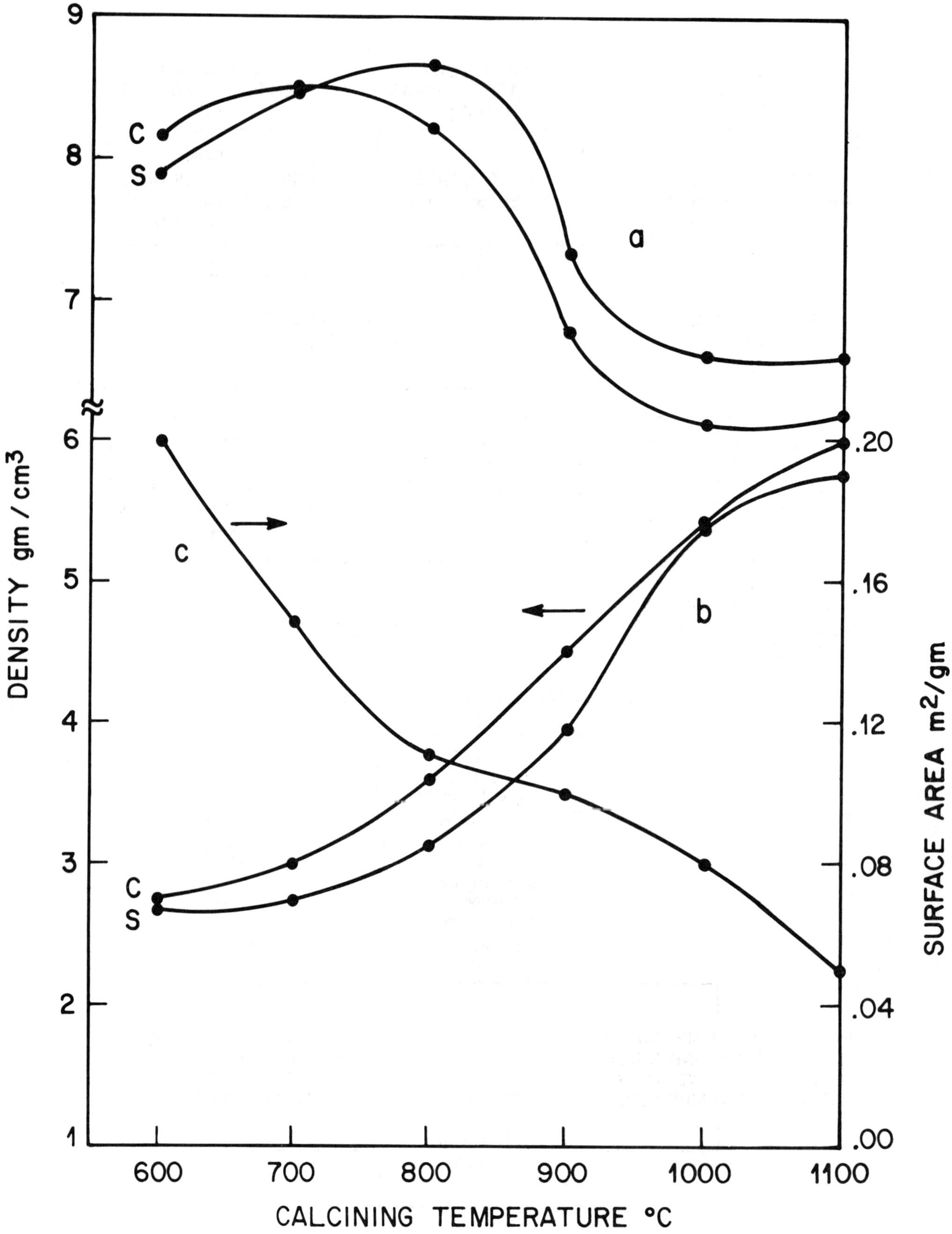
DENSITY gm / cm3
SURFACE AREA m2/gm
CALCINING TEMPERATURE °C
a
b
c
C
S
C
S
9
8
7
6
5
4
3
2
1
.20
.16
.12
.08
.04
.00
600
700
800
900
1000
1100

Fig. 2

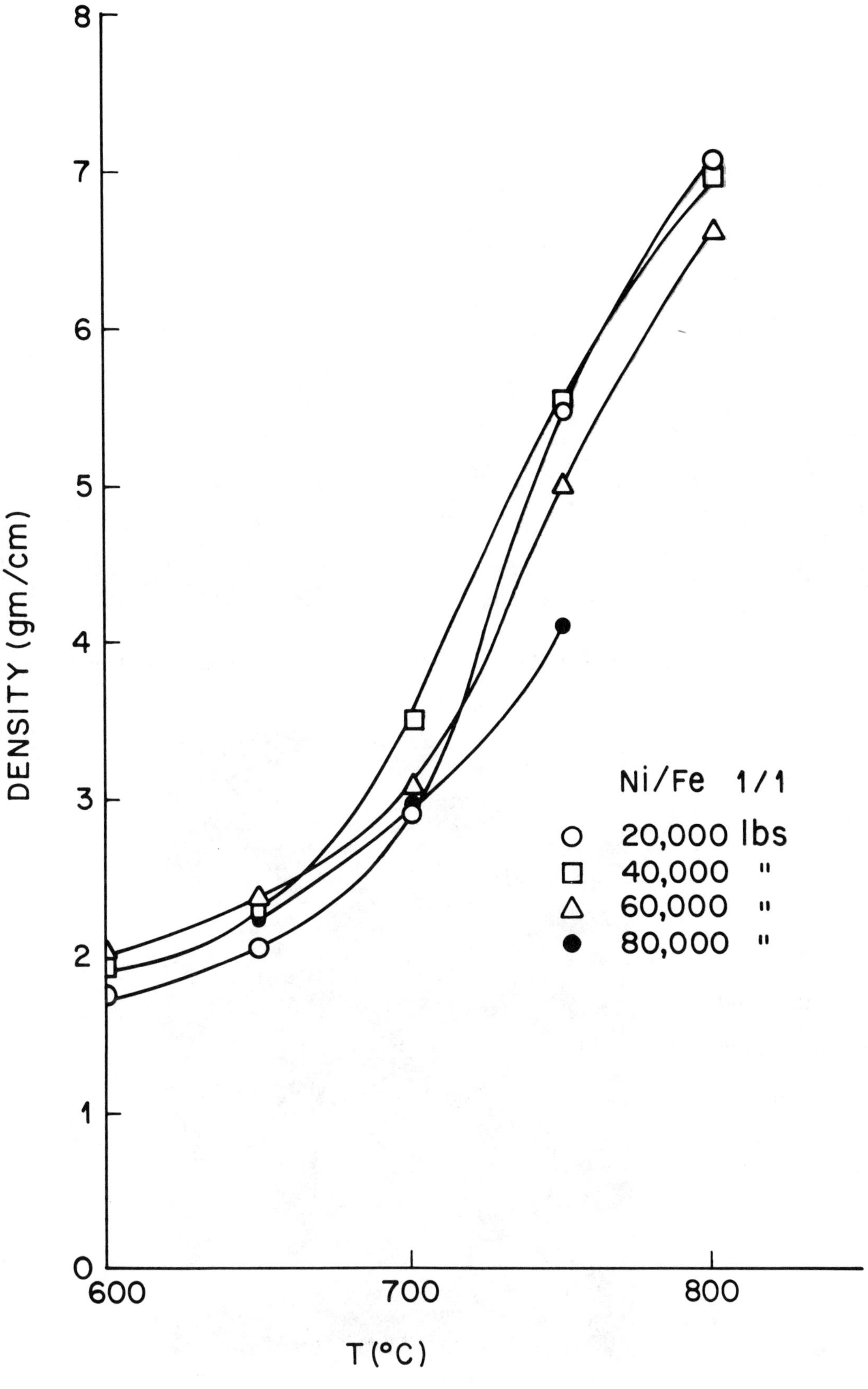

DENSITY (gm/cm)
0
1
2
3
4
5
6
7
8
600
700
800
T(°C)
Ni/Fe 1/1
20,000 lbs
40,000 "
60,000 "
80,000 "

Fig. 3

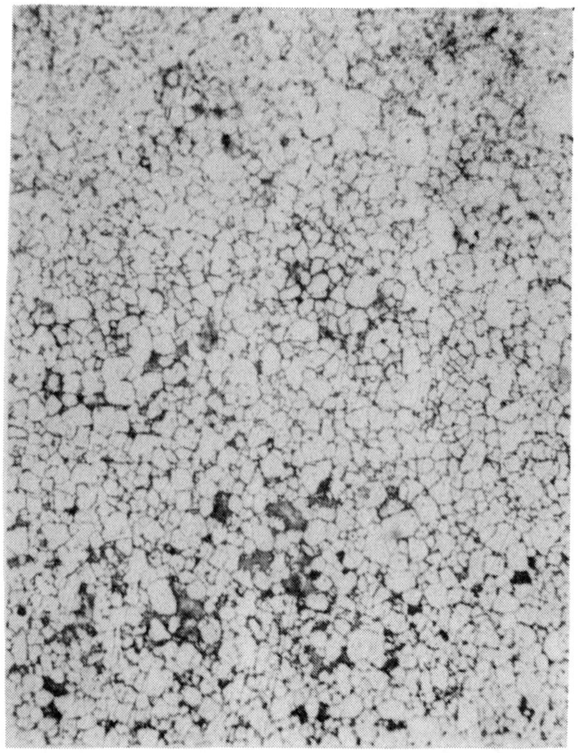

60K lbs ├──┤ 15μ

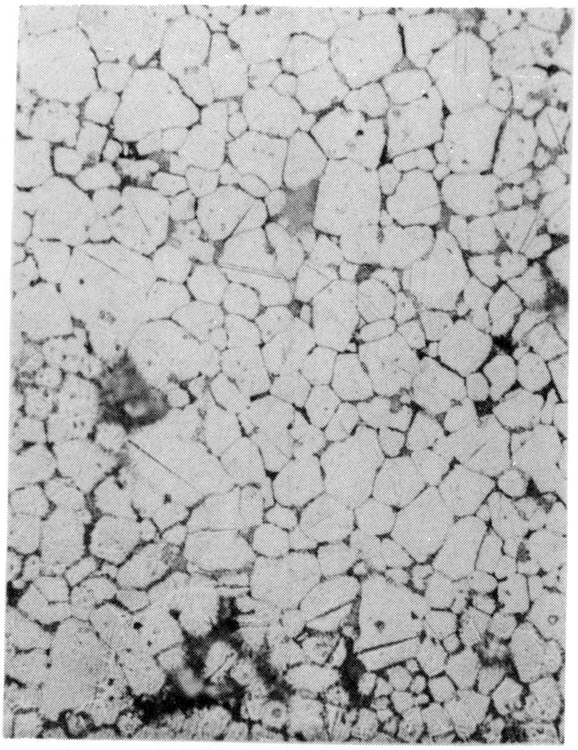

40K lbs ├──┤ 15μ

Reduced

H_2, 800°C

16 hrs

20K lbs ├──┤ 15μ

Fig. 4

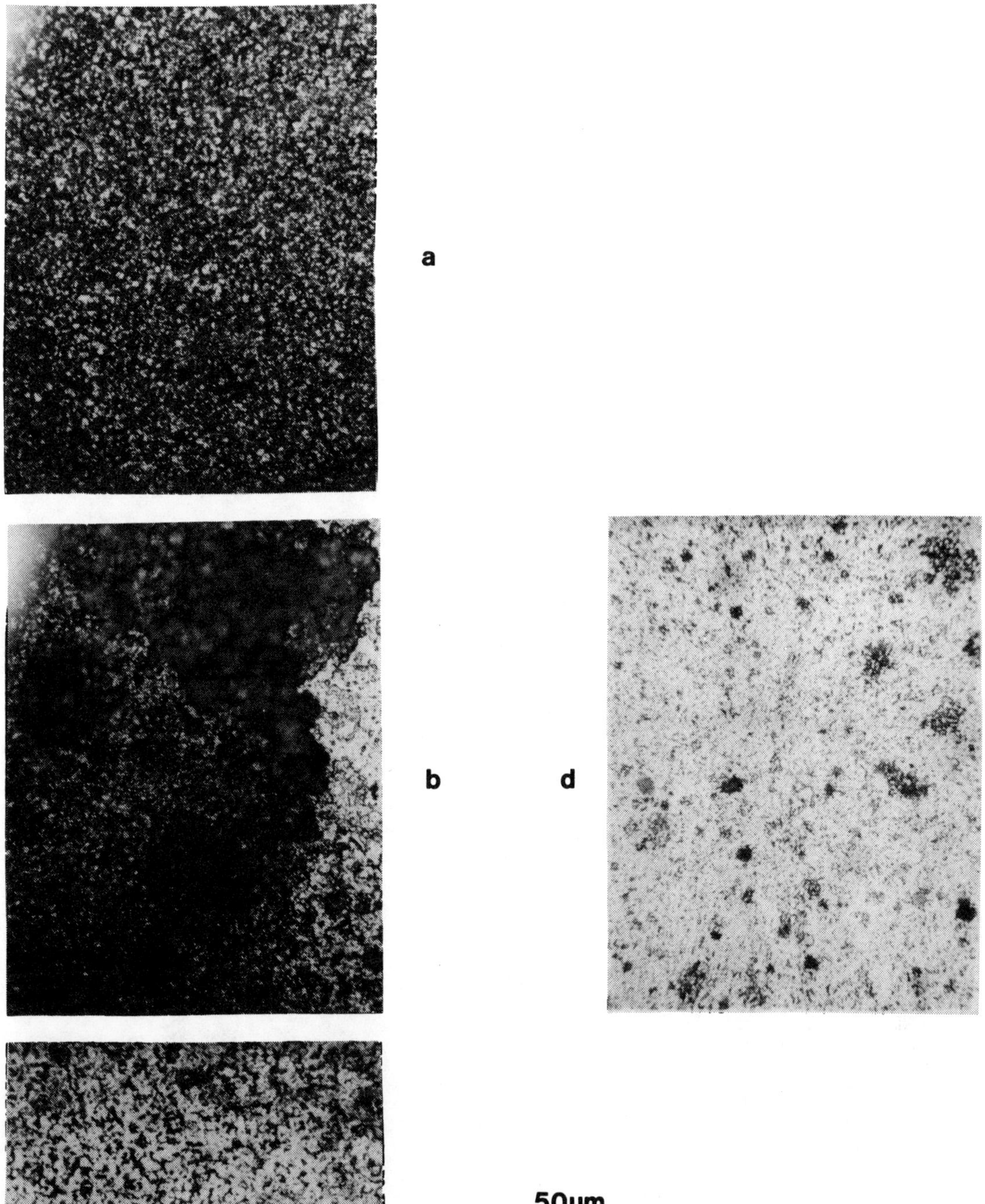

Fig. 5

THE EFFECTS OF PARTICLE SIZE AND DENSITY ON THE MECHANICAL PROPERTIES OF SINTERED IRON AND IRON/PHOSPHORUS ALLOYS

Michael J. Koczak* and Pramod Aggarwal**
Department of Materials Engineering
Drexel University
Philadelphia, PA 19103

Abstract

The effects of particle size, density and phosphorus (Fe_3P) additions on the mechanical properties of sintered iron has been investigated at two sintering temperatures e.g. 2050°F (1121°C) and 2350°F (1288°C). Evaluated mechanical properties included: yield and ultimate tensile strength ductility and hardness. Phosphorus additions, finer particle size, high densities and high temperature sintering promote strength increases. Ductility is improved primarily by finer particle fractions and higher densities. Phosphorus additions are found to increase sintering shrinkage and inhibit grain growth while providing strength increases. A transition from an interparticle to an intergranular fracture morphology is also noted with the addition of phosphorus at higher density levels, e.g. 7.2 - 7.4 g/cc. In addition, the sintering process is examined in terms of sintering temperature and diffusion coefficient of phosphorus in α and γ iron.

Background

Particle size, shape and morphology had been previously demonstrated to alter the compaction process (1-4) as well as sintered tensile and compressive response (5-18). With regard to tensile response, it was demonstrated that the tensile properties were found to be independent of particle size for sintered copper (9). In iron compacts, tensile and compressive strength increased with decreased particle size (5,6,12) and can be attributed to a finer grain size, e.g. Petch analysis, and higher impurity level with finer powders. In addition, the effects of residual porosity can clearly alter toughness, strength and ductility in ferrous materials. Modest increases in strength and dramatic increases in toughness and ductility result with decreased porosity levels (18).

---

*Associate Professor
**Graduate Student

The potent solid solution strengthening effects of small phosphorus additions in ferrous P/M alloys has been recognized initially by Lenel (19), who observed strength increases with no loss of ductility as common with carbon and manganese additions. However, the utilization of phosphorus in high performance applications has been limited due to the embrittling effects in fully dense cast and wrought ferrous products as well as limitations in cold forming (20). Previous studies have been concerned with the effect of phosphorus additions on the toughness of iron, specificially intergranular failure with varying carbon and oxygen levels (21-24). At phosphorus levels above 0.6 wt %, grain boundary segregation and precipitation have been observed (25-29). Hornbogen (27,28) and more recently Spitzig (29) have observed matrix and grain boundary precipitation in conventional fully dense alloys utilizing light microscopy techniques.

From a strengthening viewpoint phosphorus additions present an extremely effective solid solution strengthener (30). Studies have been carried out to characterize the strengthening mechanism and strain aging effects in fully cast and wrought materials (31-34). In powder processed materials Linskog (35-38) and Durdaller (39) substantiated that strengthening was accomplished with no loss of ductility.

Program of Study

The objective of the present study is to provide for a comparative analysis of particle size/mechanical property response in sintered iron and iron phosphorus alloys. The effects of particle size fractions of atomized iron powder (Hoeganaes Ancorsteel A1000B) and the influence of Fe_3P additions in atomized iron powder shall be examined over a range of densities. Mechanical properties were evaluated, e.g. ultimate tensile strength, yield strength, percent elongation and hardness are monitored. A microstructure/mechanical property correlation provides for a description of the relations of sintering temperature, phosphorus level, density and particle size upon the fracture mechanism.

Experimental Procedure

The annealed atomized iron powder and 0.7 wt % zinc stearate lubricant was mixed with 0 and 0.45 wt % P. Table I summarizes the experimental variables and property evaluations. Tensile test bars, as per MPIF standards, were compacted at 24 tsi (330.7 MPa), 35 tsi (482.3 MPa) and 50 tsi (689 MPa) with approximate green densities of 6.4, 6.8 and 7.2 g/cc. Four bars were pressed at each density level for sintering at 2050°F (1121°C) and 2350°F (1288°C) in H_2 for 30 minutes (1800 s). The dew point was maintained below -10°F (-23°C). Green and sintered densities were obtained from dimensional measurement and yield ultimate strengths and percent elongation were measured using an Instron test machine coupled with an extensometer.

In addition, iron powder was screened to particle size fractions of +100, -100+150, -150+200, -200+325, -325 mesh and zinc stearate and Fe_3P was blended with each fraction, Figures 1 and 2. A similar compaction and sintering procedure was followed in order to press and sinter tensile bars of differing densities with a 0.45 wt % phosphorus level.

TABLE I

SUMMARY OF MECHANICAL PROPERTY EVALUATIONS

COMPOSITIONS:

- A1000B IRON (FRACTIONS)
- A1000B IRON + 0.45 wt % P (FRACTIONS)
- A1000B IRON + 0.45 wt % P (PREMIX, i.e. NORMAL SIZE DISTRIBUTION)

FRACTIONS:

+ 100, -100+150, -150+200, -200+325, -325, +200, +325

DENSITIES:

- 6.4, 6.8, 7.2 g/cc

SINTERING CONDITIONS:

- 2050°F (1121°C)/30 minutes (1800 s) H_2
- 2350°F (1288°C)/30 minutes (1800 s) H_2

MECHANICAL PROPERTY EVALUATIONS:*

- YIELD STRENGTH
- ULTIMATE TENSILE STRENGTH
- PERCENT ELONGATION
- HARDNESS

* TWO SPECIMENS FOR EACH CONDITION

Results and Discussion

A. Mechanical Property Response

The compaction response for iron and the iron/phosphorus system is depicted in Figure 3. For both systems, at comparable stress levels coarse fractions provide for higher green densities. The iron/phosphorus premix, i.e. a normal size distribution, system shows the higher green density when compared to the separated fractions. Additions of 0.45% P in the form of Fe_3P marginally inhibits densification at higher densities, e.g. 6.8 - 7.2 g/cc. Following sintering at 2050°F (1121°C)/30 minutes (1800s), H_2,shrinkage of the iron/ phosphorus alloys occur particularly at the smaller size fractions, Figure 4. In contrast, the density of the iron compacts generally remains unchanged.

The yield strength and ultimate tensile strength behavior of iron and iron/phosphorus alloy following at sintering temperatures of 2050°F (1121°C) and 2350°F (1288°C) are depicted in Figures 5 to 8. A near three-fold increase in yield strength of iron compacts with 0.45 wt % was noted vis a vis pure iron, Figures 5 and 6. High temperature sintering provides for only a modest improvement in the yield strength of both iron and iron/phosphorus systems. A notable exception is the high strength exhibited by the -325 mesh fraction of iron/phosphorus at low densities.

In contrast, ultimate tensile strengths appear more responsive to high temperature sintering, Figures 7 and 8. Sintering conditions of 2350°F (1288°C) results in a substantial increase in the ultimate tensile strength particularly with the fine particle fractions. The variations in percent elongation and hardness for the iron and iron/phosphorus systems are depicted in Figures 9 and 10, respectively. At low density levels hardness values are comparable between iron and the iron/phosphorus systems and appear dominated by porosity, at higher density levels the alloying of phosphorus appears to increase hardness values above sintered iron values. The variation in percent elongation is controlled by density and porosity. Ductility improves in the iron and iron/phosphorus system with increased density and decreased particle size. Also, the premix blend appears to have the lowest ductility of iron/phosphorus blends considered. At a sintering temperature of 2350°F (1288°C)/H_2/30 minutes (1800 s), ductility increases particularly for the 0.45 wt % P additions.

Table II*

Diffusion Coefficients of P in α and γ Iron

Temperature °C (°K)	D_α (cm²/sec) P in α-Fe	D_γ (cm²/sec) P in γ-Fe	D_α/D_γ
800°C	5.7×10^{-11}	1.4×10^{-4}	4.1
900°C	1.9×10^{-10}	8.1×10^{-11}	2.4
1000°C	1.2×10^{-9}	3.5×10^{-10}	3.4
1100°C	5.8×10^{-9}	1.2×10^{-9}	4.8
1200°C	2.3×10^{-8}	3.6×10^{-9}	6.4
1300°C	7.4×10^{-8}	9.3×10^{-9}	8.0

* Based upon D_o and Q (41) provided below in fully dense Fe

	Temperature (°C)	Q(Kcal/mole)	D_o (cm²/sec)
P in α-Fe	700 - 850	40	7.1×10^{-3}
	860 - 900	55	2.9
P in γ-Fe	950 - 1200	43.7	1.0×10^{-2}
	1280 - 1350	68.8	28.3
P in δ-Fe	1370 - 1400	55	2.9

B. Sintering Response

The sintering mechanism combines liquid and solid state diffusion of P into α-ferrite phase at low temperatures, Figure 11, and a liquid phase (eutectic) formation at 1920°F (1050°C). In addition, as diffusion progresses in the Fe+ 0.45 w/o alloy, the ferrite will be stabilized and the austenite regions will diminish in size. If sintering occurs at high temperatures, e.g. ∿1300°C, in α-iron, the high rate of self diffusion, Table II leads to rounded pores and considerable shrinkage, Figure 12. Sintering in the two-phase equilibrium α/γ region can lead to compositional inhomogeneity for alloys of phosphorus contents from 0.25 - 0.55 wt % P at 2050°F (1121°C). Sintering at T ≳ 2350°F (1288°C) eliminates the two phase field and diffusion of P is accelerated in the α-ferrite structure. Conventional low temperature sintering practice at 2050°F (1121°C) places the alloy in the α/γ two phase field generating ∿45% γ-Fe and ∿55% α- Fe if complete homogeneity is achieved.

Since diffusion coefficients of phosphorus, carbon and iron in porous materials are not readily available. A comparison of volume diffusion coefficients of phosphorus in full density α-Fe and γ-Fe is provided for in Table II. It is clear that phosphorus diffusion is more difficult in γ-Fe as a result of the closely packed structure. As the temperature increases the ratio of diffusion coefficients of phosphorus in α-Fe/γ-Fe, i.e. D_α/D_γ, becomes increasingly larger, e.g. D_α/D_γ @ 1100°C = 4.8 and D_α/D_γ @ 1300°C = 8.0, Table II.

A comparison of diffusion coefficients for low temperature sintering at 1100°C in γ-Fe vs. high temperature sintering in α-Fe at 1300°C is provided for in Table III. The combined effects of an open BCC ferrite structure plus elevated temperature results in the diffusion of Fe in Fe to be 1350 times faster, P in Fe, 62; and C in Fe, 238, Table III. These effects provide for enhanced homogeneity, shrinkage and rounded porosity, Figure 12.

Table III

Volume Diffusion Coefficients in Full Density Materials

	D @ 1100°C (cm^2/sec)	D @ 1300°C (cm^2/sec)
Fe in α-Fe	1.6×10^{-9}	2.3×10^{-8}
P in α-Fe	5.8×10^{-9}	7.4×10^{-8}
C in α-Fe	5.1×10^{-5}	2.0×10^{-4}
Fe in γ-Fe	1.7×10^{-11}	3.2×10^{-10}
P in γ-Fe	1.2×10^{-9}	9.3×10^{-9}
C in γ-Fe	8.4×10^{-7}	4.0×10^{-6}

Ratio of Diffusion Coefficients for Sintering @ 1100°C and 1300°C

$$\frac{D(\text{Fe in } \alpha\text{-Fe @ } 1300°C)}{D(\text{Fe in } \gamma\text{-Fe @ } 1100°C)} = 1.35 \times 10^{+3}$$

$$\frac{D(\text{P in } \alpha\text{-Fe @ } 1300°C)}{D(\text{P in } \gamma\text{-Fe @ } 1100°C)} = 62 \qquad \frac{D(\text{C in } \alpha\text{-Fe @ } 1300°C)}{D(\text{C in } \gamma\text{-Fe @ } 1100°C)} = 238$$

Figure 12 shows the equiaxed microstructures sintered at 2050°F (1121°C) and 2350° F (1288°C) with rounded porosity more evident at higher temperatures. Also, it is noted that P addition inhibits grain growth and grain size. With increased density, the porosity is much finer and a homogeneous distribution of pores is evident. Additions of phosphorus, accelerates pore rounding as a result of the transient eutectic liquid phase. The grain size is much smaller (ASTM 6-7) in lower density (6.8 g/cc) compared to ASTM 5-6 in 7.4 g/cc density compacts. Thus, the rate of diffusion sintering and grain growth is reduced by lower density. Higher phosphorus levels appear to assist sintering, pore rounding, however it also restricts grain growth when compared to comparable iron compacts. No grain growth inhibition is noted 2350 F (1121 C).

The fracture surfaces of iron and iron/phosphorus tensile specimens were examined to assess the role of phosphorus level, density and particle size on the mode of failure. The fracture morphology of iron compacts are a result of ductile tearing at sintered neck regions at particle boundaries. The addition of 0.45 wt % P alters the fracture mode at higher densities, Figure 13. A transition occurs from a ductile interparticle fracture at low densities to an intergranular cleavage fracture at increased densities and increased phosphorus levels. Higher temperature sintering at 2350°F (1288°C), Figure 14 also appears to increase the intergranular fracture mode at high density levels and may be related to phosphorus grain boundary segregation as noted in fully dense materials.

Conclusions

1. Strength increases are noted with increased density, finer fractions, phosphorus additions and high temperature sintering.

2. Yield strength of ferrous compacts increases three fold with the additions of 0.45 wt % P to each of the atomized iron particle size fractions.

3. Ductility increases are apparent with increased density, finer powder fraction and high temperature sintering.

4. Phosphorus addition inhibits grain growth at 2050 F (1121 C) and result in intergranular fracture at high densities.

5. Analysis of diffusion coefficients indicates the beneficial effects of high temperature sintering in α-Fe.

Acknowledgment

The authors gratefully acknowledge program support from the Hoeganaes Corporation as well as the interest of the late C. Durdaller, W. Feilbach and K. Moyer.

References

1. G. Bockstiegel, Int. J. Powder Met., 2, 13 (1966).
2. J. S. Hirschhorn, Introduction to Powder Metallurgy, APMI, New York (1969).
3. R. W. Heckel, Trans. AIME, 221, 1001 (1961).
4. J. S. Hirschhorn and M. W. Garey, Int. J. Powder Met., 5, 35 (1969).
5. J. Gallina and G. Monnone, Powder Met., 11, 73 (1968).
6. G. Artisio, J. Allina, G. Mannone and E. Sgambetterra, Powder Met., 9, 89 (1969).

7. M. Etudier, Powder Met., 9, 278 (1962).
8. G. M. Krishamoorthy, Powder Met., 14, 164 (1971).
9. R. T. DeHoff and J. P. Gillard, Modern Developments in Powder Metallurgy, Vol. 4, Edited by H. H. Hausner, Plenum Press, New York, 281 (1971).
10. S. M. Kaufman and S. Mocarski, Int. J. Powder Met., 7, 3, 19 (1971).
11. J. G. Byrne and D. P. Ferriss, Powder Met. Int., 3, 202 (1971).
12. V. T. Morgan, Powder Met., 12, 72 (1963).
13. S. A. Price and J. Oakley, Powder Met., 12, 87 (1963).
14. C. D. T. Minton and G. C. Smith, Powder Met., 12, 159 (1963).
15. R. Haynes, Powder Met., 14, 65 (1971).
16. R. Haynes, Powder Met., 14, 71 (1971).
17. M. J. Koczak and A. Lawley, Powder Met. Int., 4, 4, 186 (1972).
18. H. Chung and M. J. Koczak, Effects of High Temperature Sintering on Tensile and Impact Properties of Mixed Elemental Nickel Steels, Proceedings 5th International P/M Conference, Chicago (1976).
19. F. Lenel, U. S. Patent No. 2226520 (1940).
20. G. H. Enzian, Trans. AIME, Feb. 188 (1950).
21. W. P. Rees and B. E. Hopkins, J.I.S.I., 172, 403 (1952).
22. B. E. Hopkins and M. R. Tipler, J.I.S.I., 188, 218 (1958).
23. B. E. Hopkins and M. R. Tipler, J.I.S.I., 174, 110 (1954).
24. E. D. Hondros, Proc. Roy. Soc., A286, 459 (1965).
25. M. C. Inman and M. R. Tipler, Acta Met. 6, 73 (1958).
26. P. V. Ramasubramanian and D. F. Stein, Met. Trans., 4, 1735 (1973).
27. E. R. Hornbogen, Trans. ASM 53, 569 (1961).
28. E. R. Hornbogen, Trans. ASM 55, 719 (1962).
29. W. A. Spitzig, Mat. Sci. and Eng. 16, 169 (1974).
30. W. A. Spitzig and R. J. Sober, Mat. Sci. and Eng., 20, 179 (1975).
31. C. C. Li, L. J. Dziura and W. C. Leslie, Met. Trans. 8A, 705 (1977).
32. N. P. Allen in "Iron and its Dilute Solid Solution", Ed. C. W. Spencer and F. E. Werner, Interscience, New York (1973).
33. W. A. Spitzig and R. J. Sober, Met. Trans. 8A, 651 (1977).
34. W. A. Spitzig and R. J. Sober, Met. Trans. 8A, 1585 (1977).
35. P. Lindskog, Powder Metal. 16, 32 (1973).
36. P. Lindskog and A. Carlsson, Powder Metal. Int., 4, 1 (1972).
37. P. Lindskog, Hoeganaes-AB Report S-263-01, Höganäs, Sweden.
38. P. Lindskog, U. S. Patent No. 3836355.
39. C. Durdaller, Hoeganaes Report, Riverton, N. J. (1976).
40. W. A. Spitzig, Met. Trans. 3, 1183 (1972).
41. J. Askill, Tracer Diffusion Data for Metals, Alloys and Simple Oxides, Plenum, New York (1970).

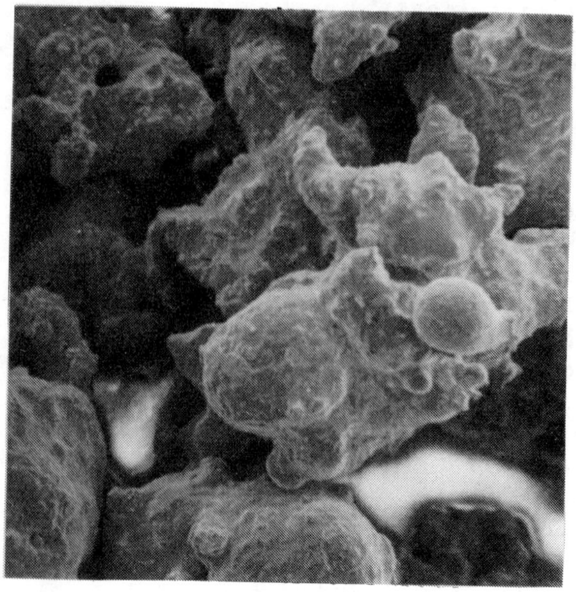

(a)

(b)

(c)

(d)

Figure 1. Atomized Iron Powder (Hoeganaes A1000B @ 100X). (a) +30, (b) -100+200, (c) -400, (d) Premix Blend.

(a) 1000X

(b) 3000X

Figure 2. Ferrophosphorus Fe_3P Alloy Addition.

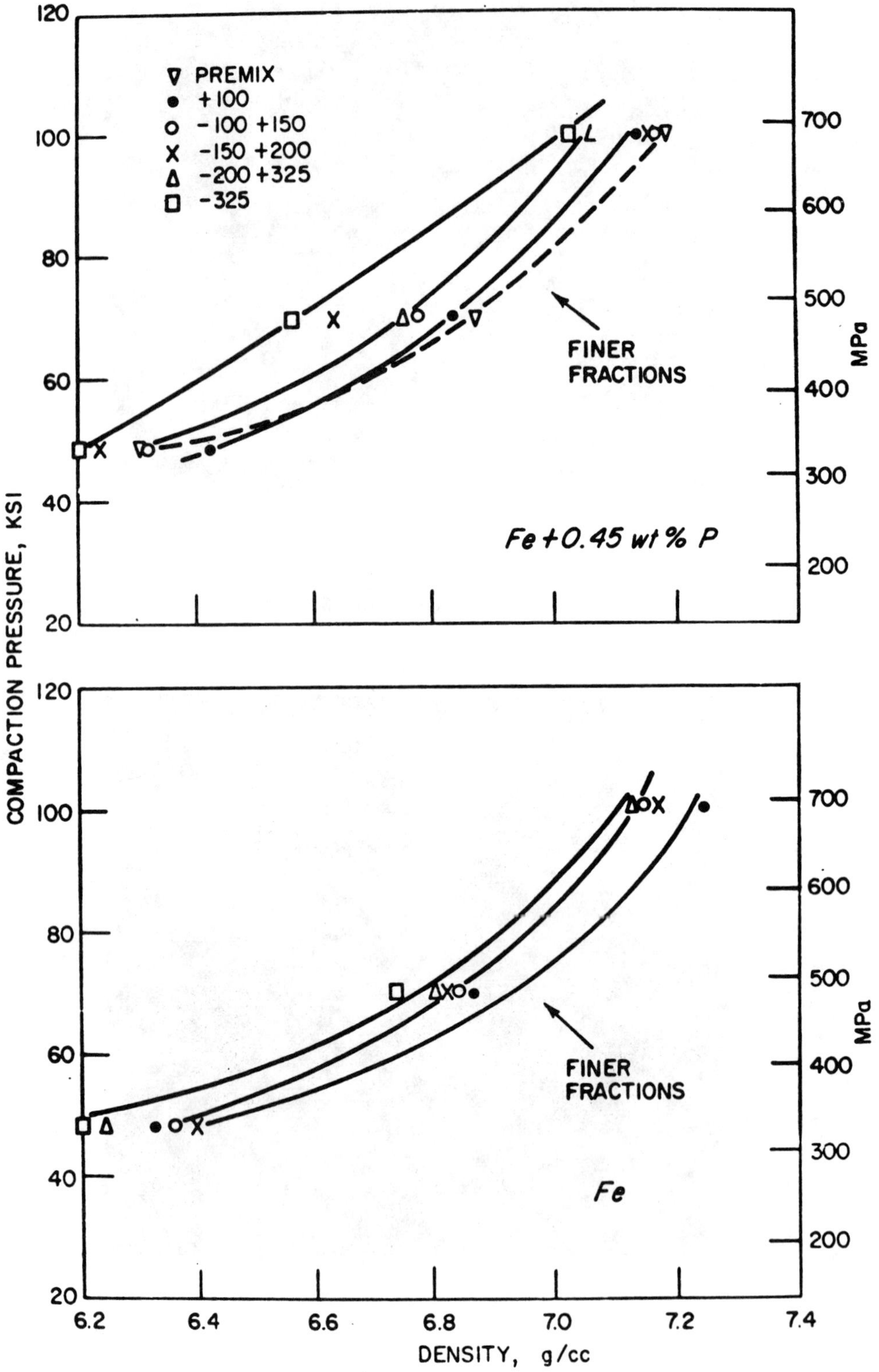

Figure 3. Green density versus compaction pressure for various particle size fractions of A1000B iron powder and A1000B iron plus 0.45 wt % P (as Fe_3P) with 0.7 wt % zinc stearate.

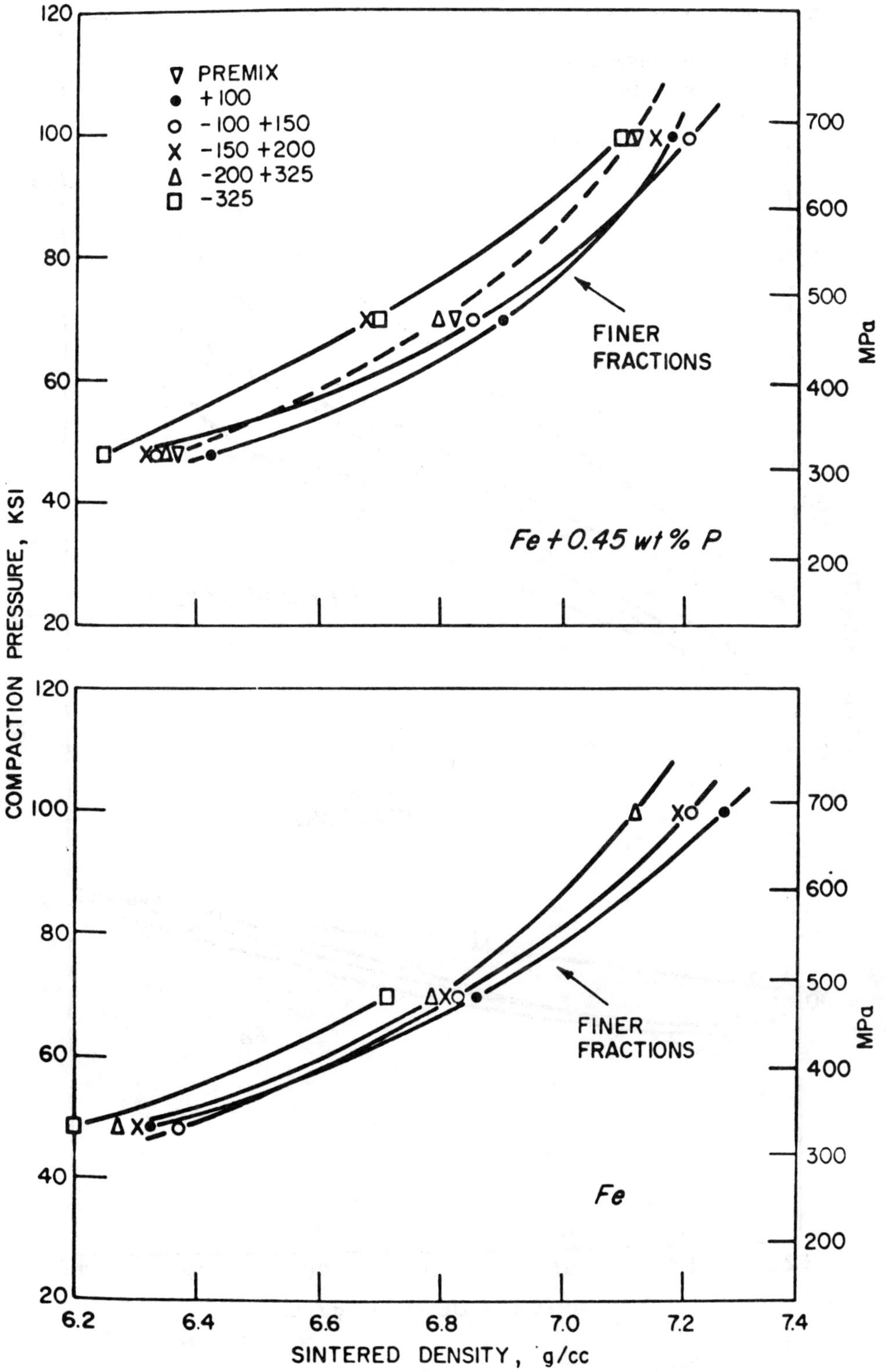

Figure 4. Sintered density at 2050°F (1121°C)/H_2/30 min (1800 s) versus compaction pressure for various particle size fractions of A1000B) iron powder and A1000B iron plus 0.45 wt % P (as Fe_3P).

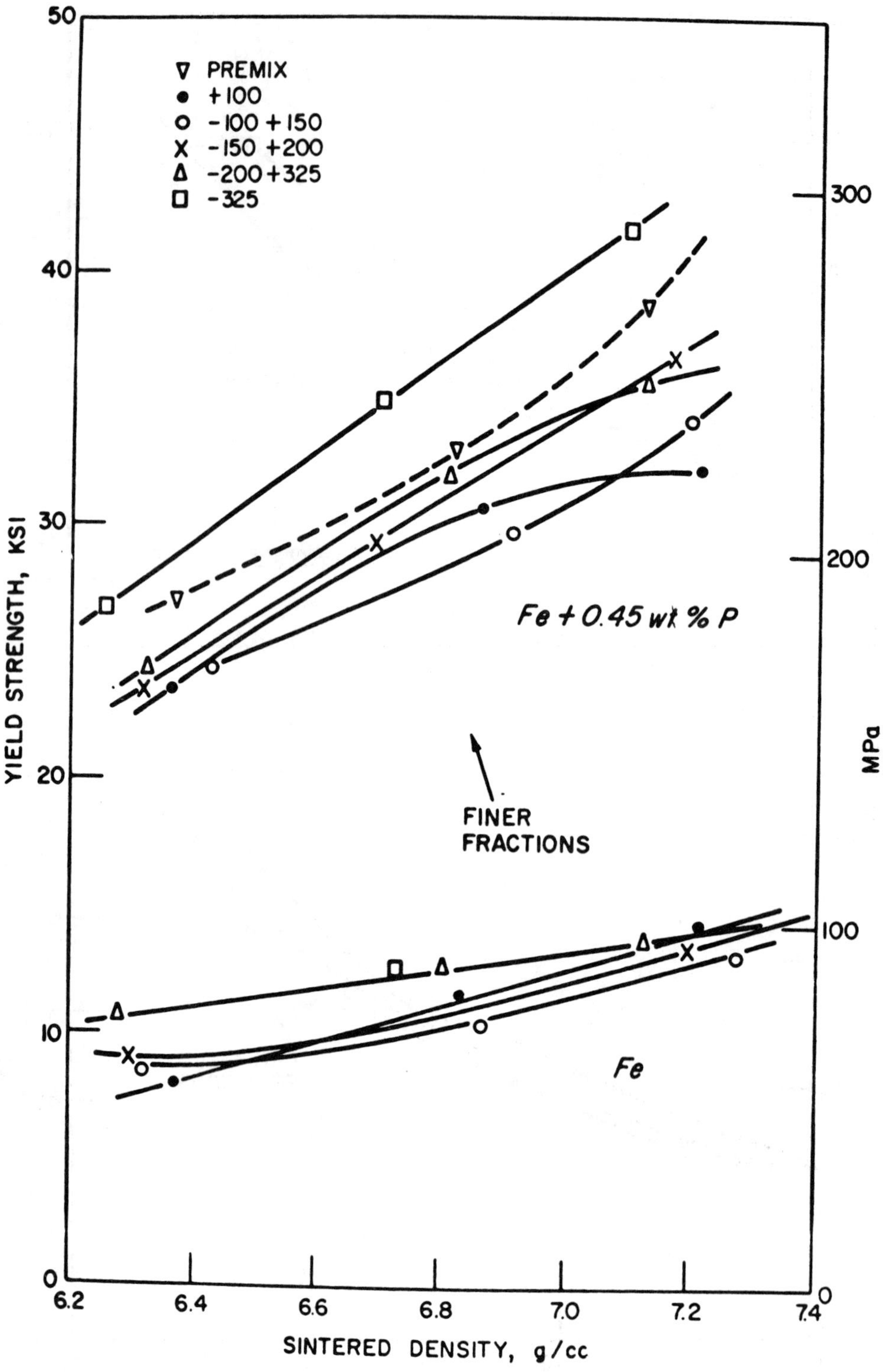

Figure 5. Average yield strength versus sintered density at 2050°F (1121°C)/ H_2/30 min (1800 s) for A1000B Fe powder and A1000B Fe + 0.45 wt % P (Fe_3P), for various size fractions.

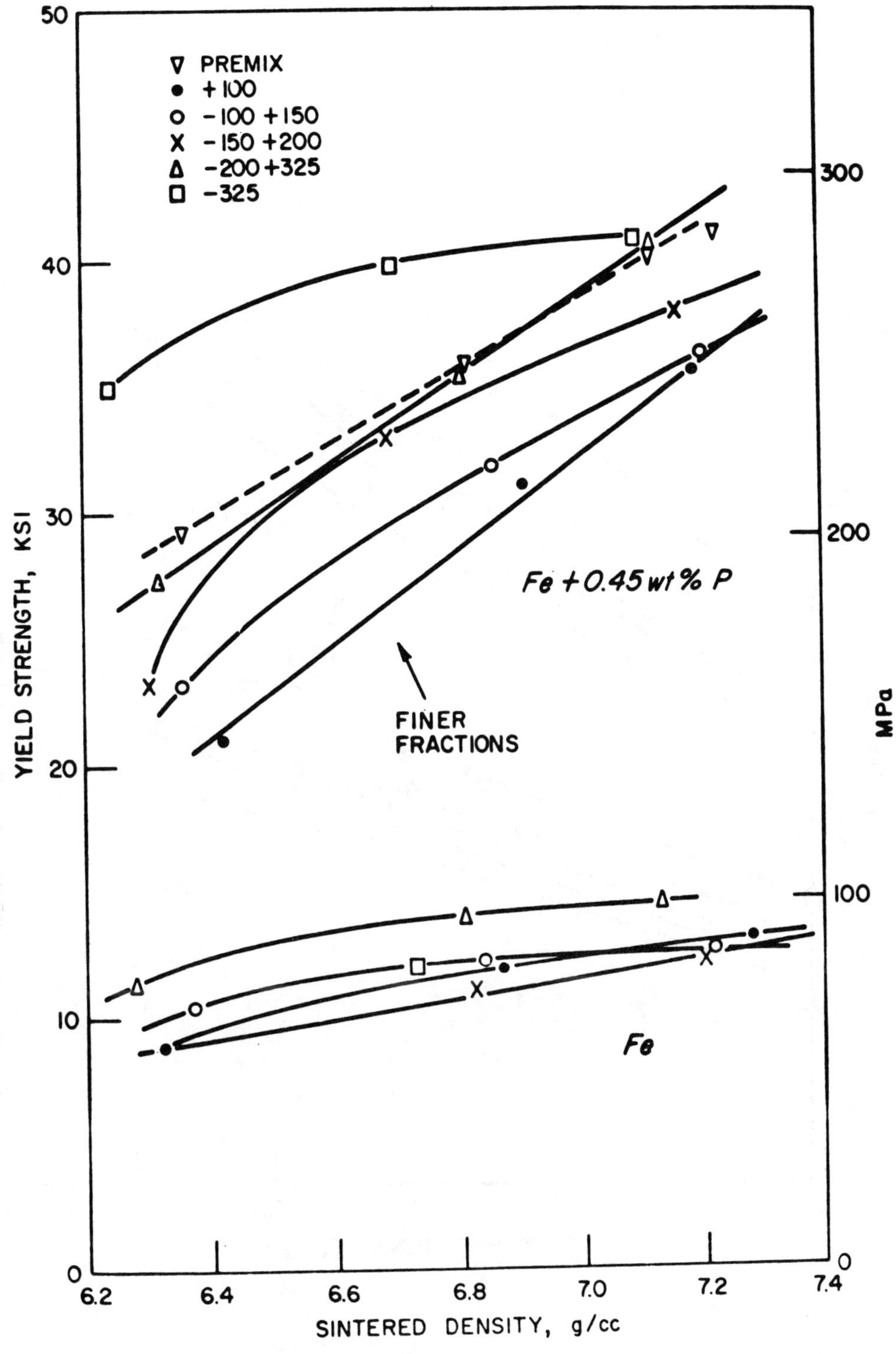

Figure 6. Average yield strength versus sintered density at 2350°F (1288°C)/ H_2/30 min (1800 s) for A1000B Fe powder and A1000B Fe + 0.45 wt % P.

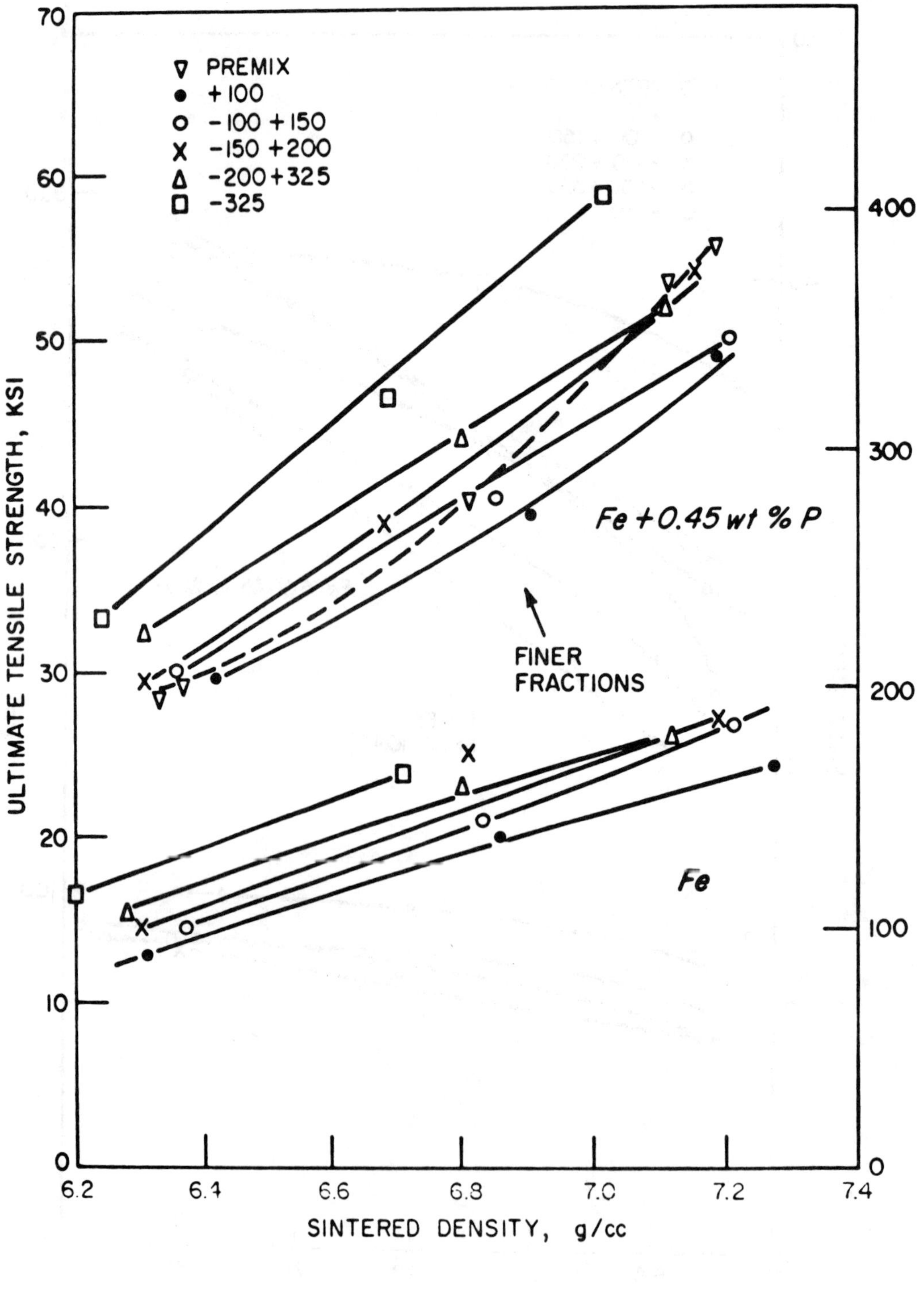

Figure 7. Average ultimate tensile strength versus sintered density at 2050°F (1121°C)/30 min (1800 s) H_2 for A1000B Fe powder and A1000B + 0.45 wt % P.

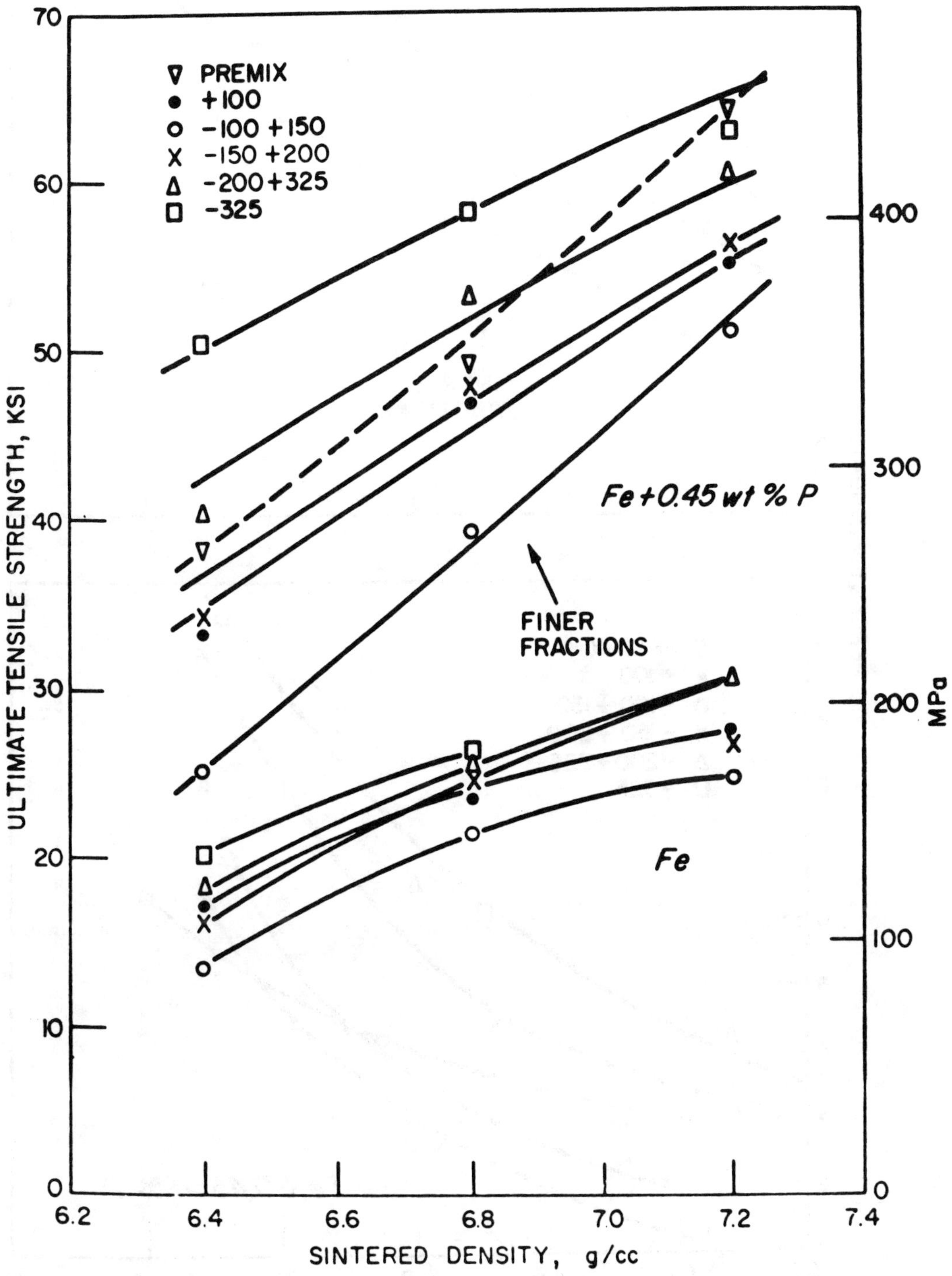

Figure 8. Average ultimate tensile strength versus sintered density at 2350°F (1288°C)/30 min (1800 s) H_2 for A1000B Fe powder and A1000B + 0.45 wt % P.

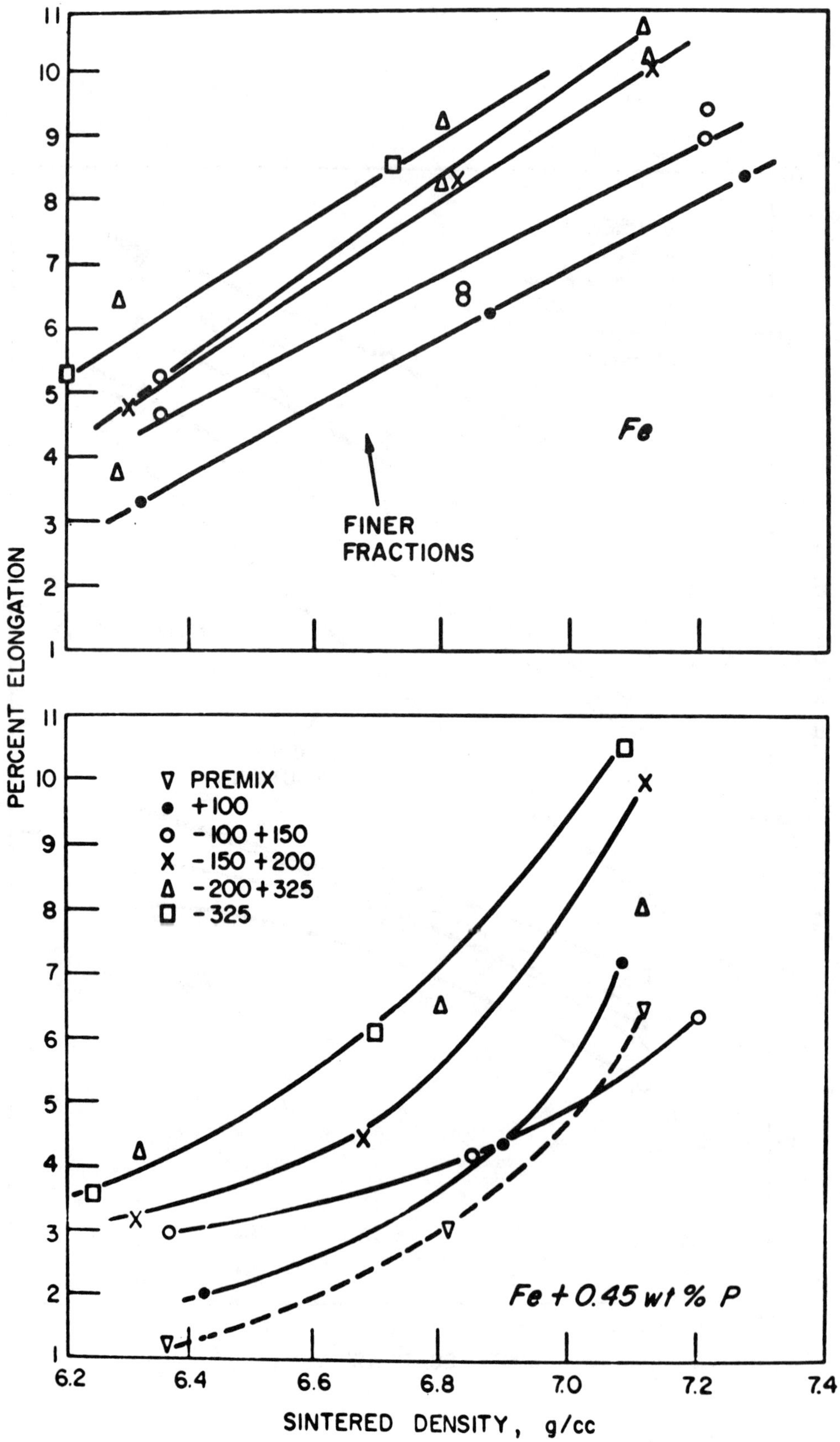

Figure 9. Average percent elongation versus sintered density, 2050°F (1121°C)/ H_2/30 min (1800 s) for various particle size fractions of A1000B Fe powder and A1000B Fe + 0.45 wt % P.

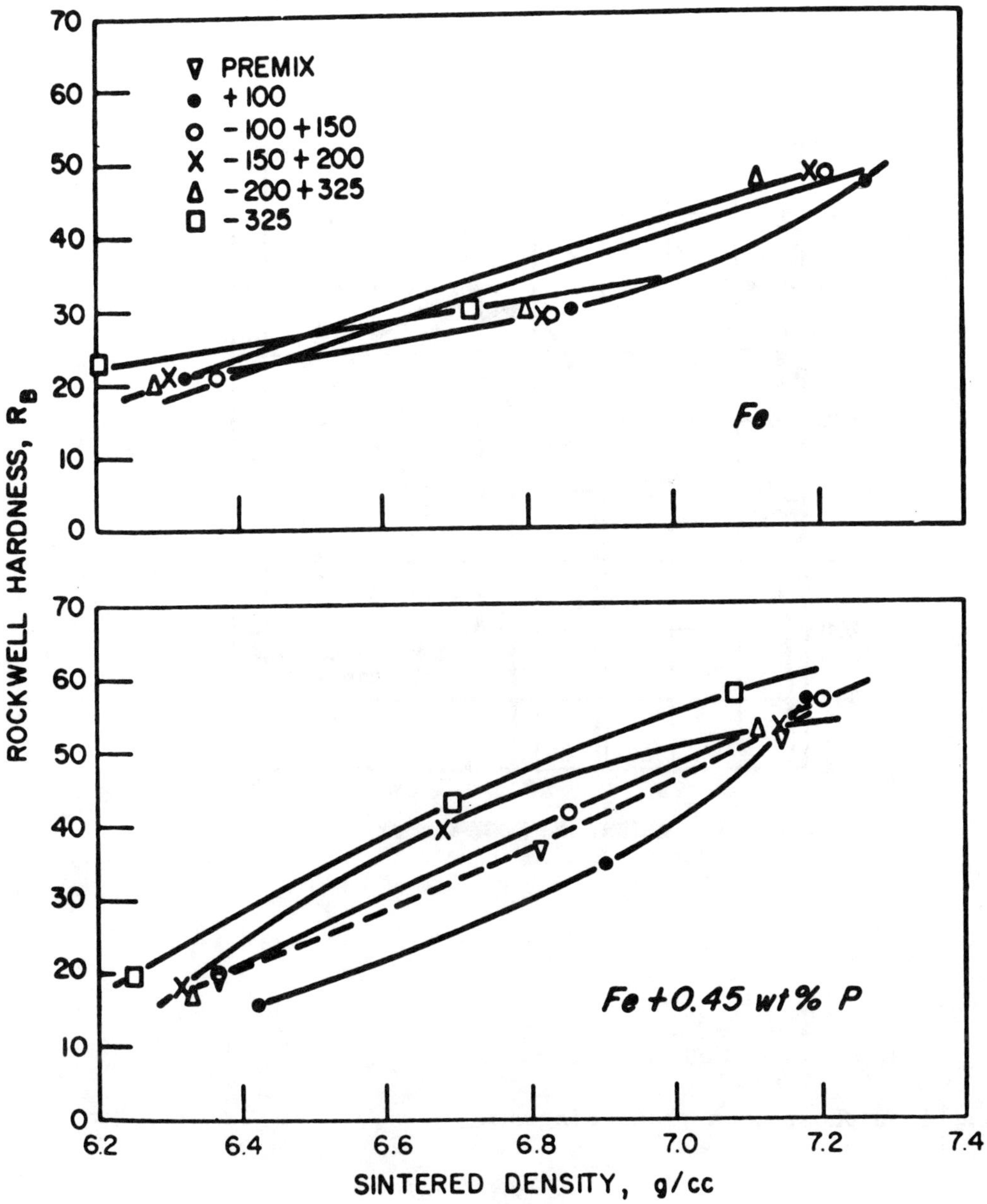

Figure 10. Hardness versus sintered density (2350°F (1288°C)/H_2/30 min (1800 s)) of compacts from various particle size fractions of A1000B Fe powder and A1000B Fe + 0.45 wt % P.

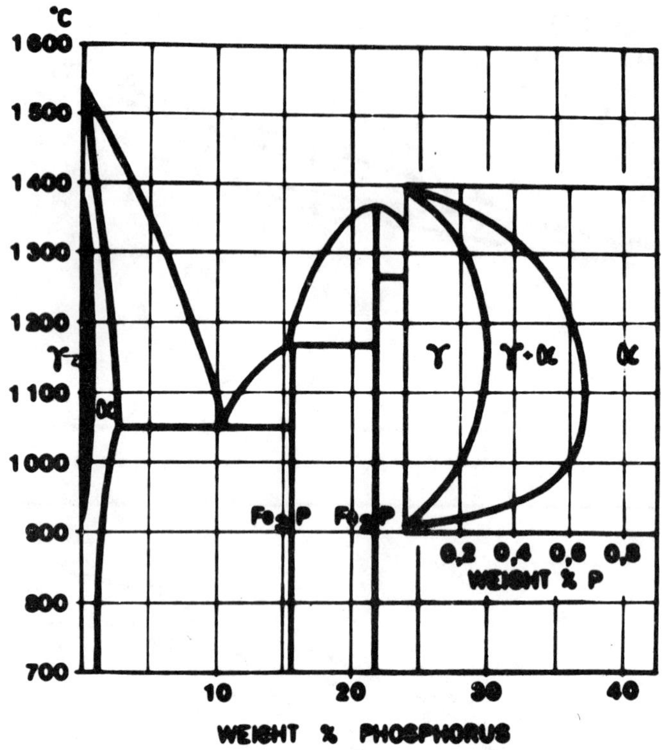

Figure 11. Iron phosphorus phase diagram.

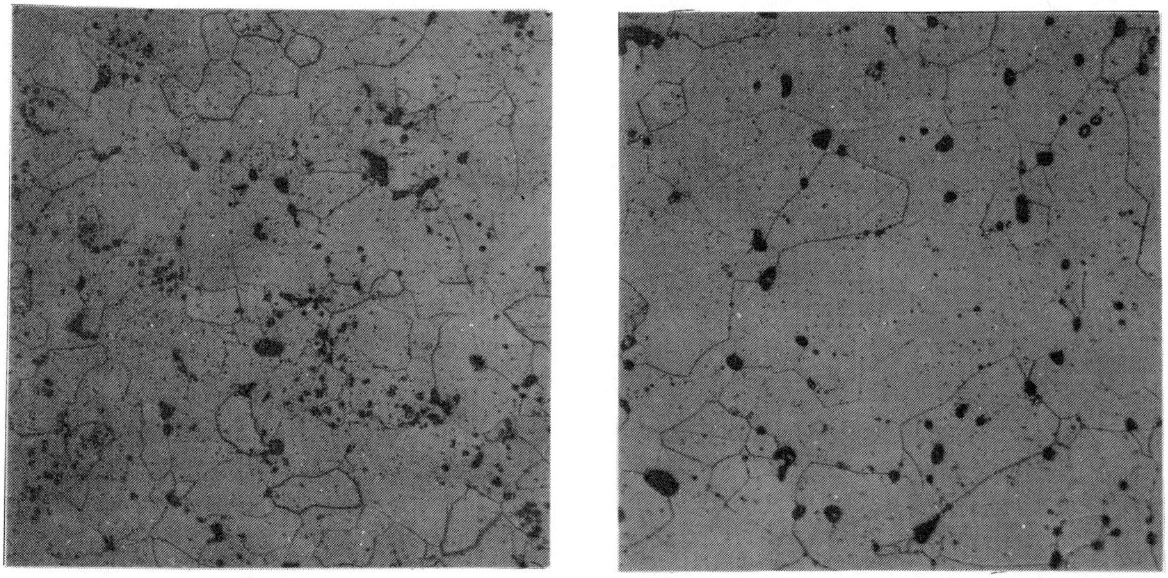

(a) 200x (b) 200x

Figure 12. Microstructure of Fe + 0.45 w/o P following sintering at (a) 2050°F (1121°C) and (b) 2350°F (1288°C).

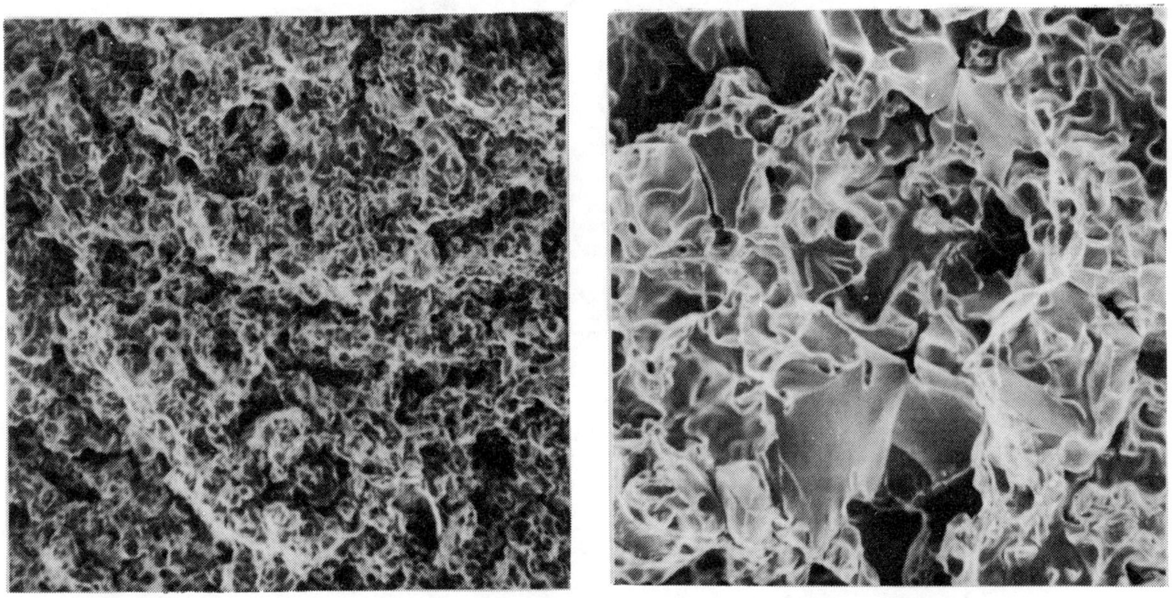

(a) 100x (b) 300x

Figure 13. Fracture surface scanning electron micrographs of A1000B + 0.45 wt % P, -100+150 fractions, sintered at 2050°F (1121°C)/30 min (1800 s) H_2 at (a) 6.8 g/cc and (b) 7.2 g/cc.

(a) 100X

(b) 100X

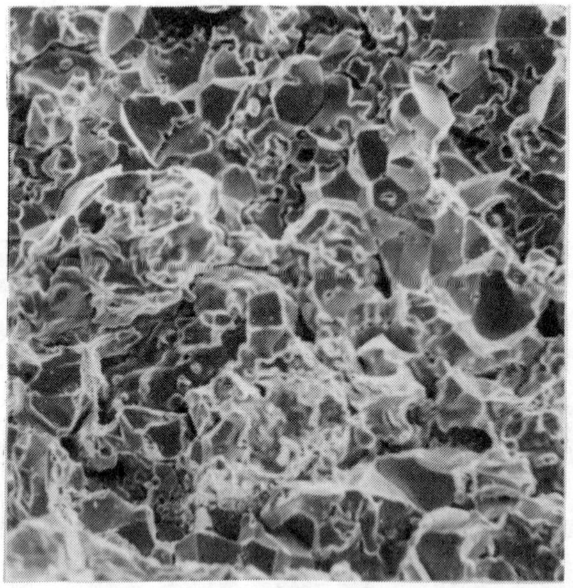

(c) 300X

Figure 14. Fracture surface scanning electron micrographs of A1000B + 0.45 wt % P, -100+150 fractions, sintered at 2350°F (1288°C)/30 min (1800 s) H_2 at (a) 6.4 g/cc, (b) 6.8 g/cc and (c) 7.2 g/cc.

PREDICTING DC PROPERTIES OF P/M MATERIALS

By
K.H. Moyer, M.J. Topolski,*
M.J. McDermott* and P. Wisniewski*
Hoeganaes Corporation
Riverton, NJ 08077

*M.J. Topolski, M.J. McDermott, and P. Wisniewski are co-op students from Drexel University. All were instrumental in the development of various portions of the technology discussed in the paper. The paper is a tribute to the innovativeness of the youth of America.

Abstract

The DC properties of P/M magnetic materials have now been well documented and many magnetic devices are currently being made. Not much has been accomplished, however, in designing P/M components to achieve optimum magnetic properties.

Recently it has been established that porosity has a decided influence on the structure sensitive magnetic properties. In fact, simply controlling the pores so that they are large and widely spaced results in improvement of the structure sensitive properties. This can be accomplished by employing coarse powder and/or high temperature sintering.

Structure sensitive properties, however, are related to many variables. Among these are grain size, chemical purity, and strains that distort the crystalline lattice. These variables must be predictable before the effect of porosity on structure sensitive properties can be understood. Predictor equations are presented for hot repressed atomized iron which contains no porosity. The effect of grain size upon this high purity soft iron is discussed and equations are presented which show how the structure sensitive properties can be predicted from the grain size of the material. An insight is also given as to how porosity alters the properties when present in P/M compacts.

In addition, initial data is presented which shows how saturation magnetization is affected by porosity. Saturation magnetization data are also presented for the various magnetic alloys that can be made by P/M technology.

Finally, state of the art properties that can be achieved by using P/M alloys for magnetic applications are summarized. These properties are compared with those of existing commercial alloys.

Introduction

Many magnetic devices which include P/M components are commercially available. Recently there has been accelerated interest in exploiting P/M for magnetic components because of weight and energy savings and general overall economy. These advantages of P/M are now widely recognized throughout the metals industry. Of course, P/M parts are different from conventional products in that they generally contain porosity although this can be removed by hot repressing to nominally full density. Powders are also different because they can be manufactured to levels of high chemical purity economically and, with careful processing, maintain their levels of purity. For instance, todays atomized high compressibility iron powder, properly sintered, has carbon content less than 0.01%, nitrogen content less than 0.002%, and sulfur content less than 0.003%. These impurity levels are much lower than found for irons economically exploited for commercial magnetic applications today. It is well known that structure sensitive magnetic properties (maximum permeability and coercive force) are dependent upon the impurity level of the material.(1) Todays improved powders, therefore, provide the possibility of becoming candidates for many new magnetic applications.

When magnetic powders are employed using a normal size distribution, despite the purity, structure sensitive properties are not different from conventional materials. The purpose of the work is to provide an insight into what can be done to improve the properties of powder compacts through control of the pores and by improvement in sintering techniques.

State of the Art

Table I compares the properties that are typical for commercial irons today with properties that may be considered typical for atomized iron compacts made from powder of a normal distribution.

Table I

Comparison of Magnetic Properties of

Atomized Iron With Commercially Avaliable Irons

Property	Commercial Wrought Irons	Commercially Available Atomized Iron (7.2 g/cm³)
Density (g/cm³)	7.86	7.2
Electrical Resistivity, (μΩcm)	10	12
Saturation Induction, (kG)	21.5	19.2
Residual Magnetism, (kG)	7.5	10.9
Coercive Force, (Oe)	1.5	2.0

It is fair to note that the properties shown for commercial wrought irons are typical of those of a 1008 steel. Those of the atomized powders at 7.2 g/cm³ resulted from compacts sintered by several commercial parts producers. The properties are not indicative of the best that can be achieved but are what might be expected if a user approached most parts producers who would sinter an atomized iron compact at 2050°F (1120°C) in dissociated ammonia for thirty minutes. It may therefore be implied that on the average the magnetic properties desired from any P/M component available today would be slightly inferior to the component manufactured from 1008 steel.

Improved Properties

Saturation

Saturation is an intrinsic property and is therefore invariant with structure. For commercially available iron it is reported as 21.5 kilogausses.(1) The saturation magnetization of hot repressed atomized iron was determined for comparison. The density of the samples was 7.68 g/cm³ (2½% porosity) after hot repressing. The saturation for this material was 20.6 kilogausses. If the loss of mass of iron owing to the porosity is accounted for, it too has the value of 21.5 kilogausses for saturation. It can therefore be stated that if pores are not present, atomized iron is equivalent to any other iron.

Saturation is important for many applications. If designers are to employ P/M compacts for their devices it is necessary to know how this property behaves when porosity is introduced. To provide this information, samples of atomized iron powder of a normal distribution were pressed to densities of 7.4 g/cm³, 6.8 g/cm³ and 6.2 g/cm³ and tested to saturation. Saturation was determined when readings of B-H were constant for increasing magnetizing field. Magnetization curves of these measurements are shown in Figure 1. It can be seen that as the density decreases or porosity increases the saturation is reduced and greater magnetizing fields are required to achieve saturation. This results because the porosity causes demagnetizing fields to occur. Larger magnetizing forces are required to overcome these opposing forces. In compacts pressed from a normal powder distribution it has been found that many small pores are present.(2,3) These small pores are extremely harmful in causing demagnetizing effects. The consequence is the shearing of the magnetizing curves such as shown in Figure 1. Therefore, if the powders employed are coarse, the demagnetizing forces resulting from the porosity will be minimized. This should cause less shearing of the curves so that magnetizing forces similar to those required to saturate the 7.7 g/cm³ hot repressed compacts will be sufficient. These magnetizing forces are similar to those that are required to saturate commercial irons. It is significant to saturate at these magnetizing forces because less current, hence no additional copper wire, would be required for the magnetizing circuit.

The data are replotted in Figure 2 to show the effect of density on saturation magnetization. It is readily seen that as the density decreases, the saturation magnetization decreases correspondingly. If the material is fully dense the normal saturation of 21.5 kilogausses (2.15 teslas) is achieved. Again saturation diminishes as density decreases. The reason for this is less iron available and the demagnetization forces caused by the porosity. It is believed that if the small pores are eliminated or reduced within

Figure 1
Magnetization Curves of
Atomized Iron at Various Densities

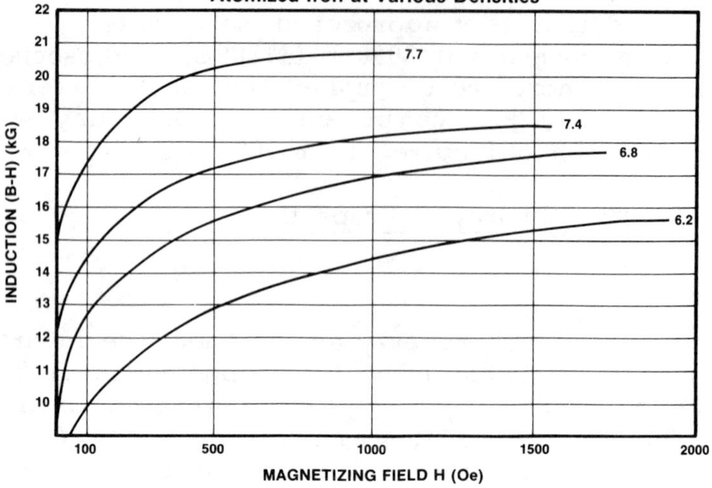

Figure 2
Density vs. Magnetization

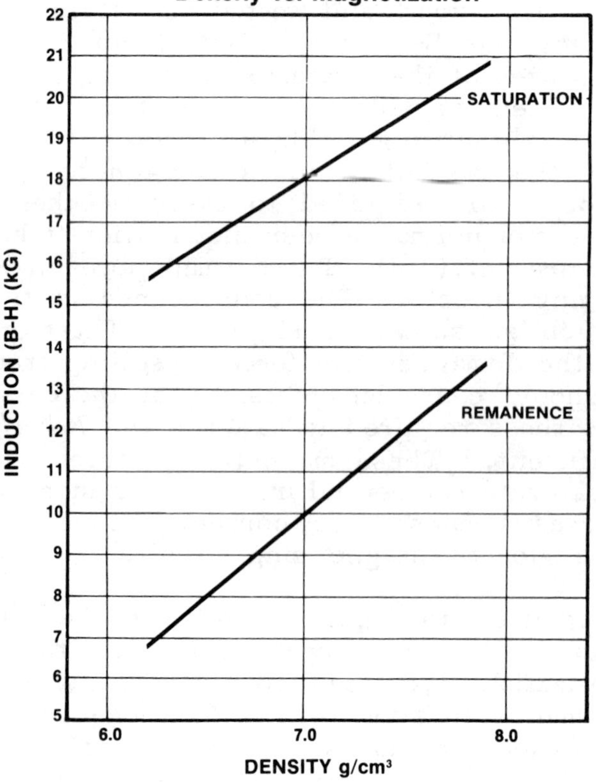

the microstructure the demagnetizing forces will be minimized and hence the saturation magnetization will reach optimum values. A compromise must be made to live with the lower saturation values if a designer wishes to employ porous P/M parts.

It was also found that the remanence decreases as the density decreases. Remanence is thought to be somewhat structure sensitive, however, in powder compacts for any given density the property has been found to be relatively invariant. This may be an advantage for designers that consider the use of P/M parts. With P/M all that need be done is to vary the density and a range of values for remanence can be achieved. This could provide an advantage for powders in the design of devices such as relays.

DC Properties of Hot Repressed Compacts

To determine the magnetic properties of hot repressed material, preforms 2 inches (51 mm) by 2 inches (51 mm) by 5 inches (127 mm) were fabricated from atomized iron powder. Prior to compaction the powder was screened into 60/100, 100/200, 200/325 and -325 mesh size fractions. These size fractions are of no concern to the properties of hot repressed material but will be used to control the size and distribution of the pores in other studies of porous compacts. The preforms made from the various size fractions were compacted to a density of 7.2 g/cm^3. Sets of the preforms according to size fraction were then sintered at 1600°F (870°C), 2050°F (1120°C), 2300°F (1260°C), 2600°F (1430°C) and 2700°F (1480°C) for 30 minutes in hydrogen. They were then heated to 1800°F (980°C) in hydrogen and hot repressed. The preforms sintered at 1600°F (870°C) were only heated to 1600°F (870°C) for hot repressing so that the grain size would not be altered appreciably. Test rings and resistivity bars were machined from each of the preforms. After machining, the specimens were resintered at their respective temperatures to remove the stresses caused by machining and to homogenize the grain structure. Test results of these rings are shown in Table II. Carbon, nitrogen, and sulfur were all at levels that are representative of pure irons. The oxygen content varied with the sintering conditions that were employed. Carbon, nitrogen, and sulfur were therefore excluded from the analysis whereas oxygen content was considered as a variable. The density of all materials was 7.85 to 7.86 g/cm^3 and therefore the material was considered to be fully dense. Resistivity of the materials was approximately 11 micro ohm centimeters. Pure irons have a reported resistivity of 10 micro ohm centimeters.(1) The materials may therefore be considered to have resistivities equivalent to pure irons.

Grain size was measured employing the intercept method. In general, grain size increased with sintering temperature. It too was considered as a variable for further analysis.

Flux densities and remanent magnetization measured from a magnetizing field of 15 Oe (1194 A/m) were invariant. If these properties are compared with conventional irons at equivalent magnetizing fields, they are similar. In fact, if the properties were compared at saturation these properties too would be equivalent since the properties are invariant and are only dependent on the material, not the structure.

So far the properties are not different from any other high purity iron. Maximum permeability and coercive force however are structure sensitive and

Table II

Magnetic Properties of Atomized Iron Cores

Property	1600 °F				2050 °F		2300 °F		
	60/100	100/200	200/325	-325	100/200	200/325	60/100	100/200	200/325
Chemical Analysis, %									
C	< 0.01	0.011	< 0.01	0.010	0.011	< 0.01	< 0.01	< 0.01	< 0.01
N	0.0015	0.0015	0.0011	< 0.0010	0.0012	0.0018	< 0.0010	< 0.0010	0.0020
O_2	0.075	0.068	0.060	0.066	0.048	0.045	0.047	0.046	0.023
S	0.003	0.002	0.003	0.006	0.002	0.004	0.002	0.002	0.004
Sintered Density g/cm^3	7.85	7.86	7.86	7.86	7.85	7.85	7.86	7.86	7.86
Resistivity $\mu\Omega$ cm	10.6	10.6	10.7	10.8	10.8	10.7	10.8	11.0	11.0
Average Intercept Length ($\overline{L}$) μm	40	59	105	89	64	58	67	70	113
Flux Density (B_{15}) kG	16.0	16.0	16.1	16.0	16.0	16.1	16.0	16.0	16.2
Residual Induction (B_r) kG	15.3	15.2	15.2	15.2	14.3	15.2	14.4	14.9	15.0
Maximum Perm. (μ_{max}) kG/Oe	4.3	4.6	5.6	4.2	5.1	4.7	5.8	5.8	6.7
Coercive Force (H_c) Oe	1.7	1.4	1.4	1.4	1.0	1.1	0.8	0.9	0.6

Table II Con't

Magnetic Properties of Atomized Iron Cores

Property	2600°F 60/100	2600°F 100/200	2600°F 200/325	2600°F -325	2700°F 60/100	2700°F 100/200	2700°F 200/325	2700°F -325
Chemical Analysis, %								
C	<0.01	<0.01	<0.01	<0.01	<0.01	<0.01	<0.01	<0.01
N	--	0.0012	0.0016	0.0012	0.0016	0.0013	0.0013	<0.0010
O_2	0.046	0.054	0.034	0.051	0.039	0.056	0.052	0.018
S	0.002	0.003	0.004	0.006	0.002	0.003	0.004	0.007
Sintered Density g/cm^3	7.86	7.85	7.86	7.86	7.86	7.86	7.86	7.86
Resistivity $\mu\Omega$ cm	10.8	10.9	10.8	11.1	11.0	10.8	11.0	11.4
Average Intercept Length ($\overline{L}$) μm	275	321	357	275	362	275	362	258
Flux Density (B_{15}) kG	16.2	16.1	16.2	16.1	16.0	16.0	16.0	16.2
Residual Induction (B_r) kG	14.2	14.9	14.9	15.0	14.7	14.9	14.9	14.3
Maximum Permeability (μ_{max}) kG/Oe	11.3	11.7	11.3	11.4	12.1	12.1	11.4	10.8
Coercive Force (H_c) Oe	0.2	0.2	0.2	0.3	0.2	0.2	0.2	0.3

therefore depend not only on the material but on the processing conditions employed. It can be seen from Table II that these properties vary with the sintering conditions employed.

Analysis of Structure Sensitive Properties

As stated, structure sensitive properties are related to the purity of the material and can be degraded by the sintering or processing conditions employed.(1) In the case of powder compacts, oxygen content can vary as a result of processing owing to the greater amount of surface area of each bonded particle. Oxygen, carbon, nitrogen, and sulfur strain the atomic lattice structure degrading the structure sensitive properties. Grain size is also important.(1) If large grains are present there is less stress on the lattice because there are fewer grain boundaries which are disordered. Indeed, best structure sensitive properties are achieved from single crystals. In this work grain size varies according to the sintering temperature employed, the higher the sintering temperature the larger the grains.

To predict how structure sensitive properties will vary in hot repressed iron a regression analysis was made with maximum permeability and coercive force as dependent variables and grain size intercept, oxygen content, and nitrogen content as independent variables. Regression analysis found that only the average grain size influenced the maximum permeability, explaining 82% of the data. The equation that was yielded was:

$$\mu_{max} = 4.6\ e^{0.003\bar{L}}$$

where μ_{max} = maximum permeability (kG/Oe)

$\bar{L}$ = average grain size intercept (μm)

Similarly regression analysis found that the coercive force was dependent on the average grain size intercept and the oxygen content. The grain size intercept explained 90% of the data and the oxygen content explained an additional 2%. Since the oxygen content explained so little of the data it was excluded from the analysis. The equation that was yielded was:

$$Hc = 1.6\ e^{-0.006\bar{L}}$$

where Hc = coercive force (Oe)

$\bar{L}$ = average grain size intercept (μm)

The predictor equation for maximum permeability is plotted in Figure 3. Included also are the individual data points that were used in the regression analysis. Similarly the predictor equation and the data points for the coercive force are plotted in Figure 4. It can be seen that reasonable agreement exists between the equations and the data points; however, the data points are bunched within two specific regions of the curve. The points falling on the left hand side of the curves all represent data points for material sintered below the delta transition temperature whereas those falling to the right hand side represent data for material sintered above the delta transition temperature.

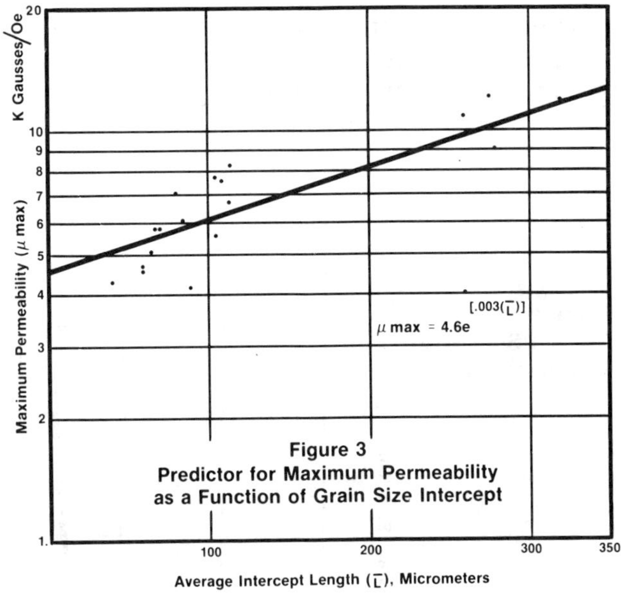

Figure 3
Predictor for Maximum Permeability as a Function of Grain Size Intercept

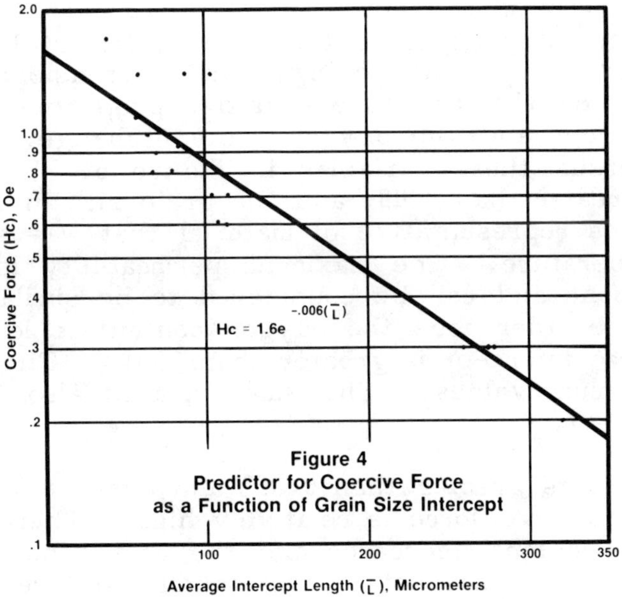

Figure 4
Predictor for Coercive Force as a Function of Grain Size Intercept

This anomaly was pursued further. To explain what is happening, a plot was made to show how the average grain size intercept varies with sintering temperature, Figure 5. It can be seen that the grains do not grow appreciably until the material is sintered above the delta transition temperature. Once the delta transition temperature is exceeded there is then marked grain growth resulting in a tripling of the average grain size intercept.

Figure 5
Variation of Grain Size With Sintering Temperature

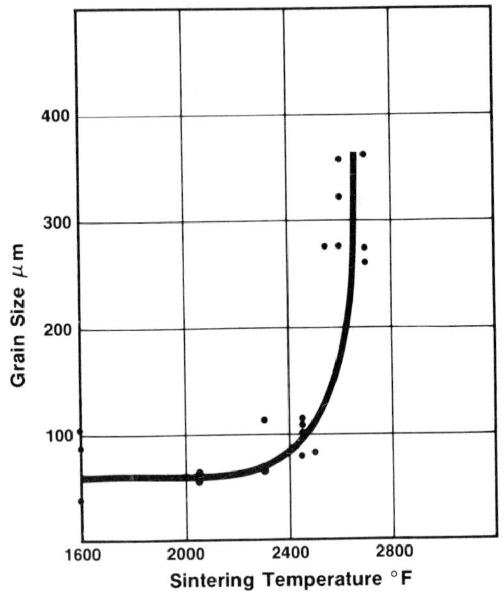

The effect of the oxygen content on the system still requires explanation. To understand what is happening, three dimensional plots were constructed to identify how the structure sensitive properties vary as the average grain size intercept and the oxygen content change. Maximum permeability is plotted versus these variables in Figure 6. Within one zone the oxygen content is greater than 0.05% and the grain size intercept is less than 110 µm . This zone is representative of material that was sintered below the delta transition temperature. The maximum permeability is at relatively low values within this zone and can be considered to be similar to that of commercial irons. In the other zone the oxygen content is less than 0.05% and the average grain size intercept is greater than 250 µm. The maximum permeability is at its optimum values. The same results also hold true for the coercive force, Figure 7.

For oxygen contents greater than 0.05% and average grain size intercept less than 110 µm the coercive force is relatively high. There is some improvement as the oxygen content decreases and the grains become larger. From Table I, it may be seen that for conventionally sintered material, 2050°F (1120°C), the coercive force is similar to most commercial irons available today. If high temperature sintering is employed, 2300°F (1260°C), the coercive force is slightly better than present materials. If the material is sintered above the delta transition temperature, the coercive force is equivalent to the purest iron that can be obtained and is relatively invariant. This is believed to occur because the oxygen content is less than 0.05% resulting

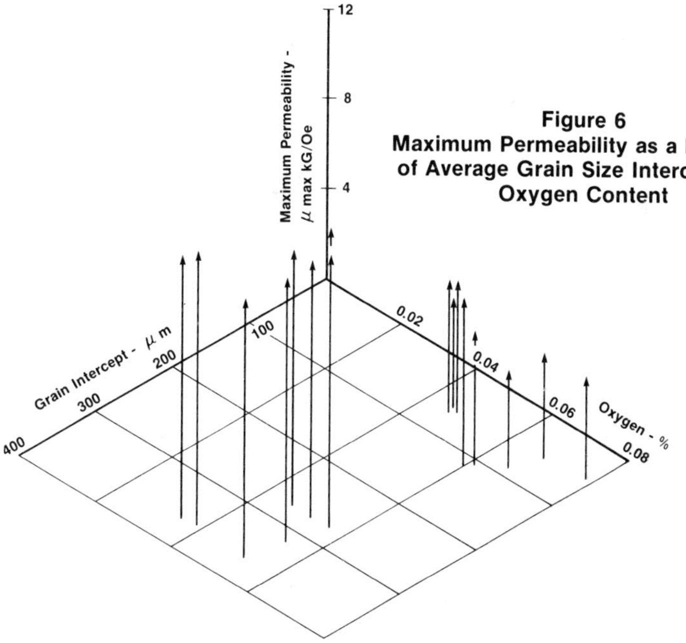

Figure 6
Maximum Permeability as a Function of Average Grain Size Intercept and Oxygen Content

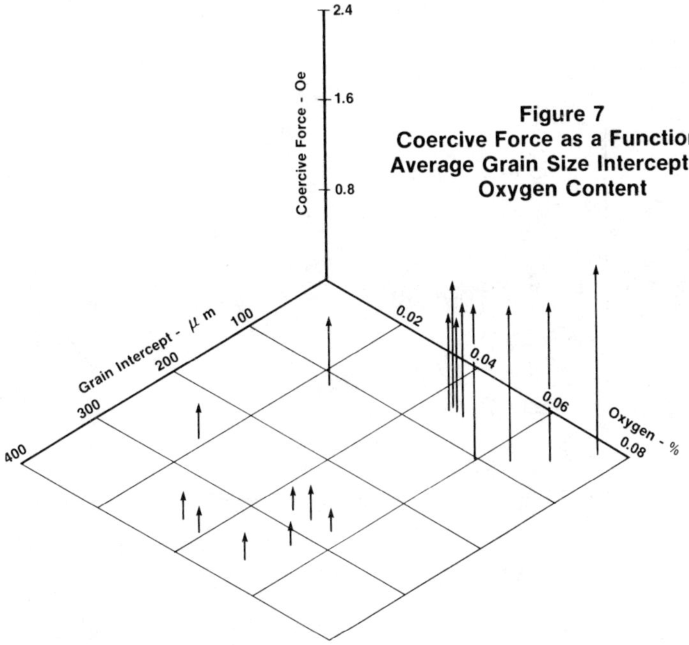

Figure 7
Coercive Force as a Function of Average Grain Size Intercept and Oxygen Content

Table III

Properties of 7.4 g/cm^3 Preforms

Property	1600°F						2050°F					
	30/60	60/100	100/200	200/325	325/400	-400	30/60	60/100	100/200	200/325	325/400	-400
Chemical Analysis												
S	.002	.002	.002	.004	.006	.005	--	--	--	--	--	--
C	<0.01	<0.01	<0.01	<0.01	<0.01	<0.01	0.019	<0.01	<0.01	<0.01	<0.01	<0.01
N	.0013	.0013	.0006	.0007	.0006	.0003	.0020	.0022	.0022	.0010	.0014	.0022
O_2	.073	.077	.065	0.053	.067	.066	.050	.045	.041	.039	.036	.044
Sintered Density g/cm^3	7.40	7.40	7.42	7.35	7.35	7.37	7.45	7.43	7.42	7.42	7.41	7.32
Resisitivity μΩ cm	11.9	11.9	11.8	12.1	12.4	12.2	11.9	12.7	12.6	12.5	12.7	13.3
Average Intercept Length ($\bar{L}$) μm	15	16	16	16	11	11	44	34	53	46	34	59
Flux Density (B_{15}) kG	13.0	13.0	13.3	13.1	13.1	13.3	13.6	13.6	13.6	13.6	13.7	13.4
Residual Induction (B_r) kG	11.7	11.9	12.4	12.1	12.3	12.3	11.7	11.6	11.0	12.0	12.1	11.7
Maximum Permeability (μ max) kG/Oe	3.0	3.0	3.3	2.8	2.5	2.4	3.8	3.6	3.8	3.6	3.0	2.4
Coercive Force (H_c) Oe	2.0	2.0	2.0	2.2	2.6	2.7	1.4	1.4	1.3	1.5	1.7	2.1
Mean Pore Area ($\bar{A}$) μm^2	134	133	89	110	60	33	191	132	147	116	67	67
Mean Free Path ($\bar{\lambda}$) μm	155	176	137	78	67	75	208	169	113	90	80	56

Table III Con't

Properties of 7.4 g/cm³ Preforms

Property	2300°F						2600°F						2700°F					
	30/60	60/100	100/200	200/325	325/400	-400	30/60	60/100	100/200	200/325	325/400	-400	30/60	60/100	100/200	200/325	325/400	-400
Chem. Anal.																		
S	.001	.001	.001	.001	.002	.003	.002	0.002	.001	.002	.002	.003	.001	.001	.002	.002	.004	.004
C	< 0.01	< 0.01	< 0.01	< 0.01	< 0.01	< 0.01	< 0.01	< 0.01	< 0.01	< 0.01	< 0.01	< 0.01	< .01	< .01	< .01	.011	< .01	< .01
N	.0016	.0018	.0020	.0012	.0014	.0024	.0014	.0014	.0011	.0015	.0010	.0014	.0024	.0012	.0015	.0019	.0015	.0024
O_2	.019	.027	.027	.023	.021	.023	.012	.020	.020	.018	.019	.027	.019	.013	.014	.017	.017	.021
Sint. Dens. g/cm³	7.42	7.44	7.41	7.47	7.36	7.36	7.48	7.55	7.56	7.53	7.53	7.62	7.51	7.55	7.57	7.61	7.69	7.68
Resistivity μΩ cm	12.3	13.1	12.4	13.2	13.2	13.2	11.8	12.0	12.2	12.2	12.3	11.7	12.3	11.8	11.9	11.9	11.8	11.8
Aver. Intercept Length ($\bar{L}$) μm	111	65	58	45	28	28	198	152	170	231	--	278	180	225	252	215	--	212
Flux Dens. (B_{15}) kG	13.3	13.7	13.8	13.9	13.7	13.9	14.2	14.6	14.8	14.6	14.4	15.0	14.3	14.5	14.6	14.8	15.1	15.0
Res. Induc. (B_r) kG	10.6	11.3	11.9	12.0	12.2	12.7	11.0	11.2	11.7	11.1	11.6	11.5	9.4	9.9	9.6	10.6	9.8	9.4
Maximum Perm. (μ_{max}) kG/Oe	4.7	4.9	4.7	4.0	3.6	3.2	9.8	8.7	8.3	6.5	6.8	6.6	9.2	9.0	8.0	7.6	7.0	6.3
Coercive Force (H_c) Oe	.9	.9	1.0	1.3	1.5	1.7	.4	.5	.5	.6	.6	.7	.4	.4	.5	.6	.6	.6
Mean Pore Area ($\bar{A}$) μm²	199	102	86	37	116	43	670	245	171	154	117	73	504	204	225	128	67	73
Mean Free Path ($\bar{\lambda}$) μm	141	143	99	117	56	62	290	279	255	212	179	225	220	249	213	232	203	276

in considerable purification and because the body centered cubic structure, coupled with less oxygen to pin grain boundaries, permits greater diffusion and hence enhanced grain growth.

Magnetic Properties of Porous Materials

Test rings were made of the various size fractions described in the previous section and sintered at 1600°F (870°C), 2050°F (1120°C), 2300°F (1260°C), 2600°F (1430°C), and 2700°F (1480°C) for 30 minutes in hydrogen. These test results are shown in Table III. It can again be seen that carbon, nitrogen, and sulfur content is equivalent to the purest commercial irons. Oxygen content decreases and average grain size intercept increases as the sintering temperature increases. The pores were also characterized in this work employing quantitative image analysis to determine the mean pore area and the mean free path between the pores. It can be seen that if powders with size fraction greater than 200 mesh are employed that the pores tend to be larger and are more widely spaced. As the size fraction becomes smaller the pores also become smaller and are spaced closer together. As the sintering temperature is increased, the pores become larger and are spaced further apart.

Flux density at 15 Oe as a function of temperature and mesh fraction are shown in Figure 8. In general it can be seen that the flux density increases as the powder becomes finer and also increases more markedly as the sintering temperature is increased. The effect of powder size and sintering on the flux density can best be seen by replotting the flux density as a function of the sintered density, Figure 9. It is seen that as the sintered density increases so does the flux density. If smaller particles are compacted there are more individual points of contact; hence, diffusion is enhanced slightly. If the sintering temperature is increased diffusion occurs more rapidly and greater densification results. It is already established that the higher the sintered density the higher the flux density will be (see section dealing with saturation). In fact, a linear relationship exists and if regression analysis is employed it may be found that:

$$B_{15} = 6.1\ (\rho) - 31.3$$

where B_{15} = flux density at 15 Oe (kG)

ρ = sintered density (g/cm^3)

Therefore, it may be surmised that fine powders and high temperature sintering enhance densification. In turn, greater densification causes the flux density to increase.

In Figures 10 and 11 however, it is seen that if fine powders are employed that the structure sensitive properties deteriorate. This is caused by the pores that are present. If fine powders are employed the pores are small and closely spaced but if coarse powders are employed the pores are larger and tend to be more widely spaced. These paramaters are plotted versus the structure sensitive properties in Figures 12 and 13. As the pores become smaller than approximately 100 square microns and as the distance between pores becomes less than approximately 100 microns the structure sensitive

Figure 8
Variation of Flux Density
With Size Fraction

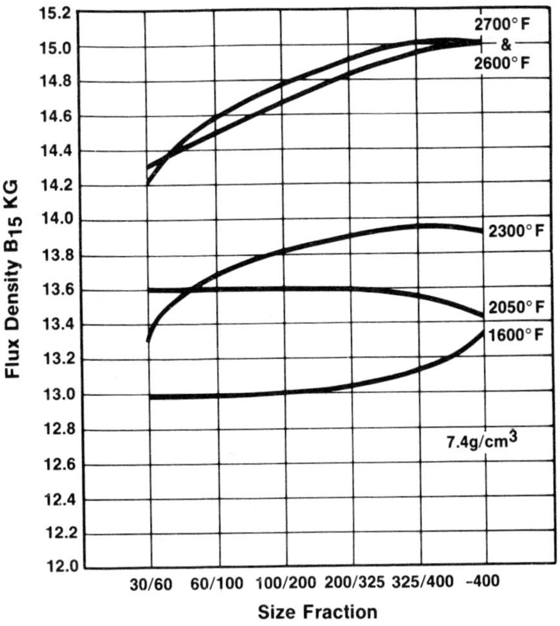

Figure 9
The Effect of Sintered Density
on the Flux Density

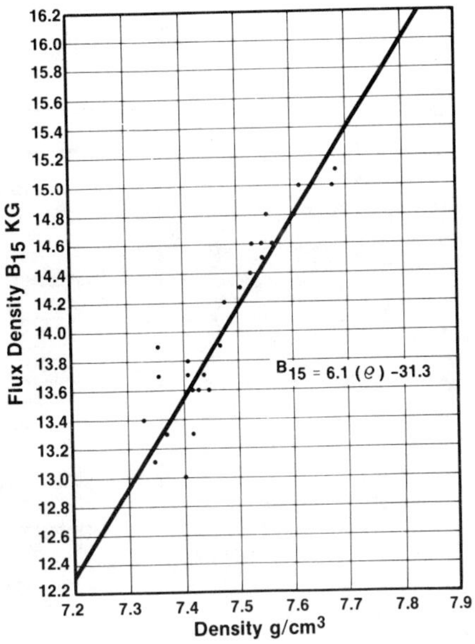

Figure 10
Variation of Maximum Permeability With Size Fraction

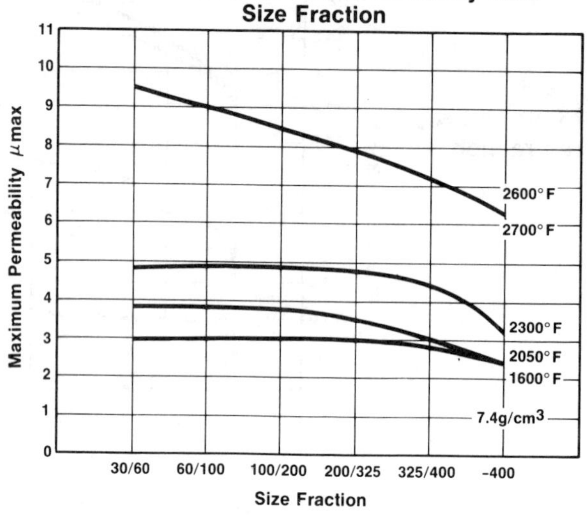

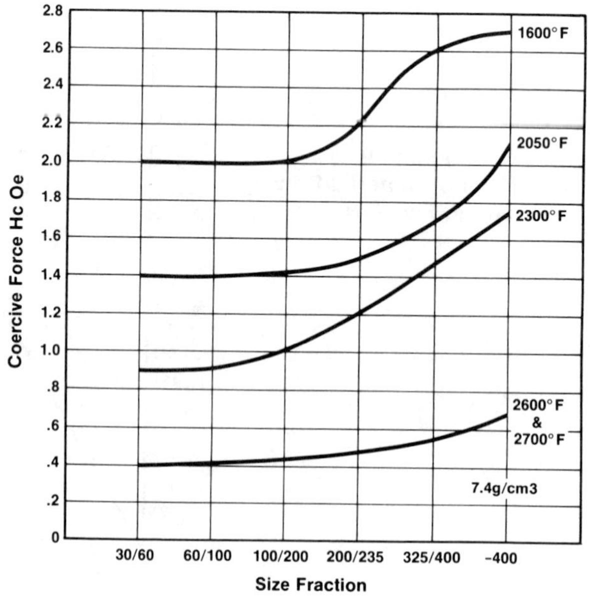

Figure 11
Variation of Coercive Force With Size Fraction

Figure 12
The Effect of Mean Pore Area on Structure Sensitive Properties

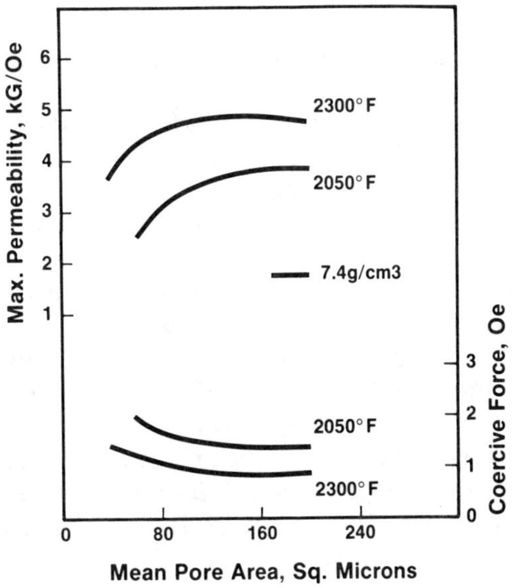

Figure 13
The Effect of Mean Free Path on Structure Sensitive Properties

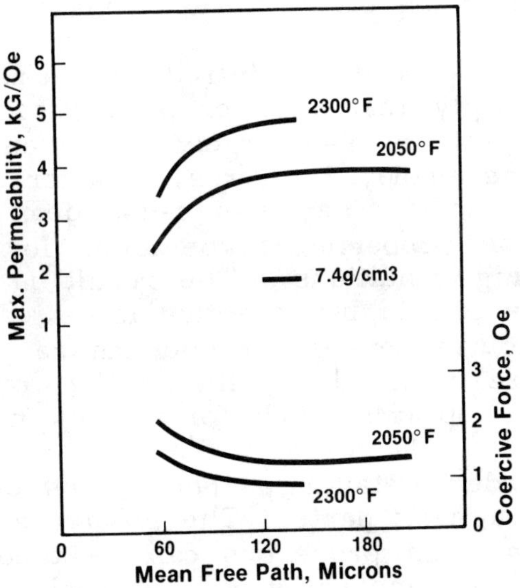

properties are degraded. Above these values the properties are at their maximum values. Also the structure sensitive properties improve markedly as the sintering temperature is increased. It has already been shown that for hot repressed material these structure sensitive properties are dependent on grain size, and high temperature sintering promotes grain growth.

The analysis of what affects structure sensitive properties is incomplete at present. The current thrust of our ongoing research program is to understand better the mechanisms that control these properties. It is also directed toward understanding the properties of more porous materials.

One additional comparison should be made of porous materials with hot repressed materials. If porosity is present, all magnetic properties are degraded to some extent depending on the powders and the processing conditions employed. Part of the degradation can be attributed to less iron available to be magnetized in porous compacts. This cannot account for the total loss in properties. Pores, like inclusions, cause demagnetizing fields to be generated. These demagnetizing fields in turn cause further degradation of the properties. At present, the extent of the degradation remains to be studied.

Conclusions

P/M parts present many new innovative possibilities for the design of magnetic devices. The properties that can be achieved depend on the powder and the processing conditions employed. When hot repressing is employed to remove porosity and conventional sintering is employed, the properties are commensurate with present commercial irons. When high temperature sintering is employed (2300°F) the properties are slightly better than commercial irons. Sintering above the delta transition temperature causes the properties to be similar to those of the purest irons.

Once porosity is introduced some compromises must be made. This is not to imply that superior properties cannot be achieved for these compacts because this is also largely dependent upon the powder and the processing conditions employed. Pores, however, reduce the amount of iron available for magnetization. They also cause demagnetizing fields to be generated which cause the properties to be lower for any given powder and/or established processing conditions. The result is lower saturation, lower maximum permeability and higher coercive force. The effect of the demagnetization fields on structure sensitive properties can be minimized. If large widely spaced pores are designed into the powder compact the structure sensitive properties will be at optimum levels for any given density.

Powder metallurgy then offers designers many opportunities to design better magnetic parts. The powder and processing conditions can be widely varied so that properties can be tailored specifically to those required for that specific application. This work just starts to provide some insight into how magnetic properties behave in porous compacts. The demagnetizing fields caused by the porosity are still not clearly understood. Also sufficient data is still unavailable to permit the designer to predict the properties he can achieve through choosing the right powder and the right processing conditions for his purpose. Hopefully, the future will provide sufficient understanding through research to render P/M parts to be employed widely for magnetic applications.

References

1. Bozorth, Richard M., *Ferromagnetism*, D. Van Nostrand Co., Inc., Princeton, New Jersey, 1951.

2. Moyer, K.H., "Porosity - Beginning to Understand Its Effects On Magnetic Properties", P/M 78 Semp 5, European Symposium on Powder Metallurgy, Stockholm, Sweden, June 4-8, 1978. Jernkontoret -Activity Group 8, Denmark, Finland, Norway, Sweden, pp 33-39.

3. Moyer, K. H., "P/M Parts for Magnetic Applications", Hoeganaes P/M Technical Conference, Philadelphia, PA, October 24-26, 1978.

P/M PLATES OF HARD-TO-WORK MATERIALS

P. Kumar and H.S. Nayar

Airco, Inc., Murray Hill, N.J. 07974

Introduction

Plate-like shapes are most often rolled or forged from castings, or they are simply cast. The yield from such processes is reduced by the defects which are normally associated with cast materials. Furthermore, the yield of hard-to-work materials is reduced by cracking during processing.

Such shapes can be fabricated using conventional P/M techniques, but some difficulties appear. Whether the powder is hot pressed[1] or cold pressed and sintered,[2] the density of plates is typically less than 90% of the theoretical density. Further, the size of the product is limited by the tonnage of the press available. Finally, the interstitial content of the product is equal to (or higher than) that of the starting powder. Unless the starting powder is very pure (and thus very expensive), the interstitial impurity of the plates is high.

Chromium and chrome alloy plates used as targets in magnetron sputtering[3] illustrate the difficulties encountered by fabricators. Such cathodes must have:

a. Low interstitial content,

b. High density,

c. Good thermal shock resistance to withstand the heat generated during sputtering,

d. Adequate mechanical strength for clamping, and

e. Non-magnetic qualities

These plates are fabricated either by casting or by conventional powder metallurgy techniques. The chromium plate is machined, brazed to a non-magnetic back-up plate, and then remachined to final shape.

This paper describes a new method, one which uses powder metallurgy and rolling practices.[4] This method overcomes the size and purity limitations without inducing casting defects in hard-to-work material.

Description of the Process

Figure 1 shows a die and punch assembly. The die has a cavity in which there are two punches. The length and width of the punches are slightly smaller than the length and width of the cavity. When a plate of another material is placed between the powder and one of the punches (as in Fig. 2), a thinner punch is used.

Openings, in the die wall like those in Fig. 2 make possible gas refinement or evacuation. Increasing the number of openings, as in Fig. 3, increases circulation in the cavity.

One punch is inserted into the die, and the cavity is partially filled with powder. After the powder is levelled, the second punch is inserted. At this point the powder's density is 55% of its theoretical density. After it is cold pressed on a hydraulic press, the powder is 65% of its theoretical density. Because the assembly is essentially free floating there is no density gradient across the thickness.

In the next step, the die is centered so that the punches stick out the same distance on both sides (Fig. 2). The punches are welded to the die using an electrode that would form a vacumn seal between the punches and the die. These welds should withstand rolling temperatures and stresses. In addition, their coefficient of expansion should be similar to that of the punch and die.

The powder can now be refined. Chromium and ferrochrome are refined by heating the assembly to 1100-1300°C and passing purified hydrogen through at 15-25 SCFH. The temperature and length of the refining phase depends on the composition of the powder, its impurities (C, O_2, N_2, S), and the desired purity. Long refining times will also help homogenize the blended powder by solid diffusion. After the powder is refined, the assembly is furnace-cooled while hydrogen is still passing through. To reduce oxidation of the outer surfaces, an inert atmosphere is maintained in the furnace.

In the next step, one tube is crimped as in Fig. 2, and gases are evacuated through the remaining tube. Heating and evacuation are continued until the outgassing rate is below 25 micrometers per minute. At that time, the assembly is cooled, and the tube is crimped and sealed. (Alternate refining and evacuation may significantly increase the purification rate, but we have not tried to do so yet.)

At this point the assembly is ready for hot fabrication. Chromium and ferroalloy are fabricated at 1100-1300°C. The first stage in fabrication, making the powder dense, can be carried out by hot pressing or by hot rolling. The objective is to obtain 90+% of theoretical density because the powder mass will be subjected to shear forces in subsequent steps.

Additional rolling increases the length of the plate and the density of the powder. In subsequent rolling passes the gap is typically 85% of the die thickness. Higher bites can also be taken provided that sufficient tonnage is available.

After the rolling pass, the assembly is cooled at 100°C per hour or less to minimize the residual stresses resulting from differences in expansion coefficients of powder, punch, and frame. However, to suppress the formation of brittle intermetallics, the cooling rate is increased in certain temperature ranges. For example, ferrochrome alloys are cooled rapidly from 850°C to 650°C. Control of rolling and subsequent cooling is critical.

At room temperature, the assembly is cut to the desired dimensions. Cutting stresses should be minimized to reduce the chances of cracking. The remnants of punches and die can be removed or retained as design requirements warrant.

Fabrication of Plates

Figures 1 and 2 give various dimensions the following nomenclature:

a. Size of die: $L_D \times W_D \times T_D$

b. Size of punch: $L_P \times W_P \times T_P$

c. Punch intrusion into die: d_i

d. Punch protrusion from die: d_p

e. Intrusion ratio: d_i/d_p

f. Thickness of the powder: T_M

g. Cold pressed density of the powder: ρ_c, expressed as a percentage of theoretical density

h. Roll diameter: G

i. Die Thickness: T_D

This nomenclature will be used to describe those elements that control the design of the die-punch assembly, namely the rolling gap ratio, intrusion ratio, protrusion ratio and the roll dimmeter.

Total protrusion (P) should be greater than $(1-\rho_C)\ T_M$ to get 100% of the theoretical density of the consolidated powder. It is practical to have a protrusion ratio of 1.2 or higher. The ratio can be calculated from the following formula:

$$p = \frac{P}{(1-\rho C)\ T_M}$$

Since too large a protrusion ratio reduces efficiency in the use of material, the ratio is best held between 1.2 and 2.0.

The density of the material can be increased during the first pass by reducing the roll gap (G in Fig. 4). If however, the gap is much smaller than the thickness of the die, the die will be substantially deformed. Furthermore, a small gap would require a higher tonnage rolling mill, and the resultant pressures would induce shear forces in the powder mass before the 90+% theoretical density is reached. A gap about the same size as the thickness of the die, or slightly smaller is desirable for the first pass. The gap ratio ($g=G/T_D$) should be about 0.95 to 1.05. The punches should be compressed flush with the die, but neither the die's thickness nor the powder core's length should be changed.

The intrusion ratio, i ($=d_i/d_p$), required to keep the punch from being forced out of the die during rolling increases as the roll diameter G decreases. Typically for a 508 mm roll diameter an intrusion ratio of 1 is satisfactory.

Tables 1, 2 and 3 describe the material used, the design of the starting assembly, and details of consolidation.

To fabricate plates 1, 2 and 3, we varied intrusion ratio and protrusion ratio, while holding other design elements more or less identical. The punches popped out of the plate 1 assembly during rolling because of the low intrusion ratio for the size of the roll (0.5 for 244 mm diameter rolls). Even though the protrusion ratio was high (3.13) the density after first rolling was only 74% because the roll gap ratio was high (1.5). The lower density of plate 3 compared to that of plate 2 was caused by the protrusion ratio of less than one. Theoretically, a protrusion ratio of one will give 100% density after the first pass provided that neither the die nor core elongates during rolling. However, plates 2 and 3 indicate that a protrusion rate of 1.2 or higher is desirable.

Plates 4 through 7 were rolled on a larger rolling mill (508mm diameter rolls). We varied the intrusion ratio and the gap ratio, while holding other design parameters more or less identical. For plate 4, the punches and weld maintained their integrity during the first pass. A comparison of plates 1 and 4 indicates that a large roll diameter and a low gap ratio help to prevent the separation of the punch from the assembly.

For plates 5, 6 and 7, the main variable was the gap ratio. Densities after the first pass were estimated from measurements of external dimensions and from radiography. As the gap ratio decreases, the density after the first pass increases.

All assemblies, 4, 5, 6 and 7, had lower oxygen content than the starting powder because they were outgassed at elevated temperatures before being rolled. The interface between chromium and stainless steel was very weak both because of the difference in coefficients of expansion of stainless steel and chromium, and because of uncontrolled cooling.

Plates 8 and 9 were similar except that plate 8 was fabricated from chromium powder and plate 9 from ferrochrome. After the first pass, one of the punches was removed to examine the core. The cores of both assemblies were crack-free and 95% dense. Fig. 5 shows micrographs of plates 8 and 9 after first and second rolling. A comparison between plates 6 and 9 shows that we obtained higher density with Inconel 601 punches than with 347 stainless steel. The higher yield strength of the Inconel at rolling temperature is probably responsible for this effect.

In plate 10, we used ferro-TiC powder made by Sintercast Division of Chromalloy Corp. The powder is an iron-base matrix with 45% TiC by colume. The assembly for the plate had only one punch with low intrusion (0.67) and protrusion (1.02) ratios. Before being rolled, the powder was refined without the punch in the assembly by being heated at 1177°C for six hours under hydrogen. The fabricated plate was crack-free. Figure 6 shows micrographs of the plate. The successful fabrication of plate 10 indicates (a) that this method can be used with a single punch; (b) that powders containing significant fractions of hard-to-work material can be formed into plates; and (c) that it is possible to reduce interstitial impurities by refining before fabricating.

Plate 11 had a starting design similar to those of plates 1-9, but before hot rolling it was hot pressed at 1190°C and 275 MPa. The estimated hot-pressed density was 86% of theoretical density. This low density points out that the largest size which can be fabricated by hot pressing is substantially limited if hot rolling is not part of the fabrication process.

The area of the plate can be increased either by taking large reductions in a few passes or by taking small reductions in several. Large reductions bow the plate unless rolls of very large diameters are used. Large reductions, also, require higher tonnage. For these reasons we preferred to take smaller reductions, and elongate the plate in several passes.

In assembling plates 12, 13 and 14, we placed plates on Monel 400, nickel, and Inconel 600 between the powder and one of the punches. Chromium and ferrochrome plates fabricated without such plates showed weak bonds between core and punch caused by the differences in coefficients of thermal expansion. We expected that by interposing a material which as a coefficient of expansion between those of chromium abd stainless steel, we could reduce the residual stresses during cooling. The coefficient of expansion for all three interposed materials is the same. We reduced the residual stresses also by slow cooling except between 871°C and 700°C where the cooling was rapid to prevent intermetallics forming.

The bond in plate 12 between Monel 400 and chromium was very strong, although the Monel had separated from the stainless punch. This separation was attributed to the incipient melting of Monel during rolling.

Plates 13 and 14 showed good bonds between the core and the interposed plate, and between the plate and the punch. However, the core itself showed a few cracks, an indication that residual stresses were not adequately eliminated and that the interfaces were almost as strong as (if not stronger than) the core.

The powder in plate 15 was refined by passing hydrogen through the welded assembly at 1230°C for 14 hours. After hot rolling, the assembly was cooled slowly. The residual stresses were further reduced by soaking the assembly at 650°C for 16 hours. At room temperature the assembly was cut to expose the core. The bond at both interfaces was along the entire cross-section and strong. Figure 7 shows the micrographs of its interface.

Figure 8 is a clear picture of the smooth surface and good bond on a sputtering cathode machined from a hot rolled plate.

Discussion

We have demonstrated that by the P/M rolling process it is possible to fabricate plate shapes from chromium, ferrochrome and ferro-TiC powders. Plates can be fabricated either of single materials or of mixtures of materials as composites with metallurgical bonding. Furthermore, the process is not limited to these materials; it can be used to fabricate such metals and alloys as molybdenum, zirconium, niobium, tantalum, and tungsten which are difficult to work because of their inherent chemical reactivity, brittleness, or high strength.

The following parameters in the process have critical effects on P/M rolling:

a. Intrusion ratio (should be ≈ 1);

b. Protrusion ratio (should be 1.2-2.0);

c. Roll diameter (a minimum size is required);

d. Rolling temperature (high temperatures are desirable);

e. Rolling schedule (gap ratio for the first pass should be ≈ 1; small reductions are preferred for subsequent passes).

In addition to these parameters the properties of the punch and die are very important. For example, materials should be more ductile than the core at rolling temperature. If a composite plate is being made, the properties of the interposed material are also important. Such material ought to have the following characteristics:

a. Its coefficient of thermal expansion should be compatible with the core and punch;

b. It should not form brittle intermetallics with either the core or the punch; and

c. It should be ductile at rolling temperature.

Other material requirements will depend on the individual application. For example, for a sputtering target neither the punch nor the interposed plate should be magnetic.

We used this process to fabricate plates longer than those reported on here. We also fabricated alloy plates using elemental blended powders. Although the homogenization was not complete, it should be possible to use fine powder (smaller than 325 mesh) to obtain satisfactory homogenization during chemical refining.

A few limitations of the process and ways to overcome them are the following:

a. Interstitial impurities are greater than in cast plates. Although hydrogen refining helps, it requires long heating at high temperatures, which can cause oxidation of the dies and punches. Alternate hydrogen refining and evacuation[5] should reduce the length of heating.

b. The assembly tends to curve in the direction of rolling as the length increases. A light final pass may circumvent this difficulty.

c. Elemental blended powder was not completely homogenized. Fine powder, or a master alloy, partially prealloyed powder, or liquid state sintering might provide better homogenization.

d. The process of making bonded or composite plates is sensitive to differences in coefficients of thermal expansion between the core and the punch.

Potential Applications

The process can fabricate plates either of one composition or composite material. It can be used to produce plates with or without metallurgical bond to other material. The size of the plate is limited by the equipment available.

We have successfully made chromium and ferrochrome plates for sputtering. We have also made ferro-TiC plates. Some other materials which can be fabricated are Cr-Cu alloy plates, Triboloy bonded to stainless steel, molybdenum or tungsten bonded to stainless steel, copper plates with graphite fiber, and beryllium and beryllium-based alloys.

Summary

This paper describes a process for fabricating plates of hard-to-work materials. Based on rolling mill and powder metallurgy technologies, the process uses materials more efficiently than casting, and produces plates characterized by uniform high density, high purity, and freedom from cracks. The process can produce plates with or without metallurgical bond to other materials. It is, finally, easily adapted to conventional rolling mills.

Acknowledgement

The authors wish to extend their appreciation to Mr. Steve Matras for carrying out the work in earlier stages of the development.

REFERENCES:

1. B.H. Hessler and J.P. Denny, "Fabrication of Beryllium from Hot Pressed Powder", Progress in Powder Metallurgy, Vol. 17, 5-11, 1961.

2. R.W. Fountain, "Chromium", p. 106, Ed. Clifford A. Hampel, published by Reinhold Publishing Corp., II Edition, 1961.

3. John S. Chapin, "The Planer Magnetron", Research/Development Vol. 25, 1, 37-40, January 1974.

4. H.S. Nayar, "Manufacture of Plates by Powder Metallurgy", US Patent No. 4, 126, 451, November 21, 1978.

5. N.C. Kothari, "Strength Properties of Type 316 Stainless Steel Specimens Sintered in Hydrogen-Vacuum", Progress in Powder Metallurgy, Vol. 11, 89-96, 1976.

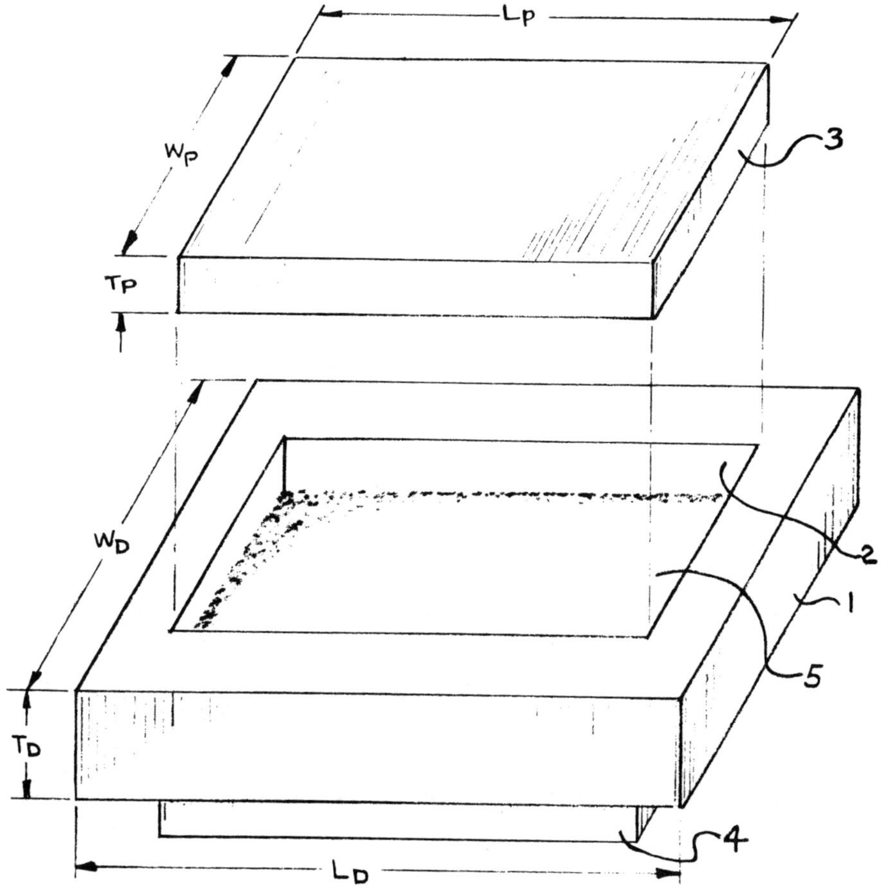

FIGURE 1: DIE AND PUNCH ASSEMBLY

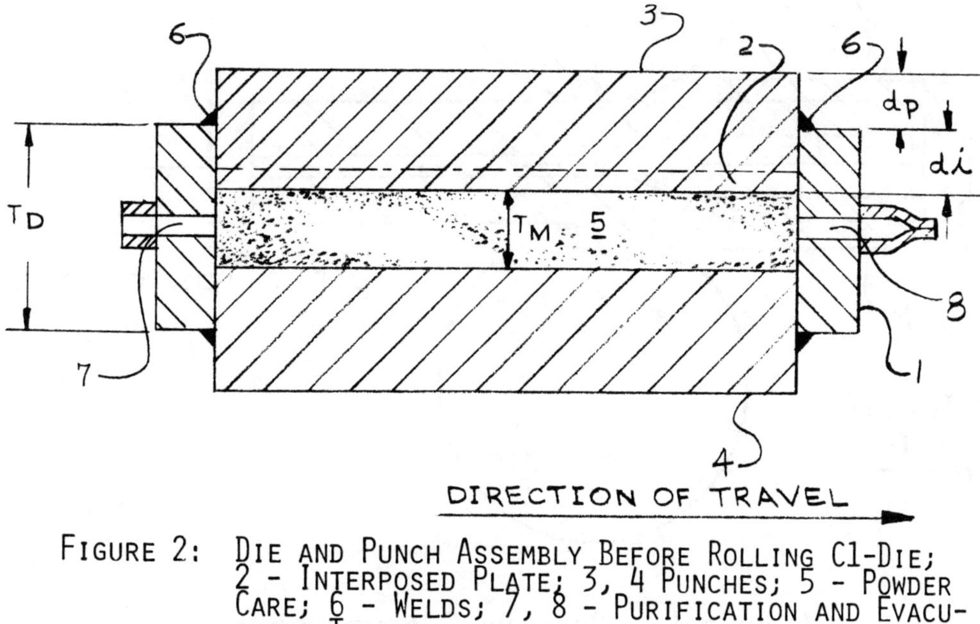

FIGURE 2: DIE AND PUNCH ASSEMBLY BEFORE ROLLING C1-DIE; 2 - INTERPOSED PLATE; 3, 4 PUNCHES; 5 - POWDER CARE; 6 - WELDS; 7, 8 - PURIFICATION AND EVACUATION TUBES

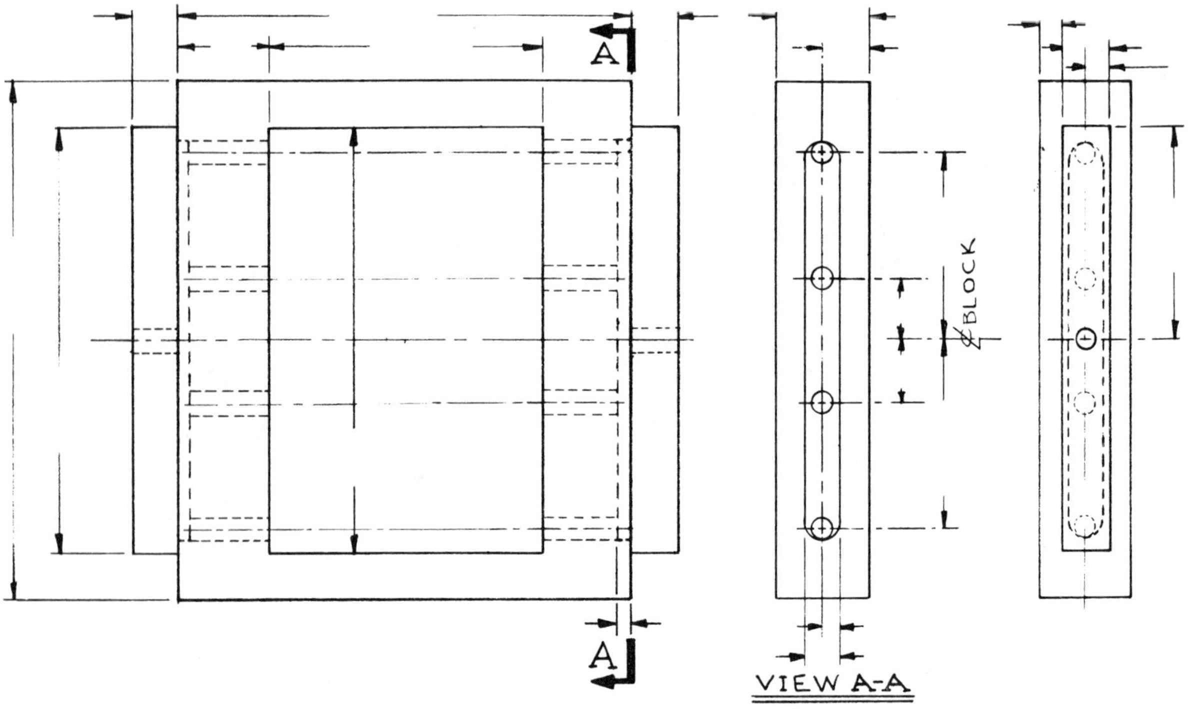

Figure 3: Modified Die Design for Improved Gas Circulation

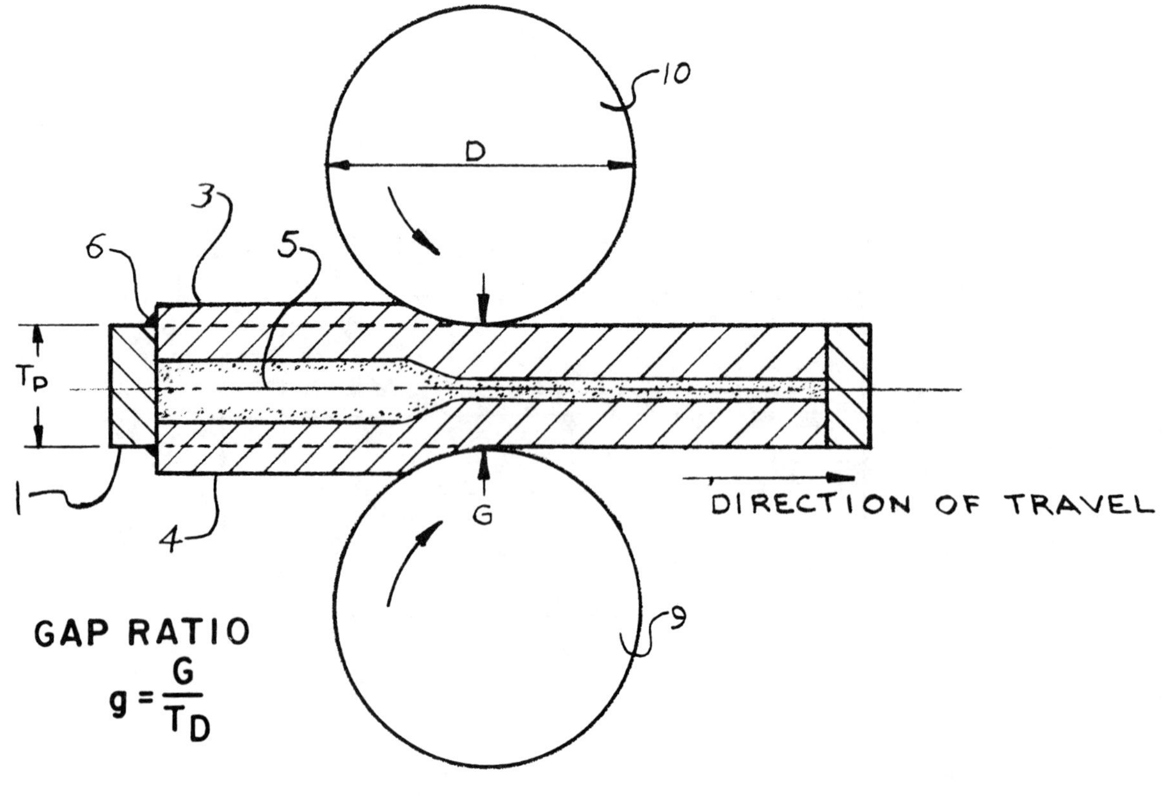

Figure 4: Rolling Parameters

PLATE 8

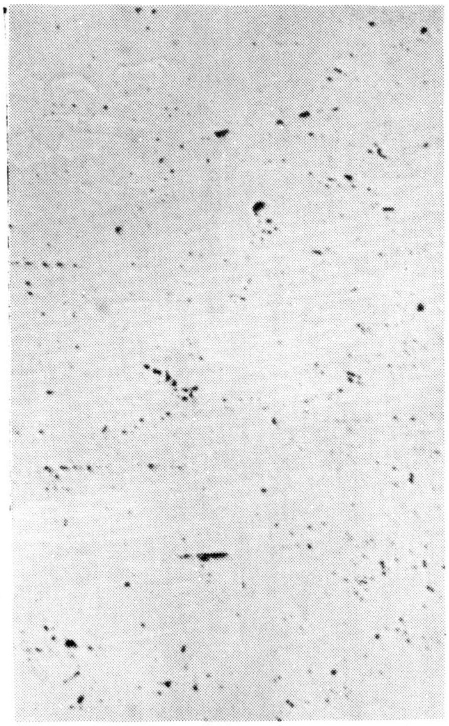

PLATE 9

After First Rolling

PLATE 8

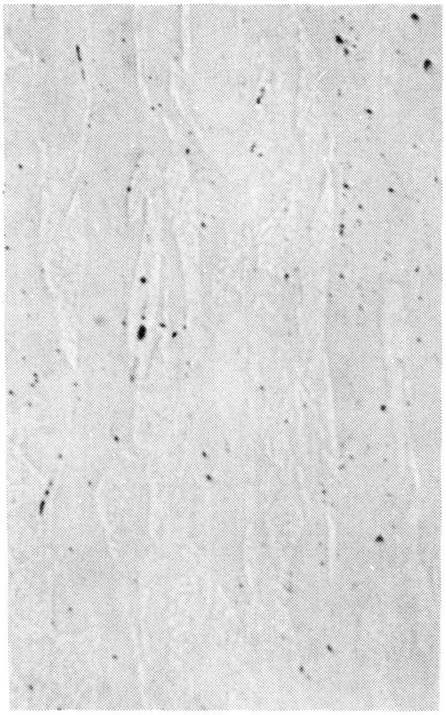

PLATE 9

After Second Rolling

FIGURE 5: MICROGRAPHS AFTER FIRST AND SECOND ROLLING

MAG 100X

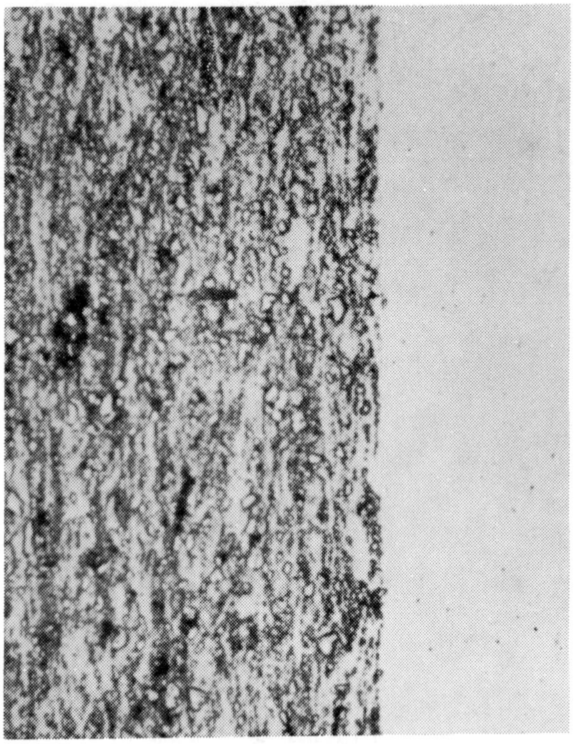

INTERFACE WITH STEEL

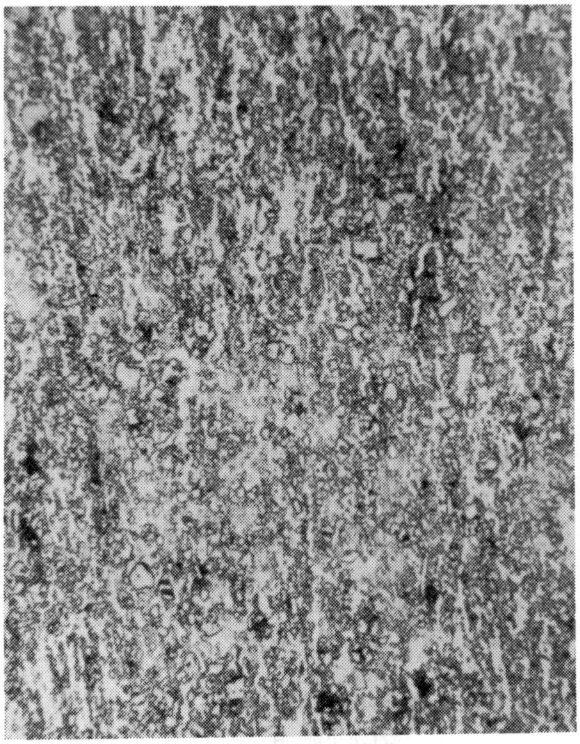

CORE

FIGURE 6: MICROGRAPH OF FERRO-TiC PLATE. MAG: 300x

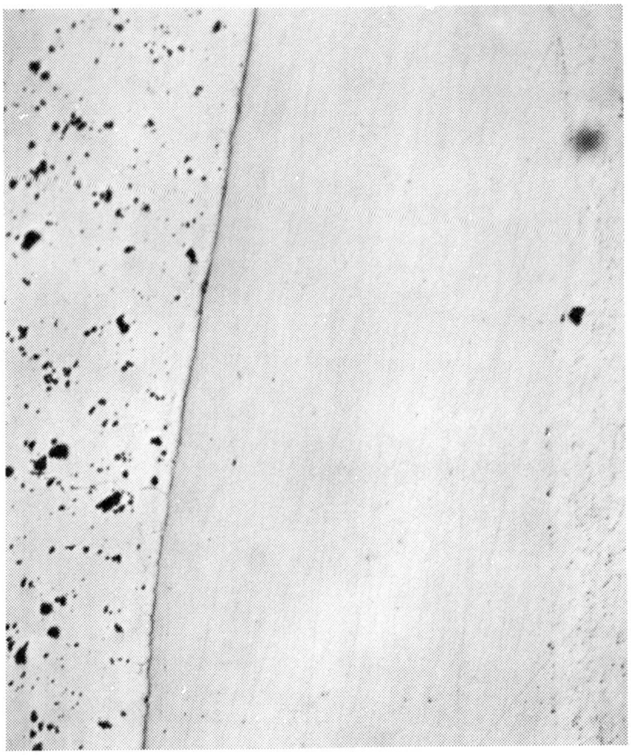

FIGURE 7: MICROGRAPH OF PLATE 15.

MAG: 50X

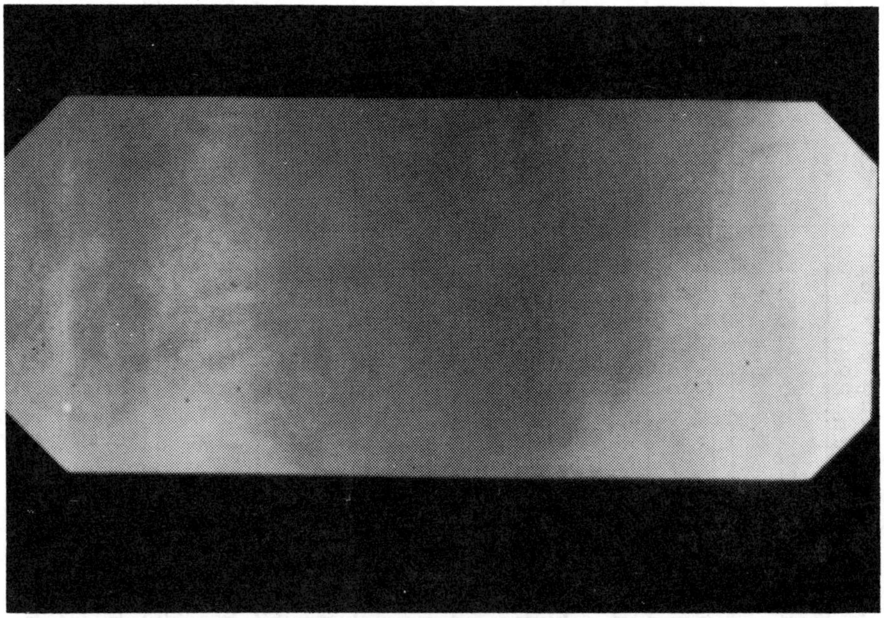

TOP SURFACE

BOND ACROSS THE INTERFACE

FIGURE 8: PHOTOGRAPH OF FINISHED CATHODE FROM PLATE 15

TABLE I: MATERIALS DETAILS FOR FABRICATING PLATES VIA P/M ROLLING

I.D.	Die	Punch	Plate between Punch & Powder	Weld Material	Powder	Chemistry of Starting Powder (in PPM)	Chemistry of Fabricated Plate (in PPM)
1	304 stainless	304 stainless	None	308 stainless	Chromium[1]	C=220; O_2=1655; N_2=88	--
2	304 stainless	304 stainless	None	308 stainless	Chromium	C=220; O_2=1655; N_2=88	--
3	304 stainless	304 stainless	None	308 stainless	Chromium	C=220; O_2=1655; N_2=88	--
4	304 stainless	347 stainless	None	308 stainless	Chromium + 11% Iron	C=170; O_2=1700; N_2=95	C=220; O_2=750; N_2=150
5	304 stainless	347 stainless	None	308 stainless	Chromium + 11% Iron	C=170; O_2=1700; N_2=95	C=230; O_2=950; N_2=350
6	304 stainless	347 stainless	None	308 stainless	Chromium + 11% Iron	C=170; O_2=1700; N_2=95	C=230; O_2=1200; N_2=155
7	304 stainless	347 stainless	None	308 stainless	Chromium + 11% Iron	C=170; O_2=1700; N_2=95	C-240; O_2=1010 N_2=200
8	304 stainless	Inconel 601	None	Inconel 182	Chromium	C=220; O_2=1655; N_2=88	--
9	304 stainless	Inconel 601	None	Inconel 182	Chromium + 11% Iron	C=170; O_2=1700; N_2=95	--

...CONTINUED...

TABLE I: MATERIALS DETAILS FOR FABRICATING PLATES VIA P/M ROLLING

(continued)

I.D.	Die	Punch	Plate between Punch & Powder	Weld Material	Powder	Chemistry of Starting Powder (in PPM)	Chemistry of Fabricated Plate (in PPM)
10[2]	304 stainless	304 stainless	None	308 stainless	Ferro-TiC	C=9.86%; O_2=18,900 N_2=1416	C=6.3%; O_2=9300; N_2=567
11	304 stainless	304 stainless	None	308 stainless	Chromium + 11% Iron	C=240; O_2=1600; N_2=100	C=280; O_2=1634; N_2=110
12	Hot rolled steel	stainless steel	Monel 400	308 stainless	Chromium	C=150; O_2=1100; N_2=120	--
13	304 stainless	347 stainless	Nickel	Inconel 182	Chromium	C=290; O_2=1700; N_2=80	C=250; O_2=1200; N_2=250
14	304 stainless	347 stainless	Inconel 600	Inconel 182	Chromium	C=290; O_2=1700; N_2=80	--
15	304 stainless	321 stainless	Inconel 600	Inconel 182	Chromium + 23% Iron	C=65; O_2=1600[3]; N_2=46	C-100; O_2=350; N_2=75

[1] Chromium powder was -60 mesh in size.

[2] Obtained from Sintercast Division of Chromalloy Corp. It contains 45% (vol.) TiC. The particle size was -325 mesh.

[3] The starting powder had the following impurities: C - 290 PPM; O_2 - 4300 PPM; N_2 - 655 PPM. It was purified outside the assembly by heating at 1190°C for 40 hours.

TABLE II: DETAILS OF ASSEMBLY FOR FABRICATING PLATES VIA P/M ROLLING

I.D.	Size of Die[1] $L_D \times W_D \times T_D$	Size of Punch[1] $L_P \times W_P \times T_P$	Size of Plate[1] between Powder & Punch $L_S \times W_S \times T_S$	Size of[1] Powder Core $L_M \times W_M \times T_M$	d_p [1]	d_i [1]	$i=\frac{d_i}{d_p}$ [2]	Fill[3] Density ρ_i (%)	Pressed[3] Density ρ_c (%)	$p=\frac{2d_p}{(1-\rho_c)T_M}$ [2]
1	216 x 177 x 32	152 x 152 x 19	--	152 x 152 x 19	12.65	6.35	0.5	51	58	3.13
2	216 x 177 x 33	152 x 152 x 13	--	152 x 152 x 14	3.6	9.1	2.54	51	60	1.28
3	216 x 177 x 33	152 x 152 x 13	--	152 x 152 x 13	2.5	10.4	4.0	51	60	1.05
4	254 x 279 x 51	152 x 178 x 19.5	--	152 x 178 x 39	12.6	6.8	0.54	53	67	1.96
5	254 x 279 x 51	152 x 178 x 19.5	--	152 x 178 x 31	9.4	10.1	1.08	55	66	1.78
6	254 x 279 x 51	152 x 178 x 19.7	--	152 x 178 x 30	9.3	10.4	1.11	55	67	1.85
7	254 x 279 x 51	152 x 178 x 19.6	--	152 x 178 x 30.1	9.1	10.5	1.15	55	67	1.84
8	241 x 267 x 51	140 x 165 x 19	--	140 x 165 x 32	9.9	9.1	0.92	55	66	1.81
9	241 x 267 x 51	140 x 165 x 19	--	140 x 165 x 32	9.3	9.7	1.05	55	69	1.89
10	216 x 178 x 33	152 x 152 x 13[4]	--	152 x 152 x 17	7.8	5.2	0.67	40	55	1.02[4]
11	266 x 241 x 53	191 x 191 x 19	--	191 x 191 x 35	9.7	9.3	0.96	54	63	1.51
12[5]	254 x 279 x 51	152 x 178 x 19.7	152 x 178 x 3	152 x 178 x 29	9.1	10.5	1.15	54	61	1.61
13	268 x 292 x 51	152 x 178 x 19 1	152 x 178 x 3	152 x 178 x 30.5	8.9	10.2	1.15	54.5	62.9	1.57

...CONTINUED...

TABLE II: DETAILS OF ASSEMBLY FOR FABRICATING PLATES VIA P/M ROLLING

(continued)

I.D.	Size of Die $L_D \times W_D \times T_D$	Size of Punch $L_P \times W_P \times T_P$	Size of Plate between Powder & Punch $L_S \times W_S \times T_S$	Size of Powder Core $L_M \times W_M \times T_M$	d_p	d_i	$i=\frac{d_i}{d_p}$	Fill Density ρi (%)	Pressed Density ρc (%)	$p=\frac{2d_p}{(1-\rho_c)T_M}$
14	268 x 292 x 51	152 x 178 x 19.1	152 x 178 x 3	152 x 178 x 30.5	8.4	10.7	1.27	54.9	64.2	1.54
15	178 x 229 x 51	102 x 178 x 19.1	102 x 178 x 3	102 x 178 x 30.5	9.8	9.5	1.01	55	62	1.64

1 Unless otherwise stated, all dimensions are in mm.

2 i and p are dimensionless ratios.

3 Densities are percentages of theoretical densities.

4 Only one punch was used.

5 The punch thickness was reduced to compensate for the interface plate in assemblies 12-15.

TABLE III: DETAILS OF ROLLING FOR FABRICATING PLATES VIA P/M ROLLING[1]

I.D.	Passes	Roll[2] Size	Roll Gap G	$g = \frac{G}{T_D}$ [4]	Force per Roll (in MPa)	Dimension of Die	Dimension of Core	Final Density of Core
1[3]	0	244	--	--	--	216 x 177 x 32	152 x 152 x 19	--
	1	244	48	1.5	--	229 x 178 x 32	178 x 152 x 13	74%
2	0	244	--	--	--	216 x 177 x 33	152 x 152 x 14	--
	1	244	33	1	--	--	--	--
	2	244	27			254 x 181 x 27	191 x 181 x 7	95%
3	0	244	--	--	--	216 x 177 x 33	152 x 152 x 13	--
	1	244	32	0.95	--	242 x 177 x 31	168 x 153 x 9[5]	--
	2	244	21	--	--	338 x 181 x 21	233 x 159 x 5	92%
4[6]	0	508	--	--	--	254 x 279 x 51	152 x 178 x 39	--
	1	508	61	1.19	689	273 x 288 x 51	179 x 184 x 28[5]	80%[5]
	2	508	27	--	4134	540 x 298 x 28	314 x 191 x 10.7	97%
5[6]	0	--	--	--	--	254 x 297 x 51	152 x 178 x 31	--
	1	508	56	1.1	689	268 x 290 x 51	167 x 180 x 19.5[5]	82%[5]
	2	508	25	--	4134	508 x 302 x 25	314 x 191 x 10	98%

...CONTINUED...

TABLE III: DETAILS OF ROLLING FOR FABRICATING PLATES VIA P/M ROLLING

(continued - page 2)

I.D.	Passes	Roll Size	Roll Gap G	$g=\frac{G}{T_D}$	Force per roll (in MPa)	Dimension of Die	Dimension of Core	Final Density of Core
6[6]	0	508	--	--	--	254 x 279 x 51	152 x 178 x 30	--
	1	508	51	1.0	1103	281 x 291 x 49	177 x 182 x 16.1	90%[5]
	2	508	22	--	5982	584 x 298 x 22	343 x 184 x 7.5	99%
7[6]	0	508	--	--	--	254 x 279 x 51	152 x 178 x 30	--
	1	--	45	0.88	1516	298 x 294 x 45	187 x 184 x 14	95%[5]
	2	--	20	--	4136	632 x 305 x 20	381 x 188 x 6	99%
8[6]	0	508	--	--	--	241 x 267 x 51	140 x 165 x 32	--
	1	--	51	1.0	--	279 x 274 x 50	165 x 170 x 19[5]	95%[5]
	2	--	25	--	--	483 x 317 x 25	290 x 186 x 9	98%
9[6]	0	508	--	--	--	241 x 267 x 51	140 x 165 x 32	--
	1	--	51	1.0	--	333 x 279 x 50	207 x 171 x 14[5]	9
	2	--	25	--	--	440 x 343 x 25	267 x 216 x 9	99%
10	0	508	--	--	--	216 x 178 x 33	152 x 152 x 17	--
	1	--	32	0.96	--	279 x 216 x 32	--	--
	2	--	17	--	--	356 x 216 x 17	4 mm thick	95%

...CONTINUED...

TABLE III: DETAILS OF ROLLING FOR FABRICATING PLATES VIA P/M ROLLING

(continued - page 3)

I.D.	Passes	Roll Size	Roll Gap G	$g=\frac{G}{T_D}$	Force per Roll (in MPa)	Dimension of Die	Dimension of Core	Final Density of Core
11	0		--	--	--	266 x 241 x 53	191 x 191 x 35	--
	1	hot pressed[7]	--	--	--	--	--	86%[5]
	1	406	45	0.83	--	350 x 317 x 43	--	--
	2	406	32	--	--	432 x 330 x 32	305 x 241 x 11	98%
12[8]	0	508	--	--	--	254 x 279 x 51	152 x 178 x 29	--
	1	--	49.4	0.97	689	292 x 305 x 49	--	--
	2		22.5	--	3447	305 x 635 x 22.5	381 x 197 x 6.8	97%
13	0	508	--	--	--	268 x 292 x 51	152 x 178 x 30.5	--
	1	--	47.1	0.92	1516	317 x 298 x 47	--	--
	2	--	21.4	--	4275	635 x 308 x 22	381 x 191 x 7.4	98%
14	0	508	--	--	--	268 x 292 x 51	152 x 178 x 7.4	--
	1	--	47.1	0.92	1516	311 x 298 x 47	--	--
	2	--	21.4	--	4137	622 x 305 x 21.4	385 x 192 x 7.1	98%
15[9,10]	0	508	--	--	--	178 x 229 x 51	102 x 178 x 30.5	--
	1	--	45	0.88	1470	229 x 238 x 45	--	--
	2	--	20.3	--	3922	483 x 254 x 21	--	99%

...CONTINUED...

TABLE III: DETAILS OF ROLLING FOR FABRICATING PLATES VIA P/M ROLLING

(continued - page 4)

NOTES:

1 Unless otherwise states, all dimensions are in mm.

2 Assemblies were evacuated to a pressure of 2-5 micrometer column of Hg, then heated to 320°C while being pumped and evacuated at temperature until the outgassing rate went down to 30-40 micrometers column of Hg/minute. At that time, the evacuation tube was crimped and sealed. The assemblies were soaked for 2 hours at 1200°C before rolling.

3 Assemblies 1, 2 and 3 were heated to 1180°C before rolling. The assemblies were air cooled after rolling.

4 g is given only for the first pass setting.

5 Estimated on the basis of external measurements and radiography.

6 The assemblies were cooled in vernaculite after rolling.

7 Prior to rolling, the assembly was hot pressed twice at 1190°C and 275 MPa (MN/m^2).

8 Plates 12-14 were cooled as per the following schedule: 1200°C/83°C per hour/871°C + 871°C/air cool/700°C + 700°C/30°C per hour/room temperature.

9 Prior to evacuating and crimping, the assembly was heated at 1230°C for 14 hours with the flow of dry hydrogen through the core.

10 The cooling schedule after rolling was 1200°C/110°C per hour/760°C + 760°C/air cool/650°C + 16 hours at 650°C + 650°C/28°C per hour/room temperature.

CHARACTERIZATION AND PROCESSIBILITY OF
HIGH SPEED TOOL STEEL POWDERS

Dr. Kishor M. Kulkarni

Glidden Metals Group of SCM Corp.
Cleveland, Ohio

ABSTRACT

New developments in near net shape fabrication of high speed tool steel components via cold press and sinter route can in many cases lead to substantial cost reduction. The recent commercial availability of high quality water atomized powders is an important element in this development. This paper describes characteristics and processibility of water atomized powders and to a limited extent compares them with similar properties of impact pulverized and gas atomized powders. The powder characteristics studied include size distribution, shape, tap and apparent densities, flow rate, hardness and oxygen content. Then, compressibility and green strength of different types of powders are compared. An attempt is made to show the interrelationship of different powder characteristics and material processibility. Some background information is provided about tool steel metallurgy, P/M mill forms, full dense sintering and powder production.

Introduction

In the past few years water atomized tool steel powders have been commercially available. These high quality powders can be compacted using existing equipment and then the compacts can be sintered to full density without the need for any post sintering operations for further densification. This paper is primarily about the characteristics of such water atomized tool steel powders. Through out the paper M-2 high speed steel (HSS) is used for illustration although most comments are practically equally applicable to other HSS powders and to a lesser extent to other non-HSS tool steels. Details of powder production and sintering are beyond the scope of this paper but these items are referred to briefly in view of the close interrelationship between powder production, characteristics and sinterability. To a limited extent the water atomized powders are compared to gas atomized powders and to those produced with an additional step of impact pulverization.

In this paper by "full dense" sintering we mean densification in excess of 98 percent of the density of wrought material or similar composition. This is of a lower limit as in most practical applications the density is significantly higher. Often density measurement cannot be relied on as an inspection method for porosity and only microscopic examination must be employed to detect if there is any porosity.

Background

Traditionally the "as sintered" density of ferrous or copper based powder metallurgy (P/M) materials rarely exceeds 90 percent of the wrought density. In some applications such as in bronze bearings and filters the applications require porosity. In others the target density is selected on the basis of a combination of product requirements, commercial feasibility and cost. Further increase in density can be obtained at added cost by techniques such as infiltration, P/M forging or other type of mechanical working and hot isostatic pressing (HIP). For various reasons so far P/M forming has not found wide applications; but the investigation effort has clearly demonstrated the sensitivity of important properties like fatigue life and impact strength to even a few percent of residual porosity. In shaped components HIP has been used primarily for nickel base superalloys for aerospace industry and for carbides as tooling components or cutting tools. By implication HIP is still too expensive for other applications. Property improvements obtained by infiltration are modest in comparison with those possible by P/M forging or HIP and infiltration is not relevent in the context of processing of tool steels.

At present most of the HSS are used in cutting tools in operations involving interrupted cuts. In such demanding applications full density is required. Yet, till recently cost effective P/M methods were not available for processing of tool steels and therefore few tool steel parts were made via P/M. Progress in technology and increasing cost for key alloying elements like Cobalt, Tungsten and Molybdenum are now providing impetus to P/M processing of tool steels.

Alloy \ Wt. %	C	W	Cr	V	Mo	Co
M-2	1	6	4	2	5	-
T-15	1.5	12	4	5	-	5

PRINCIPAL FUNCTIONS

W - HARD CARBIDES, RED HARDNESS

Cr - HARDENABILITY, HARDNESS WITH HIGH CARBON

V - VERY HARD CARBIDES, FINE GRAINS

Mo - HARD CARBIDES, RED HARDNESS, HARDENABILITY

Co - HIGH RED HARDNESS

Fig. 1 Compositions of Two Common High Speed Tool Steels and Functions of Key Alloying Elements

Particular attention is attracted to near net shape production methods in view of the increasing labor and material costs and material shortages.

HSS Metallurgy

High speed tool steels are probably the most complex amongst the common ferrous materials and there are several excellent books[1-3] and numerous technical papers devoted to their metallurgy. We shall only review some basic facts important for the remainder of this paper.

Fig. 1 shows nominal compositions of M-2 - the most common grade in U.S.A. - and of T-15 - one of the more highly alloyed grades. Functions of important alloying elements are also summarized in this figure. High hardness, its retention at high temperatures and adequate toughness are major attributes of HSS. The room temperature hardness derives from the usual martensitic reaction and from hard carbides of Cr, Mo, W and V. The high temperature or hot hardness comes because of the stability of carbides of Mo, W and V. The classic paper by Kayser and Cohen[4] clearly shows how a large portion of MC type carbides of V and much of M_6C type carbides of W and Mo remain undissolved even at the high solution treating temperature employed for HSS. Overall these alloying elements greatly reduce the kinetics of most reactions, make HSS air hardening, necessitate multiple tempering and give HSS a very high resistance to grain growth. Most HSS are solution treated at temperatures close to start of melting and for optimum properties close process control is essential during their heat treatment.

P/M Mill Forms

There is general agreement that it is desirable to have fine, uniformly distributed carbides. Unfortunately, the carbides in tool steels are prone to severe center line segregation at the ingot stage. Extensive degree of hot working is essential to break up and distribute the carbides and even then some banding and non-uniformity of carbide size persists. The carbides also cause poor hot workability. As a result of these factors conventionally produced mill forms i.e. round or rectangular bars and other products of uniform cross section, are expensive to produce and of less than optimum properties. Therefore much work[5-8] has been carried out on production of P/M mill forms and Crucible Inc. division of Colt Industries and Uddeholms (formerly Stora Kopparberg) have commercially produced P/M mill forms since early 1970's.

The P/M method in its most common form comprises production of gas atomized powder, preparation of cold isostatically pressed (CIP) compact, canning the compact and then subjecting it to HIP. Subsequent hot mechanical working may or may not be employed. The main advantages claimed of the P/M mill form are reduced distortion during heat treatment, improved grindability, finer grain size, improved toughness and improved tool performance. The products appear to be receiving growing market acceptance specially for grades such as T-15 and M-4 which are more difficult to produce by the conventional cast and hot work method. In addition, both Crucible and Uddeholms have developed[6,8] special more highly alloyed grades of tool steels which would be impossible to produce by conventional methods.

There are many applications for which the P/M mill forms will be preferable because of their above mentioned advantages. However, in a majority of the cases involving shapes they suffer from the same disadvantages as the conventional mill forms, namely, poor material usage and high machining cost. Although the basic process of gas atomization, CIP and HIP is technically feasible for near net shape making, the author believes that any method incorporating HIP is likely to be prohobitively expensive for most HSS components. Further the whole approach is aimed at end users or machine shops and not at the P/M industry. As described later, the gas atomized powder is not suitable for the usual P/M method of compaction and sintering and as far as is known such gas atomized powder is not commercially sold.

Full Dense Sintering

The cold press and full dense sinter process is a relatively recent development. The method uses water atomized powder which is compacted by either CIP or die pressing. Lubricant is unnecessary in CIP and only die wall lubrication is preferred in die pressing. (Subsequent sections will give more details of pressing and green properties). Good quality of powder and closely controlled high temperature sintering are essential.

The principal advantage of the process is its near net shape capability. This means that complex shapes can often be made with only finish grinding so that material utilization is high and machining cost can be reduced substantially. In addition, other advantages of P/M processing such as fine uniformly distributed carbides are also obtained. Successful use of full dense sintered cutting tools attests to their good toughness. Most of the work thus far has been on full dense sintering of high speed tool steels. Availability of high quality tool steel powders that can be pressed on existing equipment is an important element in the development of the technology. However, the furnace for this process has more exacting demands than normal P/M furnaces and vacuum furnaces are preferred. Also, as with any near net shape process careful attention is needed to selection of candidate components, manufacturing tolerances and tooling design.

Published literature[(9-13)] indicates that there are many companies engaged in studying full dense sintering of HSS and several patents have been issued relating to the technology. Two companies active in commercially using the process are Consolidated Metallurgical Industries, Inc. (CMI) in which SCM Corporation has a minority interest and Powdrex Inc. in England. We shall later comment more about the applications and future prospects; but, suffice to mention that we expect rapid growth in full dense technology both by this method and by other competing techniques.

Powder Production

Basically water atomization consists of melting the work material, allowing the melt to fall through a tundish and impacting the stream by water jet to form powder particles. Typically the water atomized particles are irregular in shape and consequently low in apparent density and high in green strength. As atomized HSS powder is very hard(with microhardness on Vicker's scale of around 850 HV) and has a high oxygen content of about 1,500 - 3,500 ppm. It must therefore be annealed and deoxidized to make it suitable for further processing. As for the particle shape and size distribution, some variation is possible through control of atomization conditions. Powdrex Inc. has developed a novel modification which gives powder of even greater irregularity of shape than other methods of water atomization. (SCM has licensed this powder making technology.) Unless otherwise mentioned, the characteristics of water atomized powders discussed below are those of prealloyed HSS powder produced by Powdrex Inc.

Water Atomized Powders and Their Processing

There are several powder characteristics which are customarily provided by a manufacturers along with the material. In most cases they are easy to measure and yet provide important clues about the processibility of the powder and its suitability for the intended application. Fig. 2 gives such data about a parti-

APPARENT DENSITY	2.2 G/CM^3
TAP DENSITY	3.0 G/CM^3
FLOW RATE	40.0 s/50 G.
SIEVE ANALYSIS	
+ 60 MESH	0
-60 + 80	10
-80 + 100	8
-100 + 150	16
-150 + 200	17
-200 + 325	21
-325	28

CHEMISTRY

0.85 C, 0.2 Mn, 0.2 Si, 6.1 W, 4.0 Cr, 5.3 Mo, 1.8 V, 800 ppm O_2, 150 ppm N_2

GREEN DENSITY	6.6 G/CM^3	60 TSI (827 N/mm^2) COMPACT, DIE WALL LUBE ONLY
TRANSVERSE RUPTURE GREEN STRENGTH	7600 psi (52 N/mm^2)	

Fig. 2

Example of Data For One Lot of Water Atomized and Annealed Powdrex Type M-2 Powder cular lot of M-2 powder produced by Powdrex Inc. In this section we shall discuss the various powder properties in Fig. 2 with regard to their importance and compare them with similar data for other types of powders. Much of this section is about the powder or about green compacts but some information is provided about full dense sintered items. The data were obtained with procedures in general conformity with applicable MPIF standards

Particle Shape

The as atomized particle shape is primarily a function of the cooling rate during atomization with faster cooling leading to a more irregular shape. A highly irregular shape is typical of water atomized powder and is obvious in Fig.3. Several methods have been tried to define the irregularity of the shape quantitatively; but, from a more practical view point irregularity of shape can be better judged by apparent density (A.D.) and green strength which it affects greatly as discussed later. Another technique called the Coldstream process (14) can be employed to produce fine powders from a variety of feed stock. The process depends on size reduction through a high speed impingement of the feed stock against a fixed target and would naturally lead to a far greater uniformity of particle shape than the irregularly shaped water atomized powder in Fig. 3. Also, it is well known that gas atomized powder as used for example by Crucible and Uddeholms is quite spherical and thus very regular in shape.

Sieve Analysis

The sieve analysis data in Fig. 2 is presented graphically in Fig. 4. Sieve analysis is a representation of the particle size distribution and can be used to check for lot to lot variation of the distribution. It has a direct effect on other properties like the apparent density. But, it must be noted that the sieve analysis is influenced by the particle shape so that even for the same equivalent sphere (mean volume) diameter the irregular water atomized powder may appear coarser than the more regular powders obtained with other processes. In any case, the size distribution can be varied as desired within practical limits by manipulating atomization conditions. Currently the water atomized powder is supplied to -60 mesh size and can be full dense sintered as such; although in general, finer powder gives a smaller as sintered grain size.

Apparent Density, Tap Density and Flow Rate

Apparent density is important from a practical view point because it must be considered in designing tooling to get the target as sintered dimensions. For the same reason, any given tooling can be employed satisfactorily only with powder having a particular or very similar apparent density. While all the above properties can be controlled during powder manufacture, in general, irregular powder has low A.D. and proportionately larger difference between A.D. and T.D. and a longer flow time than a more regular powder. However, for a given particle shape, finer the powder the more difficult its flow with the resulting longer flow time. Very fine powder may not flow at all when checked according to MPIF standard No. 03. As for consistency of A.D., note that A.D. follows the law of mixtures so that blending of different lots can be resorted to if unusually close uniformity is required.

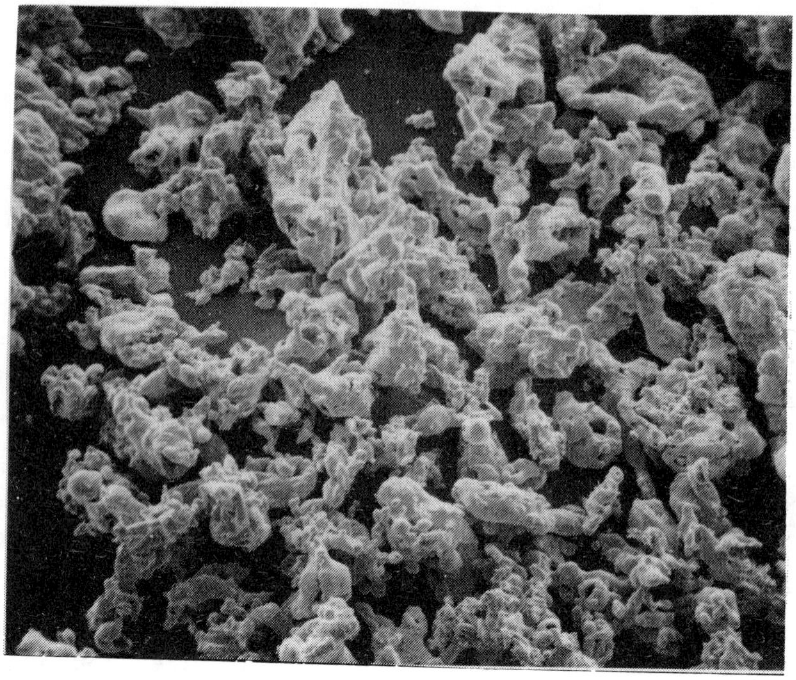

130X

260X

Fig. 3 Appearance of Water Atomized Powdrex Type M-2 Powder
(Notice the highly irregular particle shape)

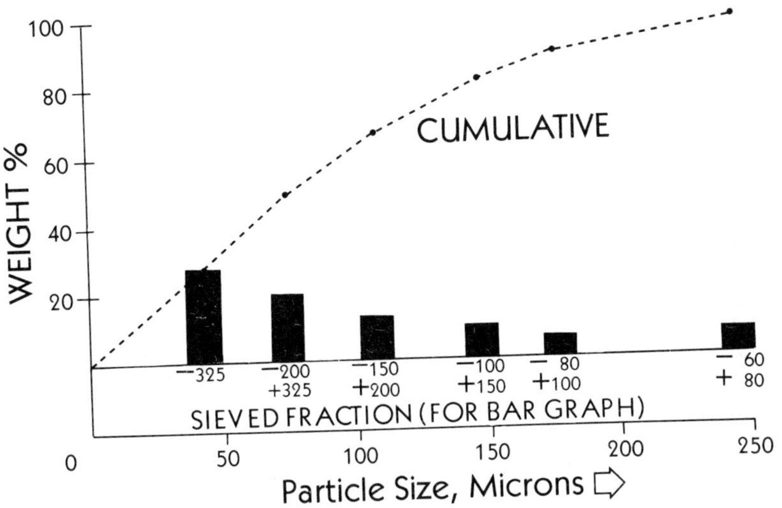

Fig. 4 Particle Size Distribution for Powdrex M-2 powder.

Compressibility and Hardness

The microhardness of annealed water atomized HSS is approximately 250 - 300 HV with some variation depending on the grade of HSS and the exact cycle used for annealing. Microhardness measurement on powder particles is tricky. A more direct indication of hardness is the pressed or green density at a given pressure, that is, compresibility. Fig. 5 relates green density to compacting pressure for samples pressed with die wall lubrication alone. Slightly higher densities result with a small quantity of admixed lubricant. It is important to point out that density values in Fig. 5 are similar to those for commonly used 410L SS powder[(15)] thus showing the excellent compressibility of the water atomized powders and the feasibility of compacting them in existing equipment.

The HSS powders can be compacted either in hard tooling or by CIP. Larger production quantities and need for closer tolerances favor die pressing where as smaller quantities but more complex shape and/or larger size suggest isostatic pressing. Elimination of lubricant and greater uniformity of density are other advantages of isostatic pressing. The upper limit for most CIP facilities is 30 tsi or 414 N/mm^2 (which is adequate for most applications) where as die pressing is often carried out at twice that pressure.

Comparison of differnt types of powders shows some interesting trends illustrated in Fig. 6. The water atomized powder has low A.D. and T.D. but the pressed density increases rapidly with compaction pressure. The increase in density continues at a high pressure of 60 tsi or 827 N/mm^2 and probably beyond. For any particular alloy the density of annealed powders at high pressure of 30 tsi (414 N/mm^2) or more is primarily dependent on the material properties. The densification at high pressure occurs by a large amount of plastic flow which is nearly independent of the initial particle shape or size distribution. The gas

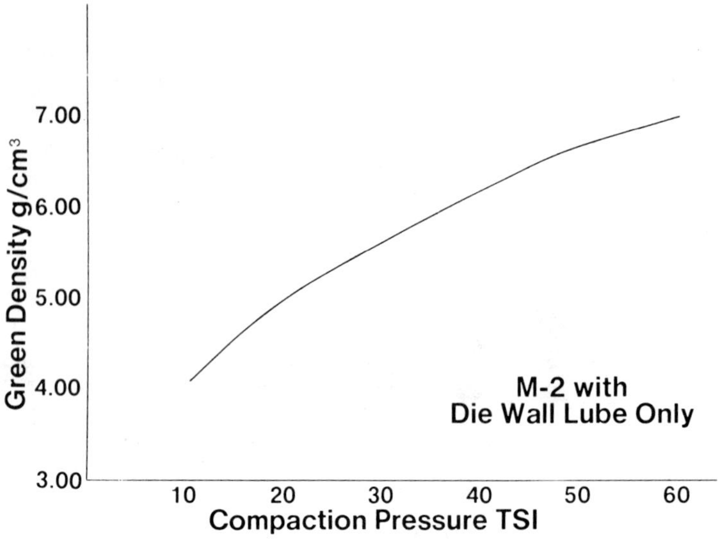

Fig. 5 Effect of Compaction Pressure on Green Density of Annealed Powdrex M-2. (1 TSI = 13.8 N/mm^2)

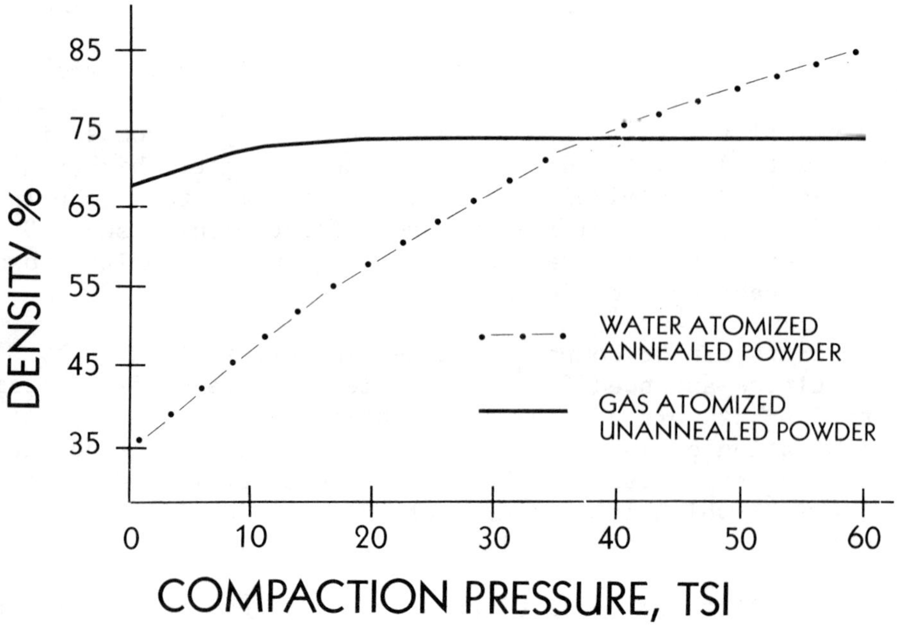

Fig. 6 Comparison of Compressibility of Water/Atomized and gas Atomized Powders (The former is the same as in Fig. 5. The curve for gas atomized powder is based on Ref. 7)

atomized spherical powders have high A.D. and T.D. but are normally compacted without annealing. Fig. 6 suggests that densification occurs at low pressures by some rearrangement of the spherical particles and some local plastic deformation at what must be initially point contacts. However, the hard particles cannot be deformed much and density becomes insensitive to higher pressures in excess of about 30 tsi or 414 N/mm^2.

Green Strength

Water atomized irregular powders have a higher green strength than spherical powders. A reasonable amount of green strength is essential in P/M industry to permit normal in process handling of the green compacts. Tests on annealed M-2 powder made with conventional (not Powdrex type) water atomization show its green strength to be similar to that of 304 L SS powder. More significantly as shown in Fig. 7 the water atomized powder made by the Powdrex process had a high green strength which is nearly three times that of powders made by other processes. This high green strength of the Powdrex type powder will be valuable in making parts with thin sections or complex details often needed on tooling components. In some cases green compacts can be machined to incorporate features that would be impossible or very expensive to achieve during compaction.

The gas atomized spherical powders normally compacted without annealing have no green strength to speak of. This factor coupled with their high hardness makes them totally unsuitable for the usual compact and sinter route used in P/M industry. The green strength depends on many different factors as shown in some careful studies by Klar and Shafer.[16] Considering the pertinent factors it is speculated that gas atomized powders, if it was annealed, would show some increase in green strength but it would still be very low compared to Powdrex

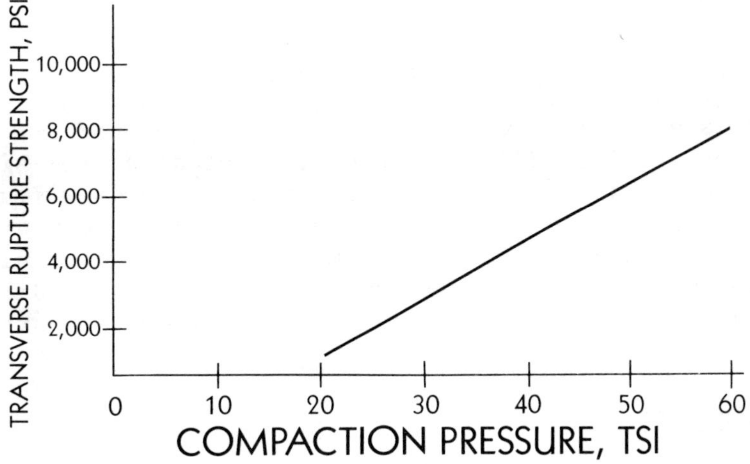

1 PSI = 0.007 N/mm^2 1 TSI = 13.8 N/mm^2

Fig. 7 Effect of Compaction Pressure on Transverse Rupture Green Strength of Annealed Powdrex M-2.

type of powder. In any case, such annealing of gas atomized powder would make it much more expensive than water atomized powder without any compensating advantages.

Chemical Analysis

The highly alloyed nature of HSS make it imperative to control chemical composition closely. Usually X-ray spectroscopy is utilized to measure most of the elements and care must be taken to follow proper procedures and use suitable standards (when available, to be secured from National Bureau of Standards). From a sintering point of view carbon and oxygen contents are the most important and to measure these, we have satisfactorily used commercially available carbon and oxygen analysers. The carbon content must be adequate for all the carbide formers like V, Mo, W and Cr and for matrix hardness but not so excessive that undesirable quantity of austenite stabilizes. Of course, carbon content can be adjusted by mixing carbon in HSS powders the same way as in the case of other ferrous powders.

Oxygen content of annealed water atomized powders is less than 1,000 ppm and is indicated for each shipment as shown in Fig. 2. The powder can be satisfactorily full dense sintered with such oxygen content which in fact decreases during sintering in the necessary non-oxidizing atmosphere so that oxygen content in sintered HSS may be similar to that in gas atomized and HIPped HSS. It is essential that the oxygen content be low to promote densification as well as to minimize the possibility of forming oxide inclusions and to avoid imparing the balanced carbon content during sintering. The caveat here is only about the oxygen content of the as received HSS powder. Our tests have shown that the HSS powder can be stored in closed containers in which they are supplied, for months without much oxygen pick up. Naturally it is desirable to use the powder quickly and as far as possible keep the storage time short. The importance of the interrelationship of C and O_2 contents is well documented.[17-19]

Dimensional Changes During Sintering

Sintered part precision depends on predictable changes in dimensions of the compact during sintering. In P/M industry consistent and low dimensional change during sintering is a highly desirable attribute of the powder. In this regard, processing of water atomized HSS powders differs greatly from sintering of other common powders. Full dense sintering of HSS powder may show linear shrinkage of the order ten percent because of the large difference between green and sintered densities. If the green density is uniform and sintering technique controlled, it would be reasonable to expect similar linear shrinkage in different directions. But, this clearly demands a good deal of expertise from the parts producer.

Sintered Microstructure

How does one judge if sintering is of good quality? One easy method is to inspect the microstructure in as sintered condition. Provided cooling is not deliberately slow the structure would be similar to but more uniform than austenitized and as quenched wrought HSS. Fig. 8 depicts schematically what to look for. The carbides should be small in size and well distributed and must not form a continuous network around the grain boundary. There should be no or very little porosity and grain size should be small. Considering the variety of grades, requirement of different applications and the heat treatments that the

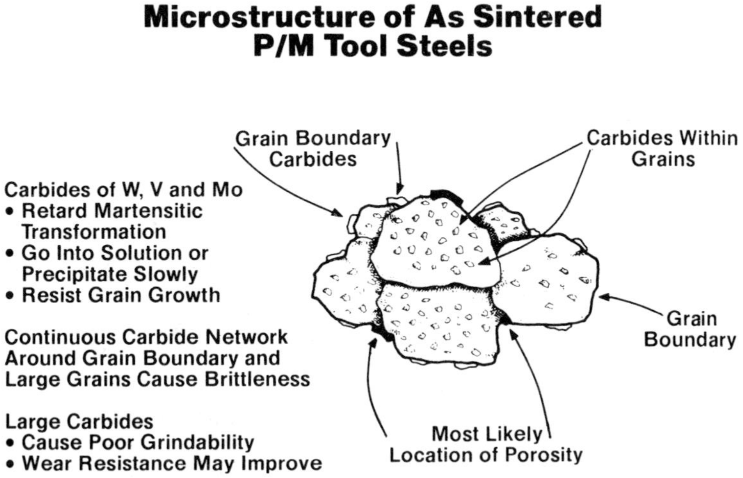

Fig. 8 Schematic of Microstructure of As Sintered P/M Tool Steels

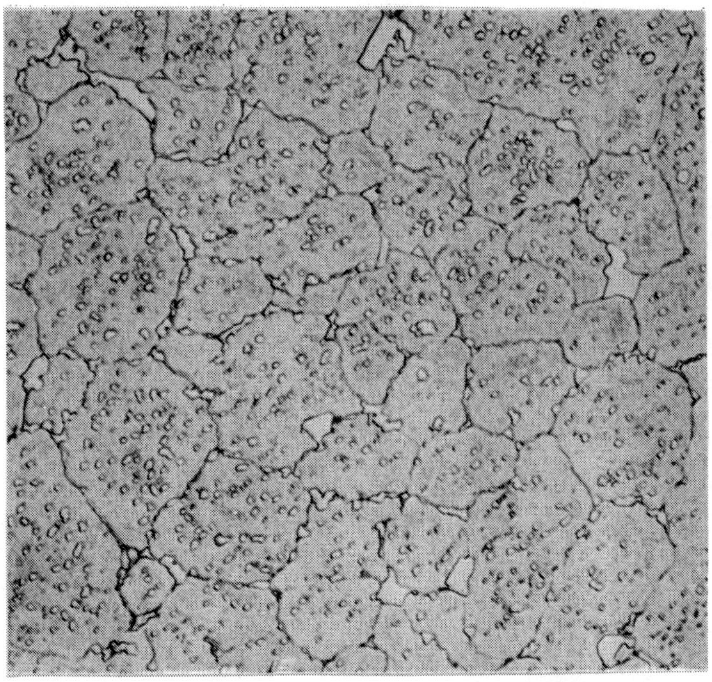

Fig. 9 Microstructure of As Full Dense Sintered M-2 Sample (Prepared from -60 mesh annealed Powdrex type powder. Magnification 500X. 5% Nital Etchant)

samples will subsequently undergo, quantitative specifications are difficult. But, these general guidelines may suffice in most cases. An example of good sintered microstructure is given in Fig. 9. Subsequent heat treatment will normally render the grain size much finer. Satisfactory performance in actual application is of course, the real acid test for the quality of the part.

Response to Conventional Processing

Let us comment briefly on how the water atomized HSS powder may respond to conventional processing. As mentioned earlier the preferred furnace is of vacuum type and sintering in the more common nitrogen bearing atmospheres may not lead to full density specially if the dew point is high. The nitrogen pick up will be considerably higher than allowed in wrought or vacuum sintered tool steels causing higher as sintered hardness as has been observed for stainless steel.[15] (It is interesting to mention that such nitrogen pick up has been studied as a source of increased hardness.[20]) Erratic results in densification and quality of microstructure are also possible because the controls on the existing furnaces may be far inferior to those built for full dense sintering. Having said all that, a conventional parts producer may still get good enough results to get a feel for the new opportunities that could develop with water atomized tool steel powders.

Applications and Future Prospects

Current Applications

Hervey[10] has discussed at length several current and potential applications of full dense sintered HSS parts. Examples of some components sintered by CMI are illustrated in Fig. 10. By and large the current applications cover metal cutting tools with one major application (not shown in Fig. 10) in a different category of wear parts viz. a cam for a diesel engine fuel injection pump. The part geometries that can be full dense sintered cover a broad spectrum and interestingly the new technology was cost effective in making relatively simple shaped T-15 indexable inserts produced by CMI since late 1978.[21] Powdrex Inc. has also reported[11] many applications of full dense sintered HSS including many of the threading dies uses in U.K. which are made from M-35 grade.

Expected Developments

P/M processing of tool steels is a rapidly developing technology and progress is expected both in work materials and in processing techniques.

A major advantage of P/M processing is the ease with which composition can be changed. Minor modifications in response to customer needs are possible right now and undoubtedly new special grades will be introduced. In fact, Powdrex has already introduced[11,12] a special grade for full dense sintering just as Crucible and Uddeholms have done[6,8] for P/M mill forms.

New applications will come not only as replacements for wrought HSS components but also for other materials. On one hand, new high hardness P/M grades may compete with carbides. At the other extreme, common grades such as M-2 processed by P/M route may replace lower alloy components either because of lower life time cost or because the increased performance demands make the lower

Approx. 3/4 Full Scale

Fig. 10 Examples of Full Dense Sintered HSS Components (Courtesy of Consolidated Metallurgical Industries Inc., Farmington Hills, MI.)

alloy material unsatisfactory. The full dense sintering technology may very likely be extended to other tool steels and to non-tool steel materials. Finally, even though this paper emphasizes full dense applications, it is recognized that not all properties are equally sensitive to residual porosity. Indeed, it is conceivable that somewhat lower density applications will develop or there will be technological break-through which will make it possible for a wide segment of conventional P/M parts producers to process HSS powders.

Acknowledgements

Thanks are due to Dr. C. I. Whitman, Director of Research, Mr. M. F. Marchitto, Manager of New Business Development, and other SCM Management for their support of this program and the permission to publish this paper. I am also indebted to many of my co-workers who have contributed to the experimental work. My special thanks to Ms. M. Cornelius for her careful typing.

References

1. G. A. Roberts, etx al., "Tool Steels", Third Edition, American Society for Metals, Metals Park, Ohio, 1962.
2. P. Payson, "The Metallurgy of Tool Steels", John Wiley & Sons, Inc., New York, N.Y., 1962.
3. R. Wilson, "Metallurgy and Heat Treatment of Tool Steels", McGraw-Hill, New York, N.Y., 1975.
4. F. Kayser and M. Cohen, "Carbides in High Speed Steel--Their Nature and Quantity", Metal Progress, Vol. 61, June 1952, pp. 79-85.
5. E. J. Dulis and T. A. Neumeyer, "New and Improved High-Speed Tool Steels by Particle Metallurgy", in Progress in Powder Metallurgy 1972, pp. 129-142, Metal Powder Industries Federation (MPIF), Princeton, N.J., 1972.
6. A. Kasak and E. J. Dulis, "Powder-Metallurgy Tool Steels", Powder Metallurgy, Vol. 21, No. 2, 1978, pp. 114-121.
7. P. Hellman, etx al., "The ASEA/STORA Process", Modern Developments in Powder Metallurgy, Vol. 4, pp. 573-582. MPIF, 1971
8. P. Hellman, "Powder Metallurgy High-Speed Tool Steels for Broaches", Technical Paper MR 78-287, Society of Manufacturing Engineers, Dearborn, Michigan, 1978.
9. R. P. Hervey, "Performed P/M Tool Steel for Cutting Tools", Technical Paper MR 78-315, Society of Manufacturing Engineers, Dearborn, Michigan, 1978.
10. R. P. Hervey, "Automotive Applications for Full Density Powder Metallurgy". Presented at the Society of Automotive Engineers, 1979 Congress & Exposition, February 27, 1979.
11. J. Smart and P. Brewin, "Complex Shapes in High Alloy Steels", Metal Powder Report, Vol. 33, No. 5, May 1978, pp. 207, 210-211.
12. "Powdrex Advances P/M High-Speed Steel Powder Production", announcement from Metal Powder Report, Vol. 33, No. 11, November, 1978, pp. 523.
13. F. L. Jagger, etx.al., "Heat-Treatment Response and Cutting Properties of Sintered Type M2 High-Speed Steel", Powder Metallurgy, Vol. 20, No. 3, 1977, pp. 151-157.
14. J. Walraedt, "The Coldstream Process: A New Powder Production Equipment", Powder Metallurgy International, Vol. 2, No. 3, August, 1970, pp. 77-80.
15. H. D. Ambs, and A. Stosuy, "The Powder Metallurgy of Stainless Steels", from Handbook of Stainless Steels by Peckner and Bernstein, Chapter 29, McGraw-Hill, New York, 1977.
16. E. Klar and W. M. Schafer, "On Green Strength and Compressibility in Metal Powder Compaction", from Modern Developments in Powder Metallurgy, Vol. No. 9, pp. 91-113. MPIF, 1977

17. P. Lindskog and S. Grek, "Reduction of Oxide Inclusions in Powder Preforms Prior to Hot Forming", from Modern Developments in Powder Metallurgy, Vol. 7, pp. 285-301. MPIF, 1974

18. A. R. Kieffer, et al., "P/M High Speed Steels Modified by Addition of Monocarbides", from Modern Developments in Powder Metallurgy, Vol. 8, pp. 519-536. MPIF, 1974

19. G. T. Brown, "Development of Alloy Systems for Powder Forming", Metals Technology, May-June 1976.

20. J. J. Dunkley and R. J. Causton, "Powder Metallurgy Extrusion of High Speed Steels and Other Alloys", as reviewed in Metals and Materials, Feb. 1977, p. 36.

21. K. Gettelman, "Powdered T-15 Inserts Fill the Carbide Gap", Modern Machine Shop, Vol. 51, No. 6, November, 1978, pp. 86-89.

GRAPHITE INFLUENCE ON P/M STEEL PROCESSING PARAMETERS

Walter V. Knopp

P/M Engineering & Consulting
Wyckoff, New Jersey

P/M steels are made by the addition of graphite to iron or alloy powders. Very little recent information has been published as to the influence of graphite on the processing parameters of P/M steels. Basic raw materials in our world are becoming more scarce and critical especially the ones produced by nature. Political conditions are also determing the availability as well as pricing. I'm sure you are already familiar with these facts.

The P/M steel area of our industry is showing a steady and healthy expansion. In recent years, attention has been focused on atomized powder for higher densitites as well as higher purities for greater strengths. The industry is also changing to lubricants that volatilize completely during sintering leaving no residue. The type of graphite addition to form a steel has been given very little attention.

Graphites are found as natural mineral deposits in Ceylon, Madagasca, West Germany, Austria, Texas, Russia, Peoples Republic of China and Mexico. The purity can vary greatly depending upon graphite origin. Graphite powders have also been obtained from carbon electrodes, moderator blocks and scrap. These materials are produced from a mixture of petroleum coke and a binder. The combination consists of graphitized coke and graphitized binder. This type of graphite does not exhibit a 100% crystallinity. A relatively new process, for the U.S.A., produces a synthetic graphite by using carefully selected petroleum coke as it's starting material. No binders are employed and graphitization is carried out above $3000^{o}C$ in an electric resistance furnace. The graphite made by this latter process is already in powder form with particles up to 7 mm in size. Controlled grinding can produce a variety of particle sizes.

Approximately 10-15 years ago, Ulf Gummeson and Athan Stosuy of Hoeganaes Corp. conducted an extensive investigation covering many different graphites from around the world as well as then available synthetic graphites. The results indicated that particle size and purity (ash content) are important for natural graphites. The synthetic materials produced low strength values regardless of purity and particle size. The reason for these poor results was not apparently clear at that time.

In dealing with graphite powders today, very little is published except a coding, approximate ash content, a particle size range and carbon content. Let us look at some of these areas. Ash contents for both natural and synthetic graphites can vary widely from less than 1% to over 15%. The impurities in synthetics are very low when compared to natural graphite. The particle size of the graphite will affect the dimensional control of the P/M part. The larger the size, the more growth will occur. The number given for carbon content is important, but it can also be misleading. Although the number may be high, complete graphization is a must.

How can the degree of graphitization and purity be determined? The use of chemical analysis, spectrographic analysis and electron diffraction. The new synthetic graphite used in this investigation has a guaranteed carbon content of 99.9% minimum. The typical trace elements of this material is as follows:

Trace Elements

Aluminum	4 PPM	Magnesium	< 1
Arsenic	< 5	Nickel	8
Calcium	43	Lead	0.5
Chromium	1	Silicon	130
Copper	2	Titanium	28
Iron	140	Vanadium	10
Gallium	< 0.5		

(1 PPM = 0.0001%)

The crystalline structure of the graphite by electron diffraction indicates the degree of graphitization. A large crystallite size (L_c and L_a) as well as an ideal interlayer spacing $\frac{c}{2}$ is indictive of this. Figure 1 is an electron diffraction pattern typical of a macrocrystalline natural graphite and an excellent synthetic graphite. Notice how distinct the pattern is.

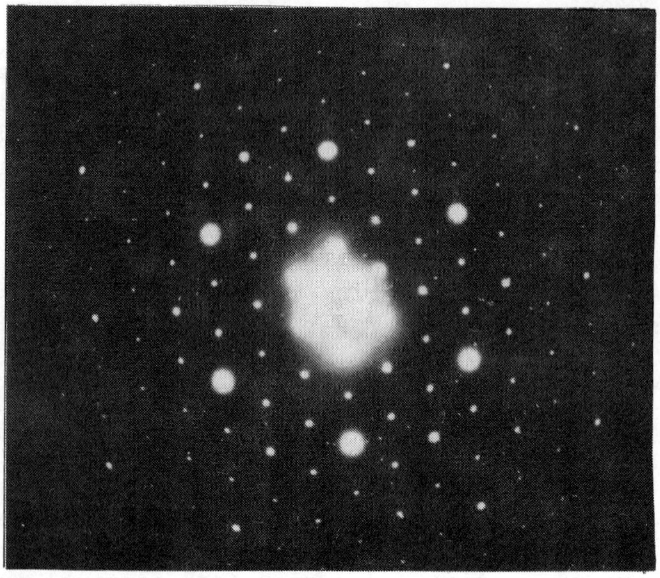

Figure 1

Figure 2 shows the pattern of the electro graphite. Notice how fuzzy this pattern is.

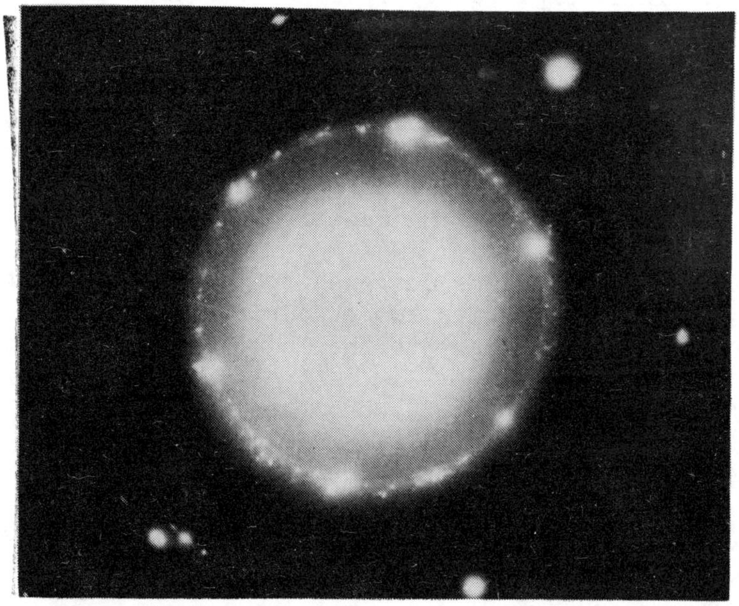

Figure 2

Figure 3 is an illustration showing the interlayer spacing $\frac{c}{2}$ and a scale of structural perfection. The theoretical interlayer spacing is 3,354 Å (Angstroms). Electro graphite is 3,380 Å with the amorphous carbon being 3,444 Å.

Figure 3

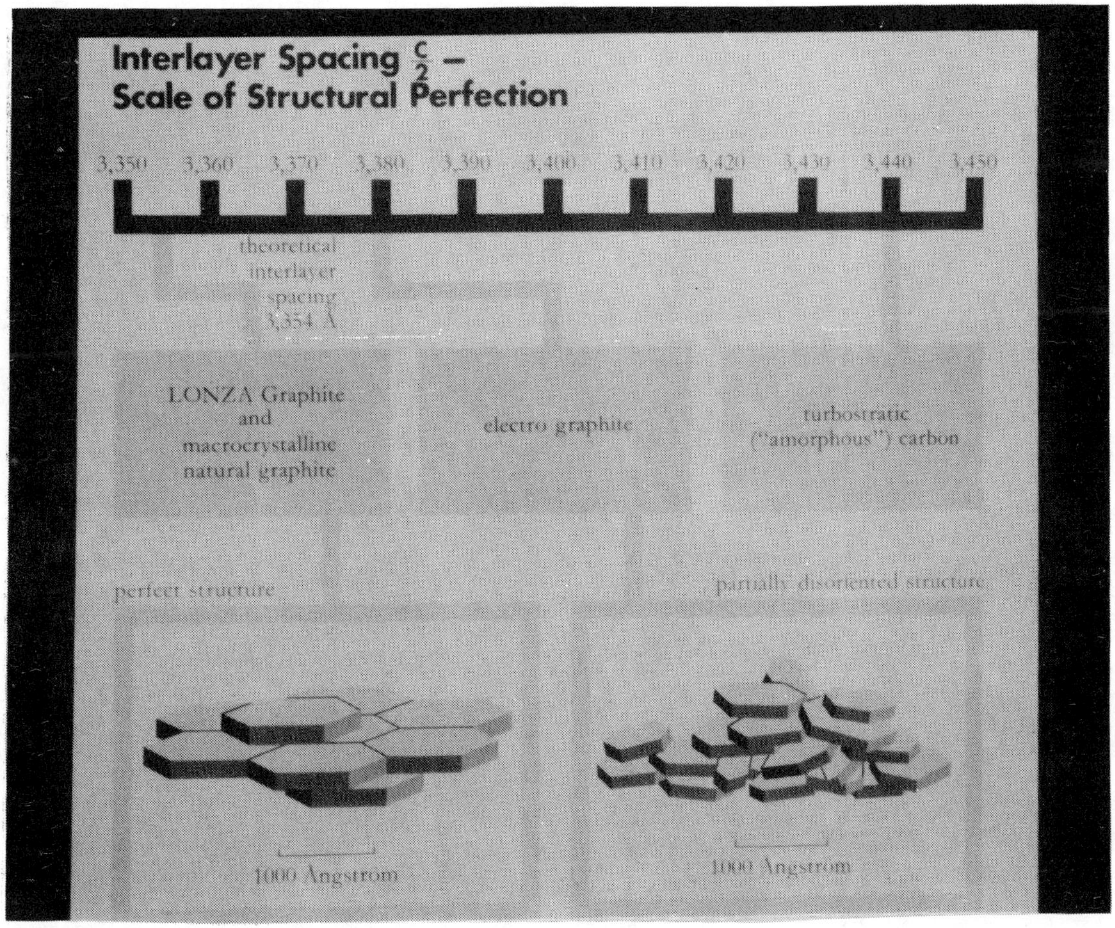

Within the last 3 to 4 years, a few companies have tested this high purity synthetic graphite. Their exploratory tests were conducted in direct comparison with the natural graphites used in the P/M industry. The results indicated that this synthetic graphite could be substituted for the natural graphite. The mechanical properties and diffusion of the graphite were comparable. Under the test conditions of one company, better strengths were obtained at a lower temperature for the synthetic graphite. However, the sintering was carried out in hydrogen and for a time of one hour. These conditions are not common in the major sector of the P/M industry.

A comparative test program was designed to compare two of the common natural graphites, designated A & B, with two grades of the synthetic graphite designated as C & D. Blends were made to the following composition:

99% Ancorsteel 1000	Hoeganaes Corp.
1% Graphite	
3/4% Acrawax "C"	Glyco Chemical, Inc.

The data covering the characteristics of the graphite powders is shown in Table I.

TABLE I

Graphite Data

Graphite	A	B	C	D
Particle Size microns				
-128 + 64	-	-	-	-
-64 + 32	1.5	-	-	-
-32 + 16	10.4	1.6	1.2	8.7
-16 + 8	37.3	11.7	9.2	35.4
-8 + 4	35.4	40.7	48.0	33.8
-4 + 2	13.4	40.0	34.1	19.0
-2	2.0	5.8	7.5	3.1
Density				
Scott, g/cc	0.123	0.098	0.096	0.147
Specific Area m^2/g	8.7	7.6	18.0	19.0
Ash Content %	4.7	2.8	0.05	< 0.1

The blends designated as A, B, C and D were compacted into MPIF Standard No. 10 tensile bars to a green density of 6.70 - 6.80 g/cc. The compacting was carried out on a automatic Kux press. The sintering was carried out using a Drever mesh belt production furnace using a Endothermic atmosphere with a Dew Point of + 30°F. A thermocouple was imbedded in one tensile bar which was placed within the test samples. Every run was made with dummy bars in front as well as behind the test samples. Figure 4 shows the temperature profile curves for the different times and temperatures. The sintering conditions were as follows:

Sintering Temperature	1832°F	1000°C
	1922°F	1050°C
	2050°F	1121°C
Sintering Time	10, 20 and 30 minutes at temperature	

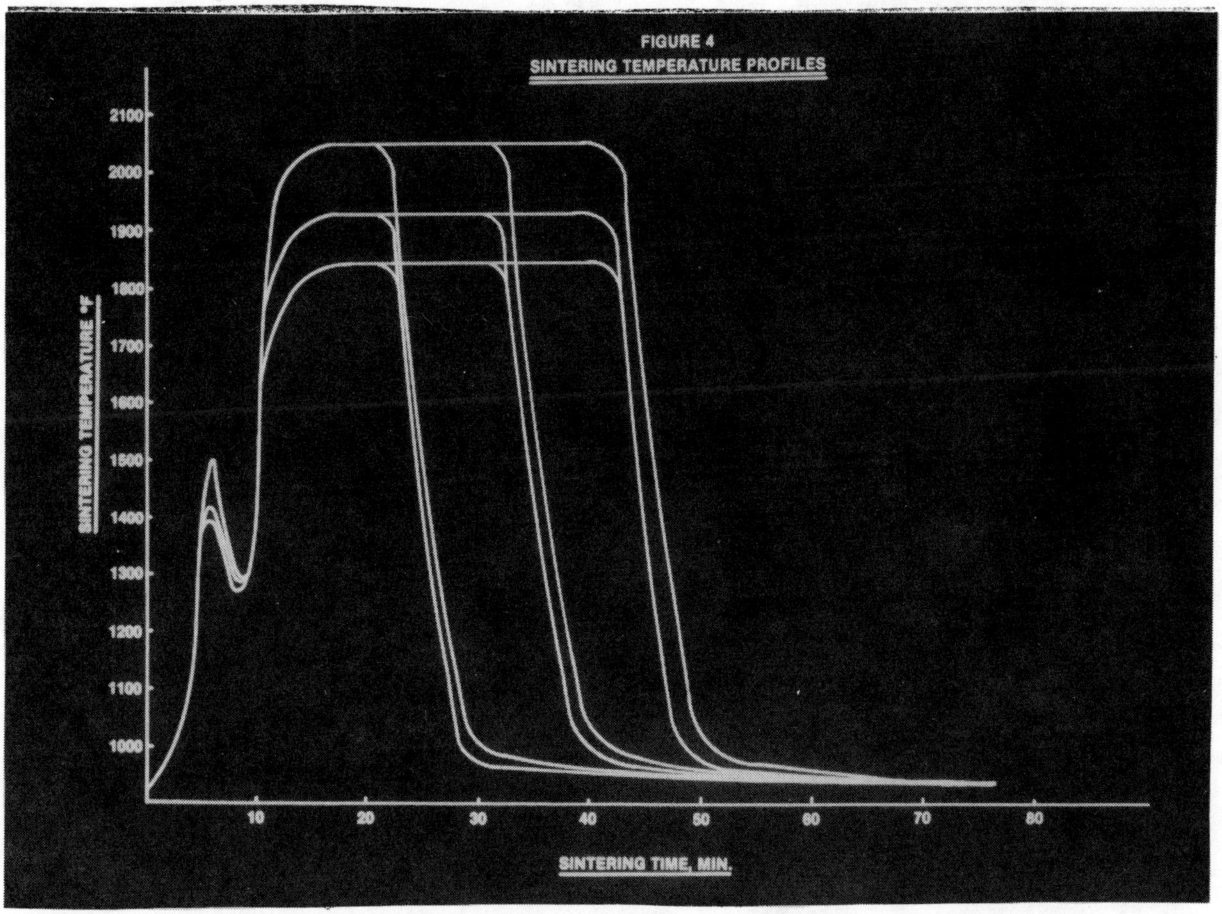

The tensile bars were weighed and measured before and after sintering to determine their densities and dimensional change from die size. The sintered densities ranged from 6.62 - 6.71 g/cc. The sintered bars were tested for their hardness, tensile strength, yield strength (0.2% offset) and elongation. The data for the three sintering temperatures are shown in Tables II, III and IV. The data shows that the resulting values for natural graphite (A & B) and synthetic graphite (C & D) are equal for all the times and temperatures of this investigation. The dimensional changes are slightly higher for the synthetic graphites but are still within 0.001"/in or less when compared with the natural graphites.

TABLE II

Sintering Temperature 1832°F (1000°C)

Ident.	Hardness Rockwell "B"	Tensile Strength psi.	Yield Strength psi.	Elong % in 1"	Dimensional Change, Die %
10 Min. Sintering Time					
A	50	28,600	26,100	3/4	+ 0.22
B	55	29,700	27,200	3/4	+ 0.25
C	54	28,900	27,400	3/4	+ 0.32
D	54	24,900	24,900	1/4	+ 0.36
20 Min. Sintering Time					
A	50	32,100	27,100	3/4	+ 0.20
B	55	32,900	28,000	3/4	+ 0.21
C	54	33,300	30,500	3/4	+ 0.29
D	52	29,700	27,800	3/4	+ 0.34
30 Min. Sintering Time					
A	45	33,300	27,000	1	+ 0.15
B	50	34,400	28,900	1	+ 0.20
C	50	33,100	30,600	1	+ 0.34
D	55	33,300	31,400	1	+ 0.34

TABLE III

Sintering Temperature 1922°F (1050°C)

Ident.	Hardness Rockwell "B"	Tensile Strength psi.	Yield Strength psi.	Elong % in 1"	Dimensional Change, Die %
10 Min. Sintering Time					
A	55	32,000	29,500	1	+ 0.25
B	57	34,100	31,300	1	+ 0.25
C	60	35,200	32,600	3/4	+ 0.30
D	58	29,200	28,500	3/4	+ 0.37
20 Min. Sintering Time					
A	52	34,900	29,900	3/4	+ 0.18
B	56	40,100	33,100	1/2	+ 0.20
C	54	38,800	35,200	1	+ 0.25
D	55	35,900	33,400	1	+ 0.26
30 Min. Sintering Time					
A	51	39,000	29,600	1½	+ 0.17
B	52	37,700	29,500	1¼	+ 0.16
C	53	37,800	33,200	1¼	+ 0.22
D	57	42,600	36,000	1¼	+ 0.24

TABLE IV

Sintering Temperature 2050°F (1121°C)

Ident.	Hardness Rockwell "B"	Tensile Strength psi.	Yield Strength psi.	Elong % in 1"	Dimensional Change, Die %
10 Min. Sintering Time					
A	57	41,700	33,200	1	+ 0.15
B	59	41,900	33,100	1	+ 0.15
C	59	39,400	36,400	1	+ 0.19
D	59	38,500	34,300	1	+ 0.23
20 Min. Sintering Time					
A	58	43,600	34,100	1-3/4	+ 0.09
B	57	44,100	37,700	1-3/4	+ 0.06
C	59	48,500	37,700	1-1/4	+ 0.11
D	57	47,200	37,600	1	+ 0.14
30 Min. Sintering Time					
A	49	45,000	33,300	1	+ 0.09
B	52	46,200	32,400	1-3/4	+ 0.03
C	54	46,300	32,200	1-3/4	+ 0.03
D	55	48,900	37,200	1-3/4	+ 0.05

The test bars were analyzed for total carbon as well as graphitic carbon. The test data for some of the tests is shown in Table V. The graphites in all cases have diffused into the iron for all the conditions investigated. The combined carbon content of 0.8% indicates that all the structures should be a eutectoid structure.

TABLE V

Carbon Analysis

Chips from Tensile Bars

Ident.	Total Carbon %	Graphitic Carbon %	Combined Carbon %	Sintering Conditions
A	0.83	0.032	0.798	1832°F 30 Min.
B	0.86	0.034	0.826	
C	0.90	0.031	0.869	
D	0.90	0.039	0.861	
A	0.81	0.022	0.788	2050°F 10 Min.
B	0.86	0.021	0.839	
C	0.86	0.019	0.841	
D	0.89	0.031	0.861	
A	0.80	0.028	0.772	2050°F 30 Min.
B	0.85	0.022	0.828	
C	0.81	0.019	0.791	
D	0.82	0.022	0.798	

Metallographic examinations were carried out on the sintered specimens. The samples were carefully polished using standard procedures and the surfaces were examined in the unetched and etched condition. The etched employed was a 40% Picral for 15-20 seconds. Figures 8 & 10 are the unetched samples of A & C sintered at 1832°F for 30 minutes. These structures indicate that the samples are not well sintered. The outlines of the original particles are still noticeable. Figures 9 & 11 show the etched structure of A & C at this same temperature. A nice pearlitic structure was obtained. The 2050°F sinter for 30 minutes is shown in Figures 12 & 13 for natural graphite A and Figures 14 & 15 for synthetic C. The unetched areas show a much better sinter and the etched areas are again of a eutectoid structure. These metallographic examinations correspond with the carbon analysis given in Table V.

What conclusions can be stated from this investigation?

1. A properly produced synthetic graphite controlled not only for its chemical purity but also for its crystalline form will produce equal results compared to natural graphite. The properties of the graphite can be reproduced and guaranteed. This uniformity should provide a more trouble free and a more consistent production of end products.

2. Even though natural and a good synthetic graphite diffus nicely, the temperature must be high enough to eliminate the original grain boundaries so that the optimum strengths are developed or time at temperature must be lengthened to accomplish the same result.

Acknowledgement

The author greatfully acknowledges the support of Lonza in this study.

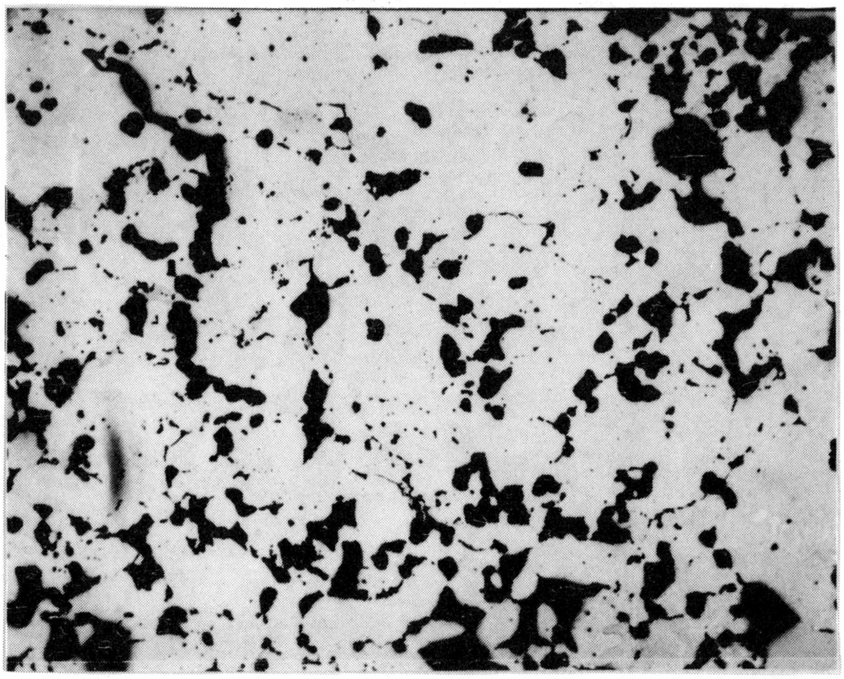

Figure 8
Graphite A 1832°F 30 Min.
Unetched 180X

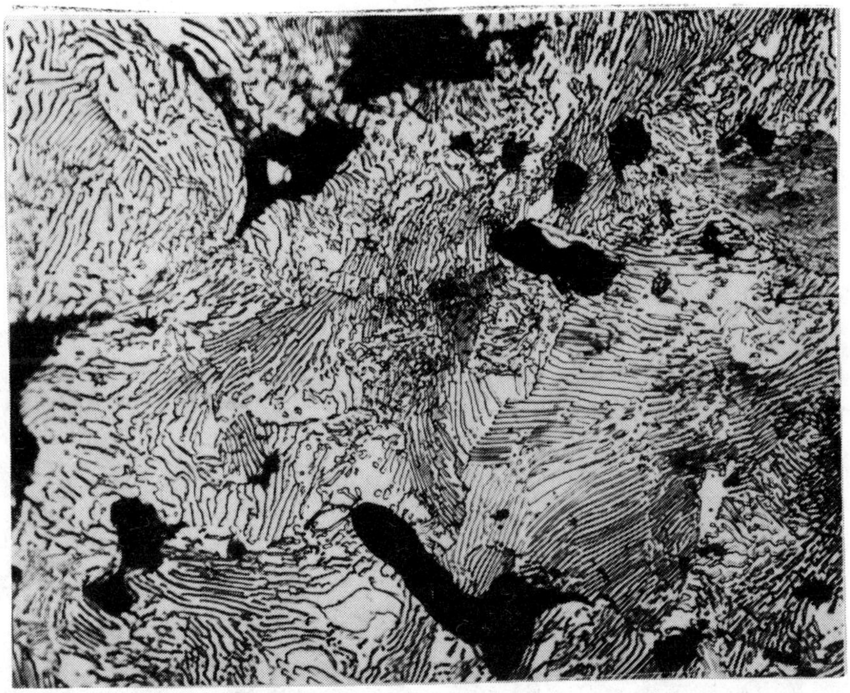

Figure 9
Graphite A 1832°F 30 Min.
Etched 5% Picral 960X

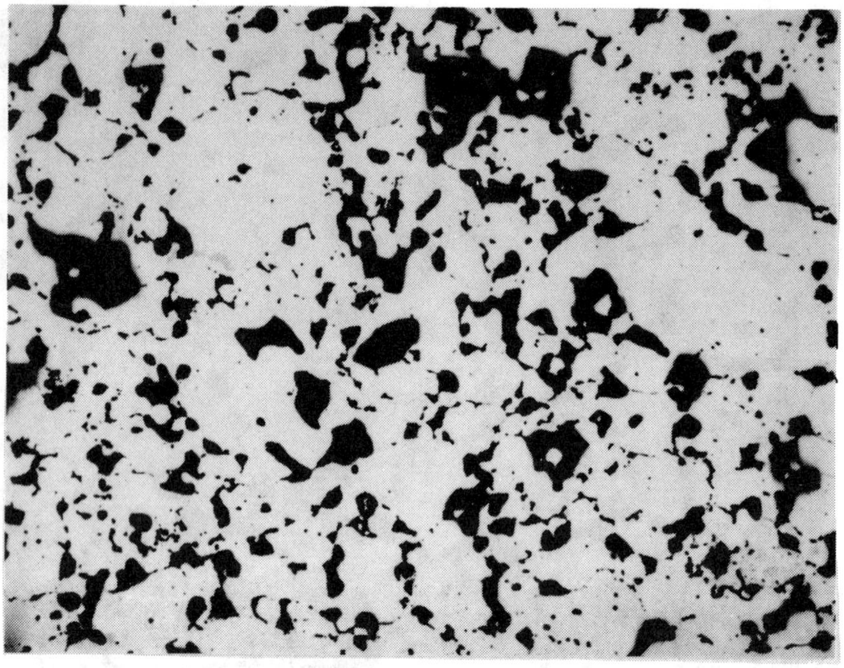

Figure 10
Graphite C 1832°F 30 Min.
Unetched 180X

Figure 11
Graphite C 1832°F 30 Min.
Etched 5% Picral 960X

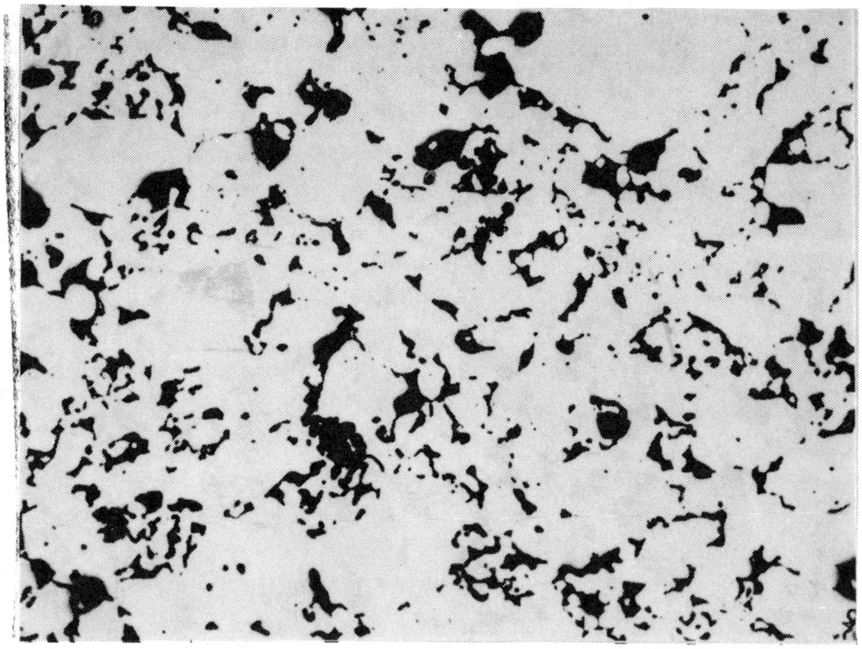

Figure 12
Graphite A 2050°F 30 Min.
Unetched 180X

Figure 13
Graphite A 2050°F 30 Min.
Etched 5% Picral 960X

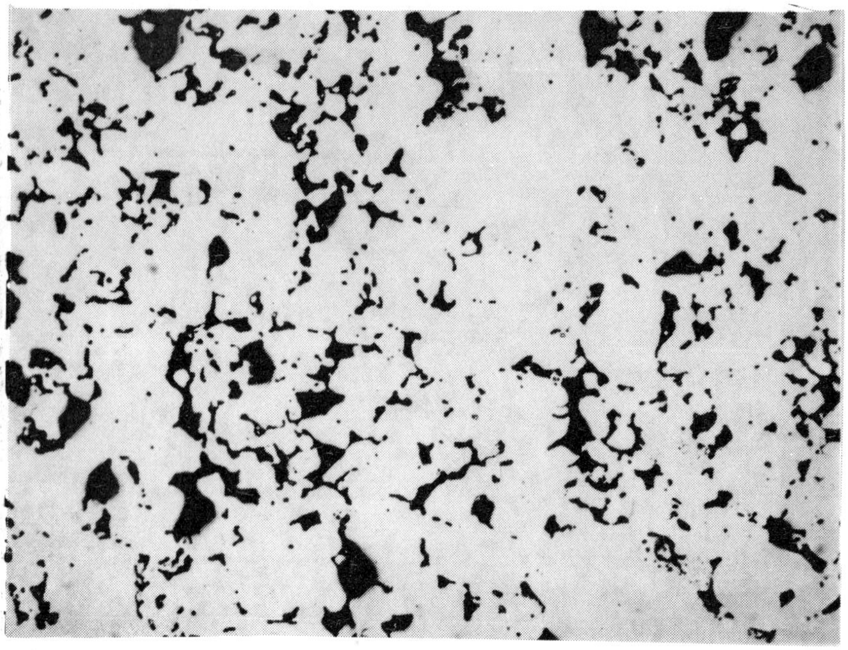

Figure 14
Graphite C 2050°F 30 Min.
Unetched 180X

Figure 15
Graphite C 2050°F 30 Min.
Etched 5% Picral 960X

POWDERS FOR PLASMA SPRAY COATINGS

David L. Houck

GTE Sylvania Incorporated
Towanda, PA 18848

Introduction

This paper deals with two techniques used in manufacturing powders for plasma spray application. Plasma spray coatings are frequently applied to various articles to impart properties such as hardness, erosion and corrosion resistance, good lubricity, or improved electrical properties. For example, in the automotive industry, molybdenum is applied to piston rings to impart wear resistance. In the aircraft industry, various parts of turbine engines are coated with tungsten carbide-cobalt alloys to prevent wear. Powders have become popular because they can be produced in a wide range of compositions.

Powders for plasma spray application should be uniform in size and composition and relatively free flowing. Flowability is necessary since the powders must be transported to and injected into the plasma flame. In general, fine powders (less than 40-micrometers average diameter) do not exhibit good flow characteristics. A narrow size distribution is important because, for example, under set flame conditions, the largest particles may not melt completely, and the smallest particles may be heated to the vaporization point. Incomplete melting obviously is detrimental to coating uniformity, whereas vaporization decreases coating efficiency. Therefore, the size ranges for plasma spray powders are typically produced such that 80% of the particles fall within a 30-micrometer diameter range, with the range of all particles within a 60-micrometer range.

The two techniques described in this paper for manufacturing powders for plasma spray applications are spray drying and plasma densification.

Spray Drying

Various agglomeration techniques exist for producing large, free flowing powder particles from fine nonflowing powder particles. One of the more sophisticated methods is spray drying. (See Figure 1.) Numerous types of spray drying systems are commercially available. (See Reference 1.) The pharmaceutical

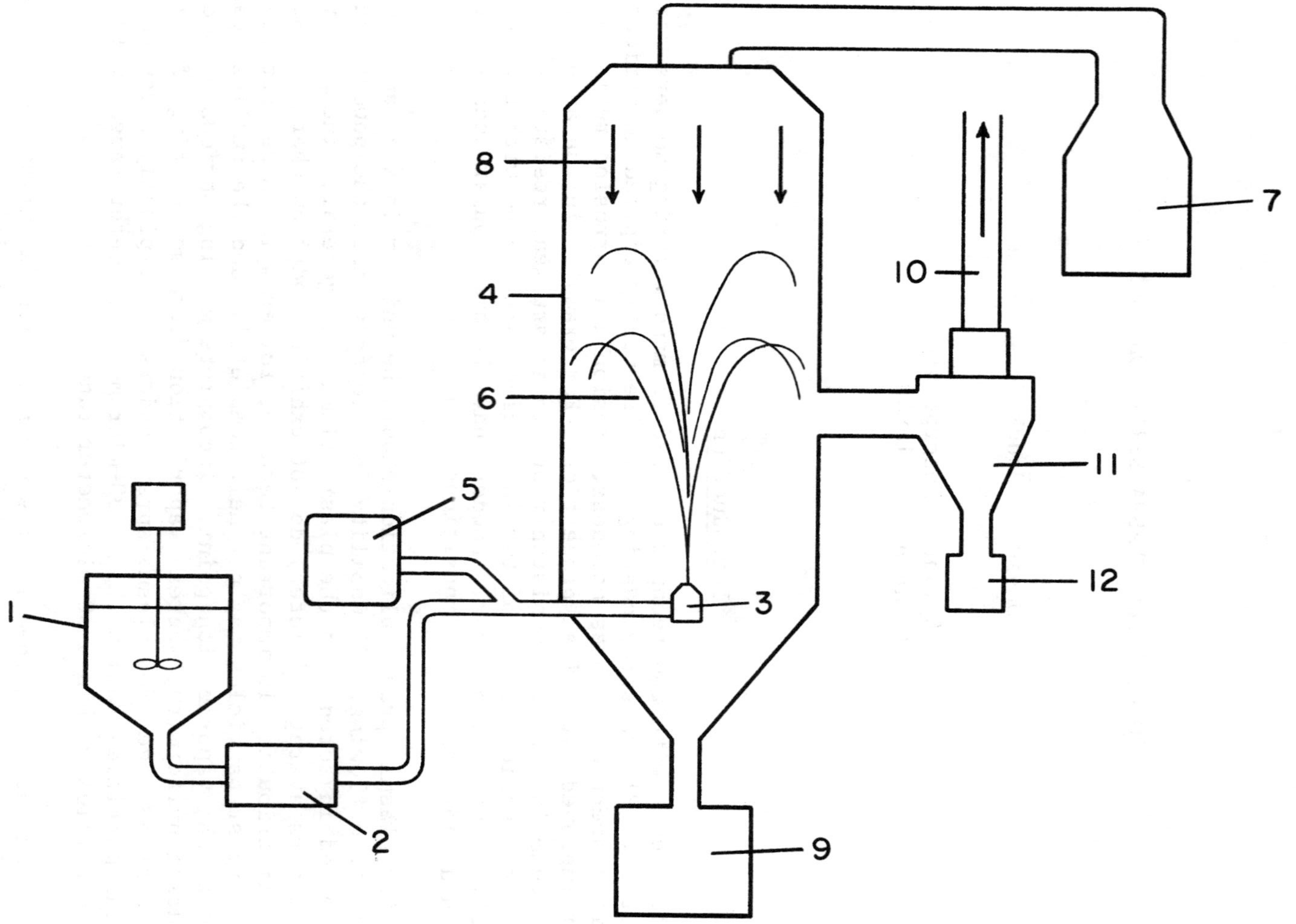

Figure 1. Schematic diagram of a spray drying system: 1/ mixing tank, 2/ pump, 3/ atomizing nozzle, 4/ spray drying chamber, 5/ compressed air source, 6/ atomized fine spray, 7/ heater, 8/ drying air passing downward, 9/ powder collector, 10/ exhausting gases, 11/ cyclone separator, 12/ fine particle collection chamber.

and food industries utilize spray drying extensively. Agglomerates are formed by atomizing a slurry of powder, binder and liquid into a drying chamber, in which the liquid is evaporated. The result is generally a spherical, agglomerated particle held together by the binder. In addition to a binder, other components may be added to the slurry to improve handling thereof or to impart a desired property to the final product.

The two most important characteristics for materials used for binders are: (1) They must be soluble in the liquid used to form the slurry and, (2) when dried, they must form a coating and/or adhere to the material being agglomerated. The resulting agglomerate must be strong enough to withstand subsequent handling without deagglomerating. Waxes and organic binders are used extensively. The following are examples of organic binders: Polyvinyl alcohol, gum arabic, carboxyl methyl, polyvinyl acetate, ethyl cellulose, polyvinyl butyral dispersions, protein colloids, acrylic resin emulsions, water soluble phenolics, wood extracts such as sodium, ammonium, potassium or propylene glycol alginates.

An organic binder has little other benefit than its ability to hold the particles together. Powder with organic material present does not always perform well in commercially available plasma spray equipment. In many cases, the binder is not strong enough to withstand handling and feeding. If it is, it often vaporizes in the flame. As a result, the operation is smoky and the binder can condense in cooler areas causing plugging or fouling of the gun, work piece or work area and creating a potential health hazard to personnel.

Some types of binder systems can be removed by various sintering procedures. Reference 2 outlines a technique for producing a molybdenum plasma spray powder. In one example, the powder was spray dried using polyvinyl alcohol as a binder. In order to remove the binder, the agglomerates were heated above its vaporization temperature (600°C for four hours in a hydrogen atmosphere). They were then heated at a temperature less than half the melting point of the material (1100°C for four and a half hours) to sinter the agglomerates by diffusion bonding. If a temperature greater than half the melting point is employed, the agglomerates themselves sinter together forming an unusable cake or sponge. The sintering of the particles restores or improves the strength which the agglomerates had prior to removal of the binder. It also causes a densification of the agglomerates, thus improving the apparent density of the powder.

For successful strengthening of the agglomerates, it is also necessary to control the relative sizes of the subparticles. In general, they should be no larger than 1/5 the agglomerate diameter, preferably 1/100 or less. If they are too large, heating under the conditions necessary to avoid appreciable caking will result in agglomerate strengths insufficient to permit handling after the sintering operation.

The technique described in Reference 2 eliminates some of the problems associated with the use of organic binders. However, some organic binder systems, upon removal by firing, leave a residue of trace contaminants. Also, for those systems which can not be removed, the binder occupies space which the powder could otherwise occupy. As a result, the apparent density of these powders is lower than desired.

Spray Drying Using Nonorganic Binders

One of the topics of this paper is a technique to spray dry plasma spray powders using a binder system which does not exhibit the problems inherent with organic binders. As outlined in References 3 and 4, this binder system consists of a soluble compound of a desired metallic constituent of the final agglomerated product. Upon heating in a reducing atmosphere, the binder decomposes to a base metal and at least one volatile product. With water as a solvent, suitable binders include, but are not limited to, ammonium complexes of metals or oxides, metal nitrates, and acetates. The following table contains examples of some particle-solvent-binder systems.

TABLE I

Examples of Particle-Solvent-Binder Systems for Spray Drying Applications

Particles	Binder*	Binder Decomposes to	Approximate Decomposition Temperature (°C)
Mo	AM	Mo, NH_3, H_2O	800 to 1000
W	AT, AMT	W, NH_3, H_2O	800 to 1000
WC-12%Co	Ammonium Complex of CoO	Co, NH_3, H_2O	800 to 1000
WC	Cobalt Nitrate	Co, NO_x	800 to 1000
WC	Cobalt Acetate	Co, CO_2	800 to 1000

*AM - Ammonium Molybdate
AT - Ammonium Tungstate
AMT - Ammonium Metatungstate

As indicated in Table I, when molybdenum comprises the particulate material to be agglomerated, a suitable binder system would be water and ammonium molybdate. Upon spray drying, the free water is driven off, leaving molybdenum agglomerates bound together by spray dried ammonium molybdate. As used herein, the term "spray dried ammonium molybdate" and "spray dried ammonium tungstate" refer to the spray dried product of the aqueous solution of the ammonium complex, since normal ammonium complexes of these metals are not known to exist in solid form.

To test ammonium molybdate as a binder system, three slurries were prepared by first dissolving appropriate quantities of MoO_3 in 28 percent NH_4OH solutions to form ammonium molybdate solutions. These solutions were then diluted with water to obtain about 2.5 gallons each. The three solutions contained amounts equivalent to 5.9, 11.2 and 20 percent MoO_3, respectively, on a weight percent solids basis. Forty pounds of molybdenum powder were then added to each solution and the solutions mixed to form a slurry.

The slurries were all spray dried under identical conditions, i.e., the solutions under continuous agitation were fed into one inlet of a two-fluid nozzle at the top of a four-foot diameter Proctor-Swartz spray drying chamber, while air was fed into the other inlet at a pressure of about 37 psi. The drying air entered the chamber at a temperature of about 230°C and exited at about 130°C. The chamber products were subjected to a standard sieve analysis. The products were then heated at 1000°C for four hours in H_2 to convert the binder to Mo. This treatment is referred to as presintering. They were then evaluated for sieve analysis, apparent density and Hall flow measurements in order to investigate strengthening and densification. The products were then sintered at 1060° for four and a half hours in H_2 and re-evaluated. Results are shown in Table II. Also shown in the table are comparative results for the same molybdenum powder processed with an organic binder.

TABLE II

Properties of Spray Dried, Presintered, and Sintered Molybdenum Powders

Lot No.	1	2	3	Organic Binder
MoO_3 (weight percent solids)	5.9	11.2	20	0
SPRAY DRIED (GREEN) PROPERTIES				
Sieve Analysis	Percent Retained			
+100 mesh	4%	8	28	9
+200	30	36	42	31
+325	32	30	17	28
-325	34	26	13	32
Apparent Density (g/cc)	2.20	2.26	2.26	1.95
PRESINTERED PROPERTIES				
Sieve Analysis	Percent Retained			
+60	1	1	5	9
+200	31	39	62	32
+325	43	40	23	25
-325	25	20	10	34
Apparent Density (g/cc)	2.26	2.26	2.32	-
Hall Flow (sec/50g)*	37	35	33	-
SINTERED PROPERTIES				
Sieve Analysis	Percent Retained			
+60	-	-	-	8
+200	34	43	53	33
+325	36	36	31	27
-325	30	31	16	32
Apparent Density (g/cc)	2.26	2.26	2.38	1.97
Hall Flow (sec/50g)*	33	35	32	45

*per ASTM B213-48

The results indicate that the apparent density of the powder produced with the inorganic binder is substantially higher than that produced with an organic binder, both after spray drying and after sintering. That is, of course, advantageous, in that the volume of material to be handled is reduced and processing time is decreased. The flow properties, as indicated by Hall flow measurements, of the sintered product are also improved over that of the organic binder product subjected to identical presintering and sintering conditions.

A scanning electron micrograph of the -170+325 fraction of the spray dried and sintered molybdenum powder is shown in Figure 2. Note that the particles are uniform and nearly spherical, thus rendering them highly flowable.

It has been observed (see Reference 5) that additions of minor amounts ($\leq$0.5%) of nickel to the molybdenum agglomerates cause the bulk density to increase during sintering at temperatures in excess of 1060°C. Figure 3 dramatically illustrates

Figure 2. Spray dried and sintered -170+325 fraction of molybdenum powder.

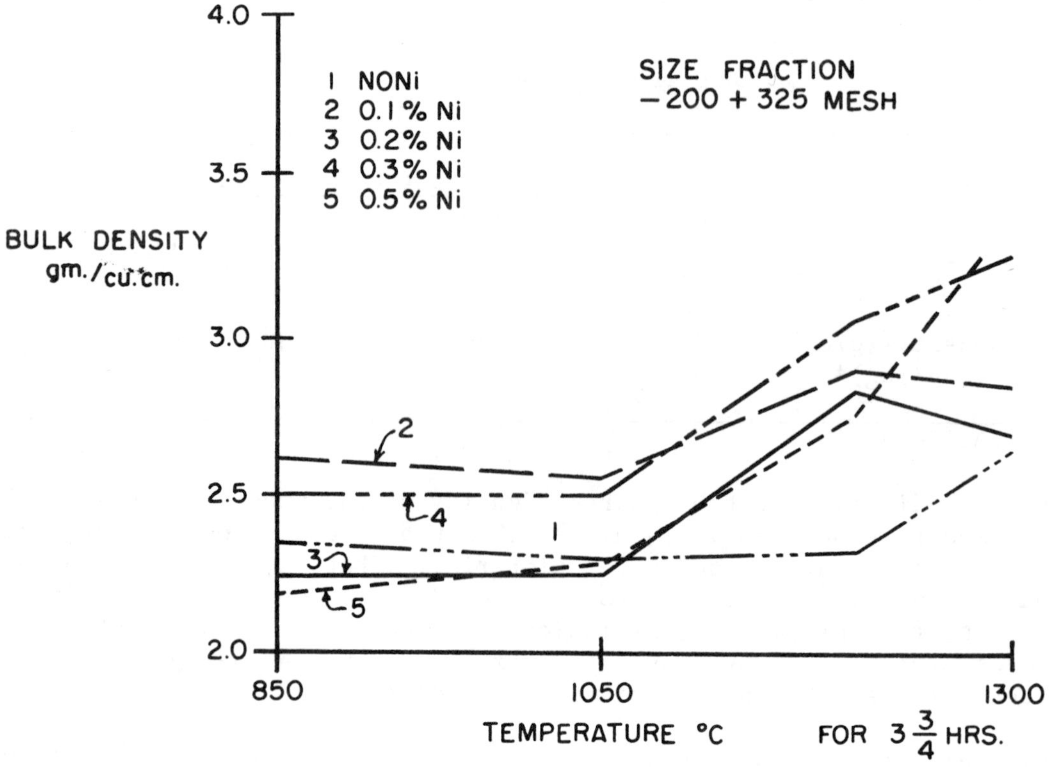

Figure 3. The effect of minor additions of nickel on the bulk density of spray dried and sintered molybdenum powder.

this effect for the -200+325 fraction molybdenum powder in which the nickel was introduced into the slurry prior to spray drying.

Plasma Densification

The second topic of this paper is a technique known as plasma densification. This process (as described in References 6 and 7) produces fine, spherical, densified particles of materials which heretofore could not be produced due to their high melting point.

The densification process consists of the following: Agglomerated powders are entrained in a carrier gas, fed through a high temperature plasma reactor, and collected. They are passed through the reactor at a flow rate such that interparticle contact and coalesence is avoided but such that at least the outer surfaces of a substantial number of the particles are melted. After melting, the particles fall through a distance sufficient to allow them to solidify and cool prior to contact with a solid surface or with each other.

Because the particles are melted while entrained in a carrier gas, the solidified particles are substantially spherical, have smooth surfaces and thus have excellent flowability. Powders made in this way may exhibit apparent densities of 40% or more of the theoretical density of a given material. In addition, the solidified particles have the same general size range as the starting material, but, depending on the porosity of the starting material, may have a smaller mean particle size, due to densification during melting. This densification is advantageous in that it leads to increased efficiency in coating operations.

If the agglomerates which are passed through the plasma flame consist of more than one type of particle, of more than one metal or ceramic or a combination thereof, these materials will react or alloy together during melting, to produce prealloyed powders or homogeneous composite particles.

Beneficial chemical reactions may also be carried out during melting. For example, by introducing hydrogen into the hot zone or by mixing carbon into the starting powder material, oxides may be reduced to low levels. Addition of carbon or boron may be used to form carbides or borides.

Plasma densification may be readily employed with any inorganic material having a melting point above 500°C including elemental metals, alloys, pure or mixed oxides, borides, carbides, or nitrides. Of particular interest for coating applications are refractory materials having a melting point above 1800°C. These include the refractory metals tungsten, molybdenum, chromium, tantalum, and niobium and their alloys. Also the cemented tungsten carbides containing up to 30% cobalt can be plasma densified.

Many types of commercially available plasma reactors can be used for plasma densification. Both induction plasma and resistance arc plasma units yield a satisfactory product. While the temperature required need only exceed the melting point of the highest melting component of the material being processed, it is preferred to have a temperature at least above the vaporization point of the lowest vaporizing component. This guarantees a relatively short residence time in the reaction zone and consequently processing of large quantities of powders can be realized.

All particles need not melt completely, since melting of the outer layer of the particle will result in some degree of sphericity, surface smoothness and densification. Furthermore, melting of only a certain fraction of the particles will nevertheless result in substantial improvement in flowability of the powder. By way of example, for plasma flame reactors having temperature capabilities between 10,000°F and 30,000°F, it has been found that a powder feed rate of from 1/2 up to 30 pounds per hour results in substantial improvement in flowability of the final product.

Induction Plasma Densification

To illustrate the plasma densification process, let us consider the technique employed to produce a molybdenum powder. Sintered agglomerates of spray dried molybdenum were separated into size fractions by screening. Each fraction was then fed separately through a commercially available induction plasma torch into a water cooled collection tank. A mixture of 126 cubic feet per hour of argon and 70 cubic feet per hour of hydrogen was fed to the plasma torch. The torch power was about 28 KVA. Nitrogen gas was fed to a powder feeder at a rate of 7 cubic feet per hour to entrain the powder and feed it through the torch. The nitrogen provides a nonreactive atmosphere as well. The results of the plasma densification process are shown in Table III.

TABLE III

Induction A/H_2 Plasma Densified Molybdenum

Feed Size	-100+200	-170+200	-200+325	-270+325
Feed Rate (lb/hr)	0.9	0.9	1.7	1.2
Torch Power (KVA)	28.2	27.4	24.5-26.3	28.8
Product				
+200	16.8	0.8	2.5	0.6
-200+325	70.1	69.7	43.4	7.9
-325+400	5.7	16.4	26.2	23.0
-400	7.4	13.1	28.0	68.5

The effect that melting has on densifying the particles is shown by the decrease in particle diameter. The -100+200 mesh feed drops 83 percent below 200 mesh. The -270+325 mesh feed decreases to 91.5 percent below 325 mesh. Measurements of apparent density show an increase from 1.8 g/cc for the spray dried feed to 5.4 g/cc for the product. The Hall flow for the feed material was 41 sec/50g whereas for the product it was 11-12 seconds.

Figure 4 is a scanning electron micrograph of the induction plasma densified molybdenum particles. Note that these particles have extremely smooth surfaces and are nearly spherical.

Resistance Arc Plasma Densification

Resistance arc plasma guns can also be used for plasma densification. A -200+325 mesh fraction of spray dried and sintered molybdenum agglomerates was fed at a rate of 1.4 lb/hr into a commercially available torch and collected in a water cooled chamber. Argon (at 0.7 cfh) was used as the powder feed carrier gas. Argon (at 28 cfh) was also used to create the plasma. The gun current and voltage settings were 500 amps and 28 volts, respectively. The resultant product had a bulk density of 5.3 g/cc and a Hall flow of 14 to 16 sec/50g. Again,

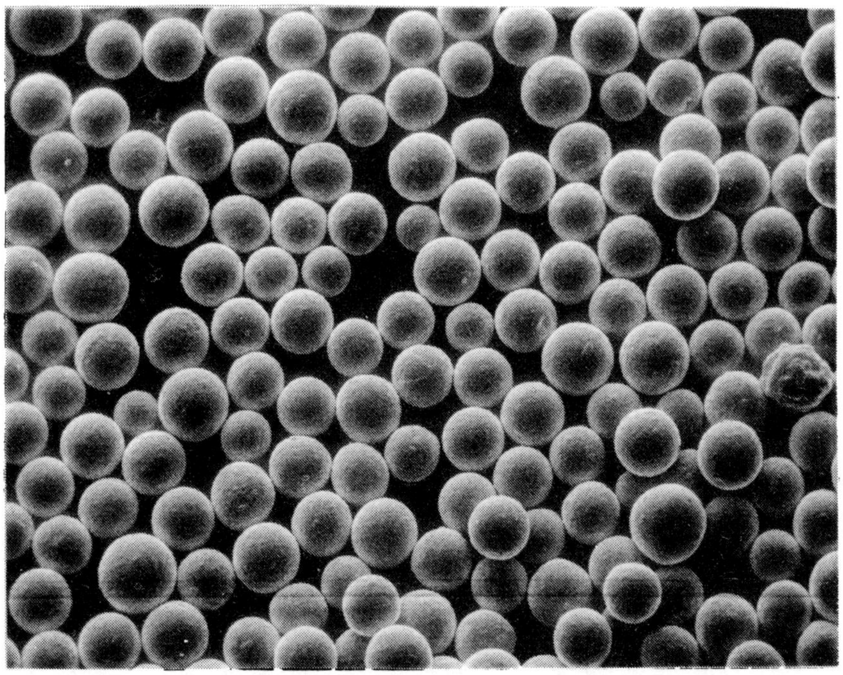

Figure 4. Scanning electron micrograph of plasma densified molybdenum particles.

due to densification, the size distribution of the final product was considerably finer than the feed material.

Conclusion

Spray drying techniques have been developed in which a soluble salt of one of the constituent materials is used as a binder. The resultant agglomerates exhibit increased bulk densities and better handling capabilities over their counterparts containing organic binders.

Agglomerates can be further densified by passing them through a plasma flame. The individual particles produced are smooth, spherical and nearly 100 percent dense. They exhibit excellent flow characteristics and when used for plasma coatings, produce high density, high efficiency deposits.

References

1. Masters, K. Spray Drying, (Leonard Hill, London, 1972)

2. U.S. Patent 3,881,911. Richard F. Cheney, David J. Port and James R. Spencer. "Free Flowing Sintered Refractory Agglomerates". Assignee - GTE Sylvania Incorporated.

3. U.S. Patent 3,973,948. John M. Laferty, Jr., Joseph E. Ritsko and David J. Port. "Free Flowing Powder and Process for Producing it". Assignee - GTE Sylvania Incorporated.

4. U.S. Patent 4,025,334. Richard F. Cheney, William D. Lafferty and George J. Long. "Tungsten Carbide-Cobalt Flame Spray Powder and Method". Assignee - GTE Sylvania Incorporated.

5. U.S. Patent 3,907,546. David J. Port and Richard F. Cheney. "Molybdenum Flame Spray Powder and Process". Assignee - GTE Sylvania Incorporated.

6. U.S. Patent 3,974,245. Richard F. Cheney, Charles L. Moscatello and Frederick J. Mower. "Process for Producing Free Flowing Powder and Product". Assignee - GTE Sylvania Incorporated.

7. U.S. Patent 3,909,241. Richard F. Cheney, Charles L. Moscatello and Frederick J. Mower. "Process for Producing Free Flowing Powder and Product". Assignee - GTE Sylvania Incorporated.

WARM PRESSED ALLOYS OF COPPER WITH TIN AND TIN-LEAD

G. Allen, S. K. Chatterjee, C. A. Mackay
International Tin Research Institute
Perivale, Greenford, England
and
W. B. Hampshire
Tin Research Institute, Inc.
Columbus, Ohio

Introduction

A simple procedure for making iron powder compacts by warm pressing a mixture of iron and tin or tin-lead alloy powders has been developed at the International Tin Research Institute.

In the work previously reported (1) the powder mixtures were pressed in a die heated to temperatures from 523K (250C) to 723K (450C) to obtain materials with a range of property values. Extension of the use of this technique to non-ferrous materials has always been intended and it is the work carried out on copper-tin and copper-solder mixtures which is reported here.

Experimental Procedure

The basis materials were -53 μm (-300 mesh) atomized copper powder (both newly purchased copper powder and some that had slightly oxidized during storage were used) and atomized tin and solder powders (with various tin-lead ratios) nominally of -53 μm (-300 mesh).

The powders were mixed in a Vee mixer for a period of six hours and the uniformity of the mixing was checked by chemical analysis. The mixtures were then compacted at 770 N/mm^2 (50 tons f/in^2) in a die maintained at 523K (250C) or 623K (350C) so that they were heated above the melting point of the tin or solder addition, into the forms of shaped tensile test pieces with dimensions conforming to those specified in BS 5600 (Part 3), 1979.

The additions investigated here were of tin and 60 Sn-40 Pb, 40 Sn-60 Pb and 15 Sn-85 Pb alloys. These additions were made at levels of 0.5, 1.5, 3.0 and 5.0 % by weight for all alloys and at 2, 4, 6, 8 and 10 % by weight for elemental tin. Scanning electron microscope examination of metallographically polished cross sections allowed analytical identification of the distribution of the various elements after compaction.

Results

The graphs in Fig. 1 illustrate the forms of the tensile strength and elongations curves for materials compacted at 523K (250C). These show that compacts with pure tin additions reach an optimum in both tensile strength and elongation between 6 and 7 percent tin. In the cases of all tin-lead alloy additions the optima occur at about 3 percent. Fig. 2 shows the results for the same properties for materials compacted at 623K (350C). Optimum tensile strength for the material with tin additions has moved to 4 percent while the elongation curve shows no comparable maximum. Materials containing tin-lead also show maxima moved towards low addition levels (1.5%) and still demonstrate associated maxima in elongation properties. Comparison of both figures shows that the tensile strength property is improved by increasing the compacting temperature with the peak strength for the pure tin addition increasing from 245 N/mm^2 (15.9 tsi) for the 7 percent tin alloy pressed at 523K (250C) to 295 N/mm^2 (19.1 tsi) for the 4 percent tin alloy pressed at 623K (350C). Increased compacting temperature improved the tensile strength of materials containing solder alloy additions even more markedly. Elongations were not so obviously affected by pressing temperature except in the case of the 60 Sn-40 Pb additions where increasing compacting temperature decreased elongation. The hardness of these materials is shown in Fig. 3 for material produced at 523K (250C) and Fig. 4 for material compacted at 623K (350C). Except in the case of pure tin additions composition has little effect on the hardness although a difference of about 10 hardness points is observed with change of compacting temperature. Fig. 3 also shows a curve labeled oxidized copper plus tin. This material was produced from copper stored in the laboratory for about 18 months which from its dark color had clearly oxidized by a natural process. Although the quantity was not measured the additional oxide did produce a substantial increase in hardness particularly at the lower levels of tin additions. The effect of increased oxide content was to decrease tensile strength and reduce the effect of compacting at different temperatures.

Metallographic Examination

Metallographic examination of specimens showed them to be essentially pore-free solid materials with a fine grained continuous copper matrix randomly interspersed with duplex lakes of the copper-tin intermetallics Cu_6Sn_5 and Cu_3Sn (Fig. 5). Microprobe analysis showed that in the samples containing solder pressed at 563K (250C), lead, apparently as the elemental metal, was distributed as fine particles or islands mainly within the copper-tin intermetallics. At the higher pressing temperature the lead islands were distributed throughout the matrix. Micrographs of samples produced from oxidized copper clearly showed

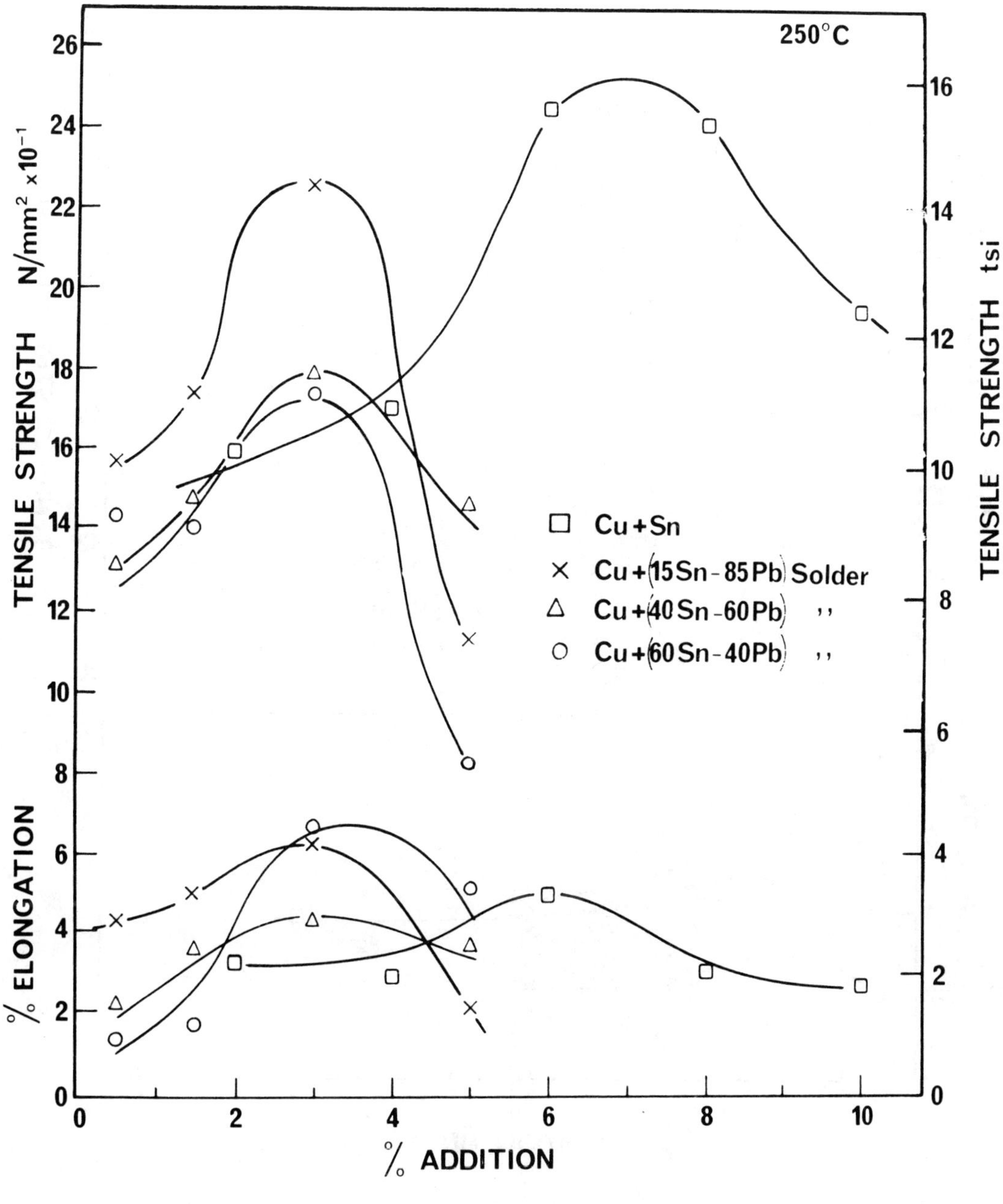

Fig. 1. Tensile strength and elongation of compacts of fresh copper with tin and three tin-lead alloys pressed at 250C.

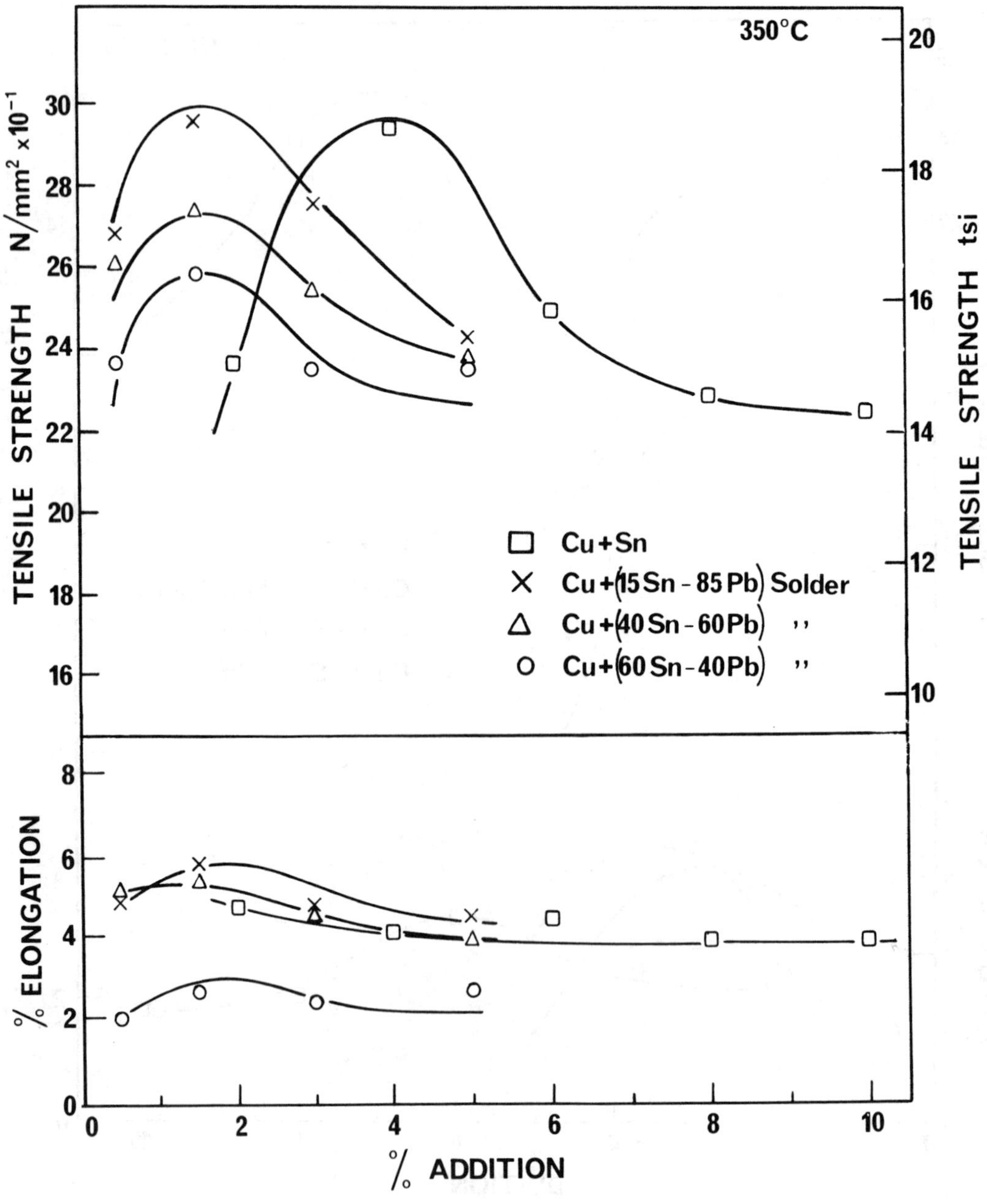

Fig. 2. Tensile strength and elongation of compacts of fresh copper with tin and three tin-lead alloys pressed at 350C.

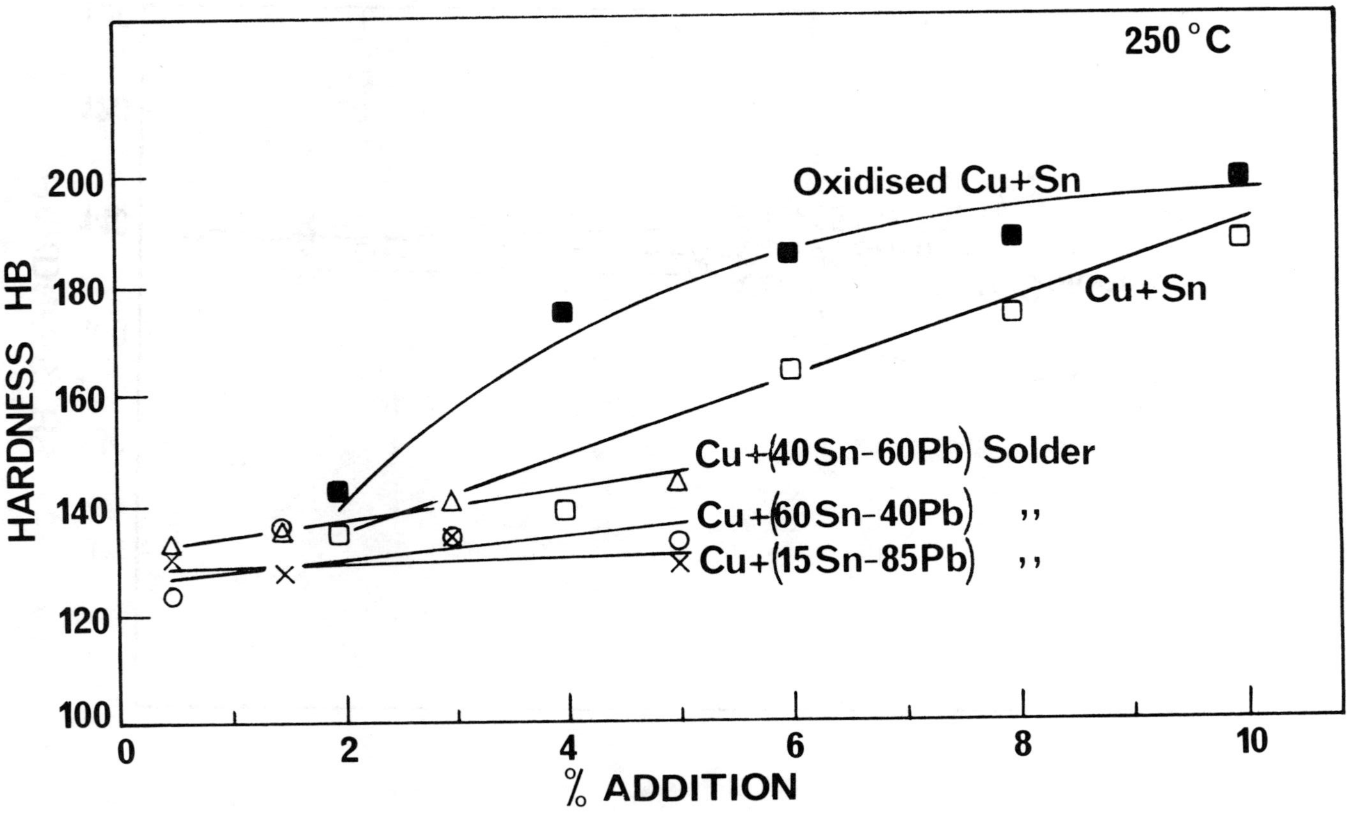

Fig. 3. Hardness of compacts of fresh copper with tin and three tin-lead alloys pressed at 250C, also showing the effect of using oxidized copper with tin.

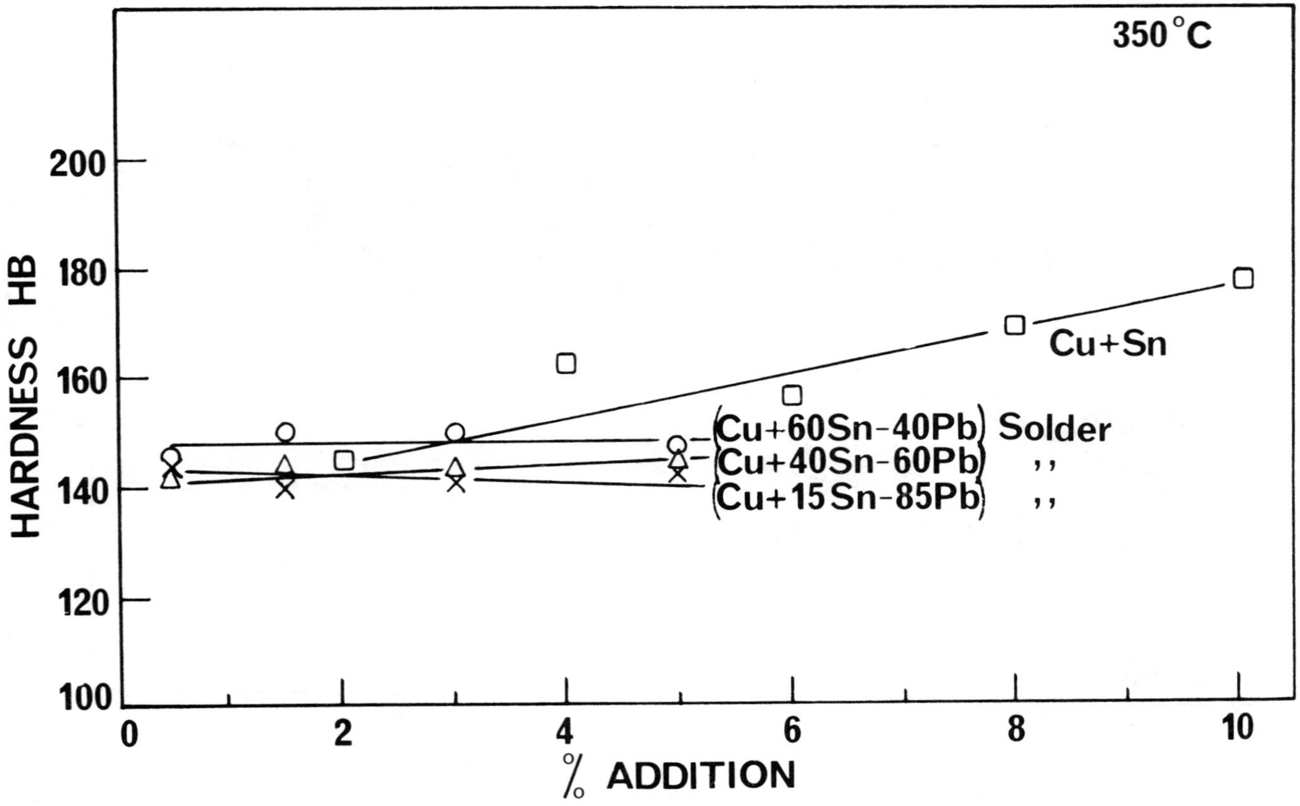

Fig. 4. Hardness of compacts of fresh copper with tin and three tin-lead alloys pressed at 350C.

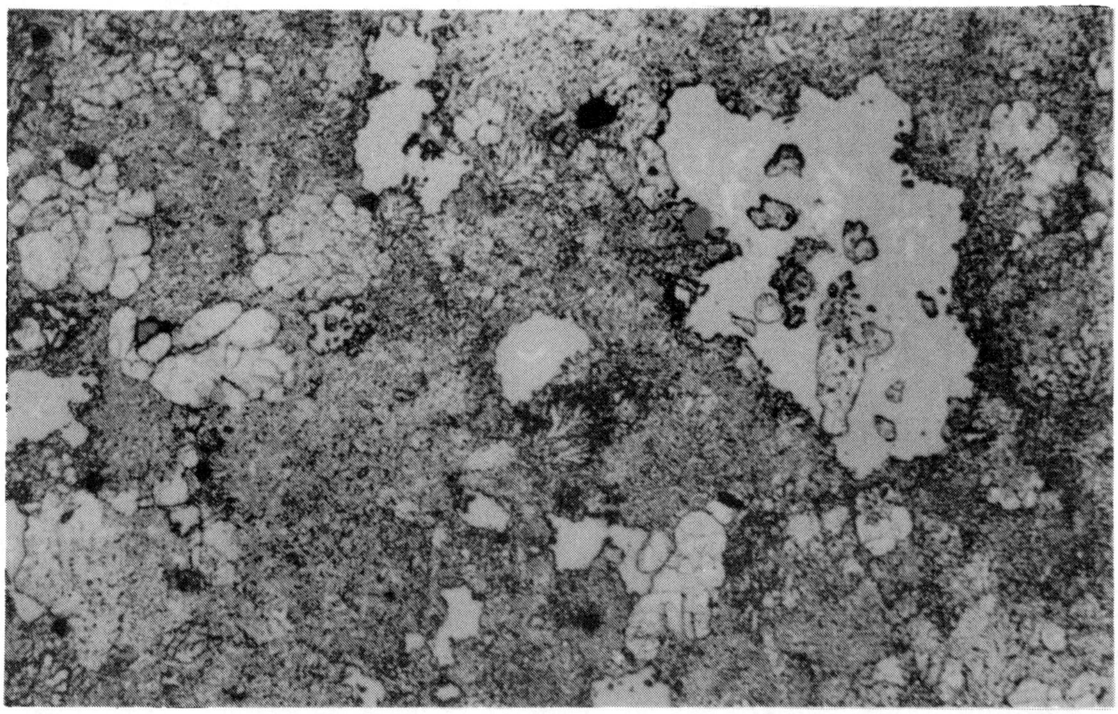

a

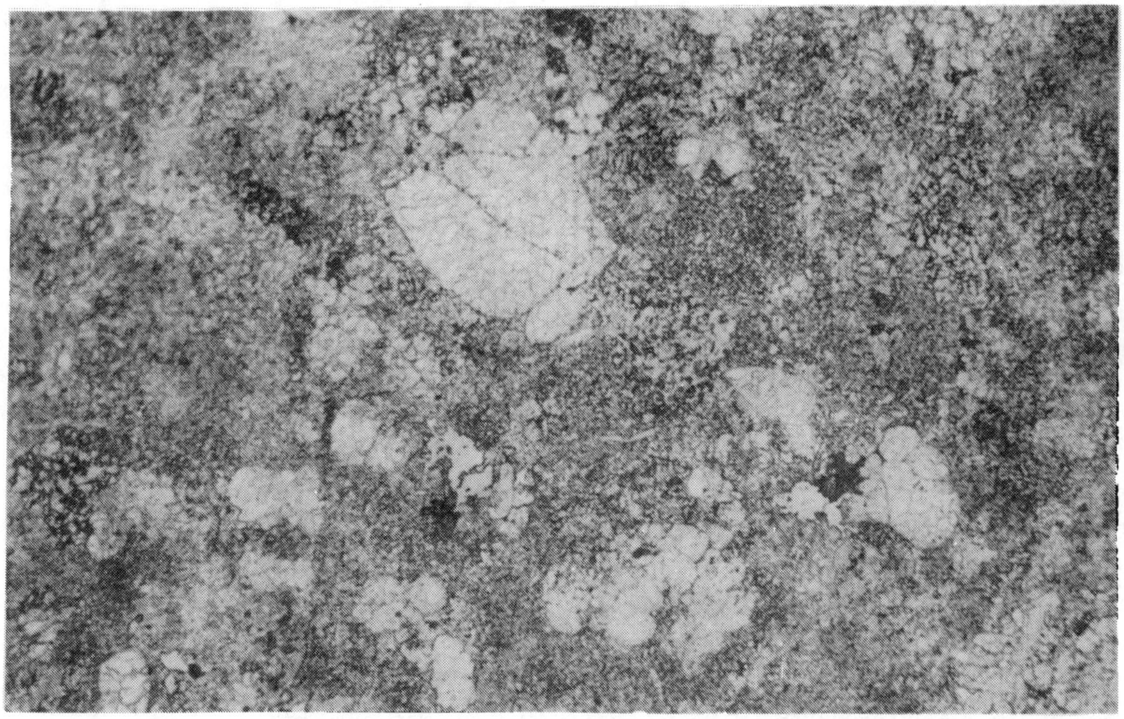

b

Fig. 5. Optical micrographs of (a) copper plus 4 percent tin (b) copper plus 1.5 percent (15 Sn-85 Pb) alloy pressed at 350C. Both micros X400 additional dark oxide films randomly distributed throughout the structure. The structures were totally different from the usual structures of both cast and sintered bronzes there being no regularity or periodicity as imposed by dendrites in a cast structure and no evidence of regular or equiaxed matrix grains as often observed in sintered bronzes. The outlines of the intermetallic regions were rough and irregular as opposed to the rounded regions found in sintered structures containing a sufficiently high tin level. As mentioned above the intermetallic regions appeared as a mixture of the compounds Cu_6Sn_5 and Cu_3Sn and not the more nearly equilibrium delta phase.

Discussion

Compared to the properties we were able to obtain during our previous work, the properties reported in this work are surprisingly good. It has been shown that copper and tin and copper and solder compacts can be produced by warm pressing (for only 2-5 seconds), without any flux, at 523K (250C) having tensile and elongation values comparable to both sintered and sandcast materials (Table 1) but with a considerable increase in hardness. The properties can be further improved by increasing the compacting temperature to 623 (350C). At this temperature the amount of tin or solder addition required to give peak strength is halved.

The mechanism for bonding proposed in our previous paper was the mutual abrasion of the harder metal powders rupturing the encapsulating oxide films so that the molten phase reacts immediately with the fresh metal surface to produce fully bonded material without the necessity for a flux. It seems likely this is only the initiation step and that the properties observed are largely due to the extremely rapid diffusion rates which are operative in the copper-tin system. Differences in hardness between iron and copper powders together with differences in ductility probably account for the difference in porosity of compacts of the two materials. Whereas in both cases the softer starting material powders are appreciably deformed during the compacting process and are thereby considerably work hardened, the ultimate hardness achieved with the copper-based materials so exceeds the full hardness of copper that it obviously includes an appreciable contribution from intermetallics which have formed. Sintering or solid state diffusion at the times and temperatures involved in the process, can not be responsible for the extremely good tensile values observed with the present materials. These must be indicative of good distribution of the liquid phase so that there is a thin film of molten metal around almost every particle. The absence of large areas of unreacted solder and the relative infrequency of large pieces of copper-tin intermetallics also suggests this. Optical examination at the highest magnification shows small blue lakes of intermetallic but is insufficient to give information on intraparticulate boundary structure. The fine detail of the structure shows a lacy network defining regions of only about 1 μm diameter. The mean copper particle size of the starting powder is probably about 27 μm with a maxima at 53 μm and a long distributive tail into the fine region. The preponderance of the fine structures after pressing must imply extensive shearing and folding of the particles probably with attendant entrapment of liquid films. This also provides ideal conditions for production of a fine dispersion of hard intermetallic phases. It is likely that the application of pressure

Composition and Treatment	Tensile Strength N/mm^2	Elongation %	Hardness HB
Warm Pressed at 250°C			
Cu + 6% Sn	245	5·0	165
Cu + 3%(15Sn85Pb) Solder	225	6·5	135
Cu + 3%(40Sn60Pb) Solder	180	4·5	140
Cu + 3%(60Sn40Pb) Solder	175	6·5	135
Warm Pressed at 350°C			
Cu + 4% Sn	295	4·0	163
Cu + 1·5%(15Sn85Pb) Solder	295	6·0	144
Cu + 1·5%(40Sn60Pb) Solder	274	5·5	140
Cu + 1·5%(60Sn40Pb) Solder	260	2·5	150
Sintered			
Cu+10% Sn	117-188	5-12	35-55
Sand Cast			
Cu+10% Sn +0·15% P	225-303	9-20	70-90

Table 1. Comparison of optimum properties of warm pressed compacts produced at each temperature with properties of sintered and sand cast 10 percent tin-bronze.

at elevated temperature modifies the copper-tin phase diagram and also the diffusion rates. Reduction of activation energies in this way would lead to enhanced reaction rates for the formation of the various phases in the copper-tin system.

The compositions for optimum properties depend on pressing temperatures although it is surprising that at both temperatures studied the percentage solder additions at which the tensile strength and elongation curves show peak values are identical irrespective of solder composition. If, as was originally surmised, the active constituent leading to bonding in this system is tin, then clearly optimum intraparticulate bonding should be achieved with pure tin additions. This neglects, however, the effects of increased fluidity, for a given temperature, of the lower melting point alloys. The results obtained however do not demonstrate this trend. Of the solder composition studied the 60 Sn-40 Pb alloy with a melting point of 460K (187C) should be the most fluid at both test temperatures and would be expected to demonstrate the best properties. In fact the best properties were obtained with the 15 Sn-85 Pb alloy which, with a liquidus temperature of 563K (290C), should not have been molten at the lower pressing temperature. Of the specimens produced at 523K (250C) those containing a 6 or 8 percent addition of pure tin showed marginally higher tensile strengths than was obtained with solder additions whereas of the alloys containing solder the high lead solder addition produced strengths markedly higher than those observed for alloys produced at this temperature with the higher tin solders.

At the higher temperature the optimum addition level for pure tin moved nearer to those for the solder additions. These latter had moved to what, on the basis of our previous work with iron powders, was surprisingly low at only 1.5%. This is explainable however if one considers the observation that in the case of copper powders there is almost zero porosity between the copper particles after pressing so that the volume of liquid required to fill any porosity is correspondingly low. It is surprising however that with our additions made as weight percentages the differences in density of the various solders used, with the consequent differences of volume of molten alloy, do not have any apparent effect on the tensile properties of the materials. At the higher temperature, except for hardness, the marginal difference in optimum properties of solder addition material compared to pure tin addition material is lost. It seems possible therefore that the formation of Cu_6Sn_5 intermetallic compounds during the pressing operation is not a predominant feature when tin-lead additions are used. The formation of a copper-tin-lead ternary eutectic could be involved and could be the reason for such good performance of these alloys. Such a reaction would produce a more ductile intraparticulate phase and help explain the extremely good ductilities observed. In addition they would not preclude the formation of the copper-tin intermetallics. Such a reaction sequence would also explain why the addition of a solder alloy which is not molten at 523K (250C) produced such a well-bonded material. Also, the effect of pressure on lower melting points could well be dependant of alloy composition and, although it is difficult to estimate the magnitude of such an effect, the melting points of lead-rich alloys could be more affected than tin-rich. Microprobe analysis of the samples showed very little evidence of the presence of lead-rich regions even in cross-sections of materials containing 15 Sn-85 Pb alloy and this too suggests a very fine dispersion not wholly consistent with the bonding reaction solely involving the tin component of the addition with lead remaining as almost pure metal. Further work is needed in this field to understand the mechanism involved but work done so far has shown that a new and unique way has been found for the manufacture of copper-base materials with excellent engineering properties.

Table 1 summarizes the optimum properties and compositions of the materials produced at both temperatures. Also included for comparison are the properties of sintered 90 copper-10 tin/bronze and also 90 copper-10 tin-0.15 phosphorus cast bronze.

As can be seen the properties obtained with pressing copper-tin and copper-solder composites at 623K (350C) were superior in strength and hardness to those of fully sintered bronze. The strength was comparable to that of sandcast bronze while the hardness of the new material was superior. Even at the lower pressing temperature the materials obtained by warm pressing gave tensile strengths and elongations approaching those of sintered bronze with hardnesses 2-3 times higher than for cast bronze. Hardness could be further enhanced if oxidized copper powder were used but this was at the expense of ductility and tensile strength.

Conclusion

It has been shown that mixed copper-tin and copper-solder powders can be warm-compacted at relatively low temperatures to produce materials with properties which suggest the possibility of their use as bearing materials. They have tensile properties which equal those of bronzes prepared by the conventional sintering and casting processes and with hardnesses which exceed them.

Free lead in copper-solder alloys can impart lubricity to the alloys which could add to the wear-resistance properties.

The process, being simple, can be used with simple die equipment and the energy consumption will be much less than with the conventional sintering process.

The oxygen content of the copper powder is not critical. This eliminates the need for a protective atmosphere during compacting. Oxidized copper powder can be used as a base material with an advantageous effect on hardness. The process as outlined here appears ideal for mass production and the manufacturing process is completed in a single operation.

Acknowledgement

The Authors are grateful to the International Tin Research Council for permission to publish this Paper.

Reference

(1) D. J. Madeley, C. A. Mackay, Powder Metallurgy International, Vol. 7, No. 4, 1975

Comparison of Endothermic (Generated) and

Nitrogen-Based Sintering Atmospheres

Prepared by: J. Scott Becker, Program Manager
P/M Applications
Applied Research and Development

INTRODUCTION

Powdered metal parts are sintered in a protective atmosphere for two critical reasons. First, the atmosphere must prevent oxidation of the parts (discoloration and blistering) while also reducing oxides already present in the powder before sintering. Secondly, the atmosphere must provide an acceptable carbon potential in order to prevent excessive carbon loss from the green parts. Several generated atmospheres meet these basic requirements. Recently, nitrogen-based atmospheres have also been found to meet these requirements. This paper compares these atmospheres and the properties of parts sintered in them. It also discusses operating considerations, the problems associated with lubricant removal, as well as recommended nitrogen-based blends for various operations.

REQUIREMENTS OF A SINTERING ATMOSPHERE

OXIDE REDUCTION

Most commonly used iron and copper powders have a hydrogen loss of .1% to 1.5%. This hydrogen loss is a measure of the amount of oxide contained in the powder as received by the parts fabricator. Therefore, it is necessary that the sintering atmosphere not allow further oxidation, and that it reduce oxides already present in the green parts. The importance of reducing these oxides relates to the actual sintering mechanism. This mechanism is based upon the diffusion of metal atoms between the individual powder particles, and this diffusion is severely limited by the presence of oxides on the surface of the particles. The result of poor metal atom diffusion is low sintered hardness, strength, and ductility.

In order that the sintering atmosphere both prevent oxidation and discoloration and reduce existing oxides, it must be composed of a mixture of more than one gas. An ideal gas for preventing oxygen and moisture from entering the furnace and causing oxidation is nitrogen. Nitrogen molecules are about the same weight as oxygen molecules (heavier than H_2O molecules) making nitrogen a very effective blanketing or inerting gas. However, nitrogen will not reduce oxides. The ideal gas for reducing oxides is hydrogen, which reacts with oxides to form water vapor (raising the dew point). However, since hydrogen molecules are very light, hydrogen is not a good blanketing gas. The ideal sintering atmosphere must, therefore, contain both hydrogen and nitrogen.

CARBON CONTROL

Since a suitable sintering atmosphere must also control carbon loss from the green parts, carbon potential of the atmosphere is crucial. Simply stated, carbon potential is a measure of the amount of carbon in the sintered part that will be in equilibrium with a given atmosphere. Carbon potential is normally measured as the CO concentration squared over the CO_2 concentration:

$$[CO]^2/[CO_2]$$

Factors which critically affect carbon potential at a given temperature are CO concentration, CO_2 concentration and H_2O content (dew point). Dew point strongly affects carbon potential through the following reaction:

$$C_{(part)} + H_2O_{(atm)} \rightleftarrows CO_{(atm)} + H_{2(atm)}$$

The important atmosphere components and their impact on carbon potential at a given temperature are as follows:

TABLE I

Atmosphere Component	Relative Impact of Conc. Increase	Magnitude of Change
CO	Increases Carbon Potential	Medium
CO_2	Decreases Carbon Potential	Large
H_2O	Decreases Carbon Potential	Large
H_2	Increases Carbon Potential	Small
N_2	Neutral	-
O_2	Decreases Carbon Potential	Large
CH_4	Increases Carbon Potential	Large

To summarize this table, the following atmosphere components promote a higher carbon potential - CO, H_2, CH_4; while these components promote a lower carbon potential - CO_2, H_2O, O_2.

For several practical reasons, the total carbon in the green compact is normally greater than the desired combined carbon in the finished part. A more or less typical practice is to start with a powder blend of 1% carbon and produce a finished part of .6 to .8% carbon. This means that some carbon is intentionally sacrificed. Most of this lost carbon combines with the oxides present in the powder to form CO. This carbon loss occurs at any atmosphere carbon potential since the reaction occurs inside the parts. The objective of controlling the sintering atmosphere carbon potential is to prevent further loss of carbon to the atmosphere.

Returning to the practical consideration of what gases are required in the sintering atmosphere, one concludes that either CH_4 or CO or both, are necessary to provide a suitable carbon potential and CO_2, H_2O and O_2 must be avoided. Since different parts are designed with different finished carbon contents, the required amounts of CO and/or CH_4 vary. Also, many parts contain no carbon and therefore require no carbon potential in the sintering atmosphere. Examples are nonferrous metals and most stainless steels.

The ideal sintering atmosphere for iron-carbon parts is, therefore, a blend of hydrogen, nitrogen and methane (CH_4) and/or carbon monoxide. In the furnace methane reacts with water vapor as follows:

$$CH_4 + H_2O \rightarrow CO + 3H_2$$

Therefore, the presence of CH_4 also produces CO.

COMPARISON OF AVAILABLE SINTERING ATMOSPHERES

COMPOSITIONS

There are four common sintering atmospheres available today. These are endothermic gas, exothermic gas, dissociated ammonia and nitrogen-based atmospheres. The following table lists the typical composition of the input atmospheres.

TABLE II

Typical Components	Atmosphere Composition			
	Endothermic	Dissociated Ammonia	Exothermic	Nitrogen Based
N_2	39%	25%	70-98%	97%
H_2	39%	75%	2-20%	2%
CO	21%	-	2-10%	-
CO_2	.2%	-	1-6%	-
H_2O (Dew Point)	-5° to +45°F	-20° to -60°F	-10 to -50°F	-90°F
O_2	10 to 150 ppm	10 to 35 ppm	10 to 150 ppm	5 ppm
CH_4	.8%	-	10 to 200 ppm	1%
NH_3	-	5 to 100 ppm	-	-
Specific Gravity	.61	.30	.89	.95

As can be seen, all common sintering atmospheres contain H_2 and N_2 and all but dissociated ammonia also contain either CH_4 or CO or both. All atmospheres also contain a certain amount of impurities like H_2O, O_2 and CO_2 which cause oxidation and reduce carbon potential.

Determining the best atmosphere for a particular sintering application consists of applying the various principles discussed above in conjunction with other important factors like cost, availability, consistency, and reliability of supply.

Regardless of whether the nitrogen comes from a generator or from an industrial gas supply, it is always the least expensive component of the blend (derived ultimately from the air) and also the best component for inerting the furnace. Therefore, it is logical to utilize an atmosphere composed mostly of nitrogen with only those additions of other components necessary to achieve the desired results.

Hydrogen, regardless of how it is supplied, is normally the most expensive component of the atmosphere and also the most explosive and energy intensive. Therefore, it is logical to utilize an atmosphere composed of as little hydrogen as is necessary to achieve the required reduction of oxides and protection of the parts.

Several hydrogen sources are available for sintering. These include:

1. Pure industrial hydrogen stored as a gas or liquid.
2. Dissociated ammonia.
3. Alcohols.
4. Endothermic gas.

The choice between the first three hydrogen sources is largely a matter of relative economics and personal preference. In most cases, if dissociated ammonia is already available, it is preferred.

Utilizing endothermic gas as the hydrogen source is somewhat different in that it represents blending a combustion generated gas with a nitrogen base. Ideal ratios for this blend as derived from the table below are from 15% to

25% "endo," balance nitrogen. This blending (often referred to as endo dilution) has the advantages of higher combustibles, especially CO, and generally lower cost, but has the disadvantages of reduced purity and continuing (although greatly reduced) dependence on a small combustion plant (generator) and varying raw material supply (natural gas quality). Typical data for endo dilution are contained in Table III. The relative sooting levels are interesting to note. For reasons not fully understood, sooting increases between 60% to 30% endo, balance nitrogen, but decreases substantially at or below 20% endo.

TABLE III

ENDOTHERMIC DILUTION*

	Percent Endo in Blend			
Component	100%	60%	30%	20%
H_2 (%)	35.5	24.5	13.8	10.8
CO (%)	18.3	10.7	6.7	5.5
D.P. (°F)	+18	-8	-35	-40
O_2 (%)	.47	.28	.15	.12
R_B	68-75	70-80	70-78	73-80
SOOT (Index)	1.0	1.2	2.0	1.0
CARBON (%)	.6	.55-.6	.5-.55	.60

*TOTAL FLOW 1500 SCF/HR 18" FURNACE

CONSTANT CH_4 ADDITION OF 12 SCFH

Carbon bearing components (CO, CH_4) are relatively expensive, combustible and energy intensive, and tend to cause sooting. Therefore, it is desirable to limit their content to that level necessary to provide the required carbon potential.

Obviously, it is highly desirable to minimize the amount of all impurities in the sintering atmosphere. In addition to oxidation and decarburization caused by H_2O, O_2 and CO_2, raw or uncracked ammonia is also detrimental as it causes severe nitriding of the parts.

SELECTION

Nitrogen-Based Atmospheres typically composed of 1% to 4% hydrogen and 0% to 1.5% natural gas, balance nitrogen have found growing acceptance for sintering because they combine the following characteristics:

1. Sufficient hydrogen to efficiently reduce oxides present in the furnace while providing an excellent reducing potential (hydrogen-to-water vapor ratio).

2. Sufficient CH_4 to provide a carbon potential of from 0 to .8% carbon in the sintered part.

3. Extremely low concentrations of H_2O and O_2 and no measurable CO_2 or raw ammonia.

4. An atmosphere comprised primarily of nitrogen is very efficient in keeping air out of the furnace. Also, since nitrogen molecules are heavier than water molecules, nitrogen is very efficient at sweeping water, produced by the reaction $H_2 + 1/2\ O_2 \rightleftarrows H_2O$, out of the furnace. This removal of the product of the reaction is essential if the reduction of oxides is to proceed efficiently.

The generated atmospheres, of course, also meet the requirements for a good sintering atmosphere when properly applied. On points one and two, these atmospheres are generally every bit as good as nitrogen-based atmospheres. However, nitrogen-based atmospheres are substantially purer than generated atmosphere, as was seen in Table II. Also, nitrogen-based atmospheres are substantially denser than most generated atmospheres and this increased density provides the advantages previously described.

Up to this point, it has been assumed that any given atmosphere composition, whether generated or nitrogen-based, is constant over time. This is, in fact, true for nitrogen-based atmospheres which are created by mixing pure components in fixed proportions. This is not quite true of generated atmospheres which fluctuate in composition depending upon the following variables.

1. Generator/dissociator condition (especially catalyst condition).

2. Relative humidity and temperature of air.

3. Purity of feedstock (i.e., moisture in ammonia or peakshaving of natural gas).

4. Temperature of generator reaction chamber.

Variations in atmosphere composition can have several negative effects on the quality and consistency of sintered products. The two most important areas of concern are carbon control and size control. Variations in generator output, especially dew point and uncracked hydrocarbon levels, can substantially vary furnace carbon potential and thus finished part carbon levels, hardness and crush strength. Variations in dew point and carbon potential also strongly influence final part size (growth). Therefore, since nitrogen-based atmospheres are more consistent and provide more reliable carbon and size control, quality problems are often alleviated.

An additional benefit of nitrogen-based atmospheres is that they are essentially custom blended at the furnace. This enables the operator to intentionally alter the composition to affect final part size or carbon level. This measure of control can often be of great value when sintering parts subject to tight size and composition specifications.

MECHANICAL PROPERTIES OF SINTERED PARTS

Many factors combine to determine the mechanical properties of sintered parts. The sintering atmosphere represents only one of many variables which include powder composition, green density, etc. The major ways in which atmosphere can influence finished part mechanical properties are:

1. Reduction of internal oxides (on powder when compacted) which influences the rate of diffusion during sintering and degree of bonding.

2. Carbon potential which influences as-sintered carbon content, hardness, and strength.

3. Impurity levels which influence strength and ductility of sintered parts.

Typical mechanical data for parts processed in nitrogen-based atmospheres is given below. Generally, mechanical properties are about the same for any given part processed in endothermic verses nitrogen-based atmosphere. Occasionally a somewhat higher or lower hardness, at a given carbon content, is observed with the nitrogen blend. When the hardness is lower, it is generally the result of a somewhat coarser pearlite structure caused by a difference in cooling rate between nitrogen and endo. When the hardness is higher, it can normally be traced to a somewhat better sinter, probably due to reduced impurity levels in nitrogen. As is often the case, when the hardness is about the same in both atmospheres, it is probably the result of a balancing between these two influences.

TABLE IV

TYPICAL PHYSICAL PROPERTIES - NITROGEN SINTERED

Powder	Density g/cc	Transverse Rupture Strength KSI	Apparent Hardness R_B
Ancorsteel 1000 1% C Premixed	6.7	95.2	68
Ancorsteel 1000 .6% C Premixed	6.7	77.4	45
Ancor MH-100 1% C Premixed	6.4	74.9	50

Powder	Density g/cc	Tensile Strength KSI	Apparent Hardness R_h
Aluminum 201 AB	2.4	20.0	-
Bronze 90/10	6.6	14.2	-
Brass 90/10	7.4	19.8	56

OPERATING CONSIDERATIONS IN ATMOSPHERE SELECTION

When operating considerations are examined, several substantial advantages of nitrogen-based atmospheres become apparent. Many of these advantages are common to all heat treating operations. These include:

1. Reduced maintenance and downtime.

2. Improved process flexibility and dependability.

3. Improved safety.

4. Greatly reduced natural gas consumption and dependence.

5. Low capital investment.

Several operating advantages are more specific to the sintering process. These include:

1. Improved belt and muffle life (made possible by a more pure atmosphere).

2. Reduced scrap losses due to off-size parts (as described previously).

3. Reduced furnace conditioning time in unmuffled furnaces (due to greater atmosphere density and purity).

LUBRICANT REMOVAL

Powder metal blends normally contain about 1% lubricant. This lubricant may be acro wax or a stearate of zinc or lithium or other similar compound, but all lubricants are complex molecules containing carbon, oxygen and hydrogen. The lubricant is required in order that the compacted green part can effectively be ejected from the mold after compaction. Without a lubricant, the green parts would not eject from the mold intact and undistorted.

The removal of this lubricant from the green part is an important first step in the sintering process. The lubricant is volatilized at a temperature in the range of 800°F to 1400°F. Presintering is sometimes performed as a separate operation in a presintering furnace, but normally the presentering and sintering functions are performed in separate sections of the same continuous furnace.

Once the lubricant has volatilized from the green part it must be swept from the furnace by the furnace atmosphere. Otherwise the lubricant forms soot on the sintered parts. Since most sintering operations utilize one continuous furnace with the same atmosphere passing through both the sintering and presintering zones of the furnace, the ability of an atmosphere to sweep lubricant from the furnace is important. The atmosphere considerations for sintering and presintering are not the same, however.

To efficiently sweep lubricant from the furnace, it is advantageous that the atmosphere be as dense as possible, for the same reasons that density improves water vapor removal. This favors a high percentage of nitrogen and is consistent with the requirements of sintering. A problem arises, however, because lubricant removal is substantially improved by the presence of certain gaseous impurities. Since lubricants are essentially carbon chain molecules, the presence of CO_2, O_2 and H_2O promotes the breakdown of the complex lubricant molecules into smaller molecules. These smaller molecules are more easily swept from the furnace and thus there is less tendency to form soot. The dilemma presented by the necessity of lubricant removal is that these same gaseous impurities which promote lubricant breakdown during presintering also promote decarburization during sintering.

One answer to the problem of lubricant removal is to introduce one atmosphere into the sintering section and another into the presintering section of the furnace. When utilizing generated atmospheres, this can be accomplished by introducing an endothermic atmosphere into the sintering zone and an exothermic atmosphere into the presintering zone. With nitrogen-based atmospheres this can be accomplished by introducing a dry hydrogen/nitrogen/ methane blend into the sintering zone and an intentionally humidified (wet) nitrogen/hydrogen/ methane blend in the presintering zone. Of course, when presintering and sintering are performed in separate furnaces there is no problem utilizing two different atmospheres, and some combination furnaces are designed to allow different atmospheres. However, most sintering furnaces are designed for only one atmosphere and this atmosphere must be a compromise between the above requirements.

Endothermic atmosphere provides a reasonable compromise between these requirements in most applications, and this is one of the reasons endo has been the standard atmosphere for sintering iron-carbon parts. Conventional, dry nitrogen-based blends also provide a reasonable compromise for presintering and sintering low and medium density parts. However, sooting problems can arise when sintering high density (6.9 g/cc or greater) parts in a conventional hydrogen/nitrogen/methane blend. Recent developmental work has shown, however, that this problem can be overcome.

The answer to sooting problems when sintering high density P/M parts in nitrogen-based atmospheres is dew point. As mentioned previously, intentionally humidified hydrogen/nitrogen/methane works well for presintering, and when dew point and methane levels are properly controlled a reasonable balance between delubing and sintering requirements can be achieved in the same furnace. It should be noted that the actual dew point in the presintering zone is no higher for the humidified atmosphere than for the dry atmosphere. This indicates that the water has fully reacted with the methane in the atmosphere and the lubricant coming off the parts in the furnace. Comparative data for humidified and dry blends are shown below.

TABLE V

ATMOSPHERE COMPOSITION IN THE PRESINTERING ZONE

	Component %				DEW POINT °F
	H_2	CO	CO_2	CH_4	
Humidified H-N with Natural Gas Addition	4.1	.95	.003	.68	-75
Dry H-N with 1% Natural Gas Addition	.73	3.2	.03	.45	-60

(Balance Nitrogen)

TABLE VI

AVERAGE ROCKWELL B HARDNESS

	POWDER BLEND			
	IRON-.8% CARBON		IRON-.8% CARBON-2% COPPER	
Density, g/cc →	7.0	6.6	7.0	6.6
Humidified H-N with Natural Gas Addition	75	63	88	77
Dry H-N with 5% Natural Gas Addition	63	63	90	79

Nitrogen is humidified by the simple procedure of bubbling it through water. With a few relatively simple control features a bubbler can produce a controlled dew point of anywhere from -60°F to +80°F.

As would be expected, either air (O_2) or CO_2 can also be used instead of water. Dew point is preferred, however, because it is inexpensive, and requires less critical control. Also, the reaction of H_2O with methane produces beneficial hydrogen. The problems associated in trying to make controlled additions of air into a furnace are obvious.

In summary, the following statements apply to soot control in nitrogen-based sintering:

1. For low and medium density parts there is no increased tendency to soot.

2. For high density parts humidified nitrogen blends are often necessary to control sooting.

3. When possible, it is advantageous to put dry nitrogen blends in the sintering zone of the furnace and humidified nitrogen blends in the delubing zone. Some mixing of atmospheres can not, of course, be avoided; but the better the separation, the closer to the ideal sintering environment is achieved.

RECOMMENDED NITROGEN-BASED SINTERING BLENDS

In light of all the previously discussed considerations, the following approximate nitrogen-based blends are recommended for the materials listed:

TABLE VII

RECOMMENDED H-N BLENDS

	H_2 (%)	N_2 (%)	Humidified	CH_4 (%)
Iron .5% C or Less	2	97.5	-	.5
Iron more than .5%C	2	97	-	1
Iron - High Density	2	97	+20°F	1
Iron - Infiltrate	3	96	-	1
400 Stainless	10	90	-	-
300 Stainless	30	70	-	-
Nickel Alloys	10	90	-	-
Copper & Brass	2	98	-	-
Aluminum*	-	100	-	-

*Normally all aluminum must be sintered in 100% nitrogen for certain process and safety considerations.

The process of establishing the exact optimum blend composition and atmosphere flow rate is largely a matter of applying past experience to existing conditions. However, because every furnace (even those of the same model) has its own unique characteristic, no two furnaces behave exactly the same in any atmosphere. Determining the ideal atmosphere composition and flow rate is best accomplished by utilizing gas analytical equipment to monitor furnace conditions. The example below illustrates how optimum hydrogen level of 2% and total atmosphere flow rate of 2200 SCFH was determined in a copper sintering application, by analyzing trace oxygen levels in the furnace.

TABLE VIII

SINTERING COPPER

(24" Convocool)

TRIAL	H_2 (%)	FLOW RATE (SCFH)	HOT ZONE O_2 (ppm)	COOLING ZONE O_2 (ppm)	PRODUCT SURFACE
1	5	2700	3.5	800	Highly Oxidized
2*	5	2700	4.0	30	Clean
3	2	2700	3.5	35	Clean
4	2	2050	2.0	55	Light Oxide
5	2	2500	3.0	37	Clean
6	2	2200	1.5	45	Clean

*Furnace Exhaust Modification

Another factor that sometimes influences atmosphere composition and/or flow rate is furnace conditioning. When a furnace (especially an unmuffled brick furnace) is converted from a relatively wet atmosphere like endo or exothermic to dry nitrogen the drying out process often takes an extended period of time. During this drying period, the atmosphere cannot be optimized since more hydrogen and total flow are required to counteract the adverse effects of the previous atmosphere. The example below shows how long this process can take, and how the hydrogen level can be reduced as the furnace dries out. This example is somewhat extreme, but normally a brick lined furnace does require at least 2-3 weeks to remove the adverse effects of operating under a wet atmosphere.

TABLE IX

FURNACE CONDITIONING

MONTH	HOT ZONE D.P. (°F)	PPM H_2O	REQUIRED H_2 LEVEL (%)
Aug.	+35	7000	10
Oct.	+20	3400	8
Dec.	+10	2200	6
Feb.	+ 5	1600	5

$[H_2] / [H_2O] = 100{,}000 / 7{,}000 = 14.3$

30" Roller Hearth, Total Flow 2000 SCFH

Material Cu-Pb Alloy Bonded to Steel

CONCLUSIONS

The intent of this paper is to discuss the requirements of a suitable sintering atmosphere and to show how both conventional generated atmospheres and nitrogen-based atmospheres meet these requirements. Since both atmosphere systems provide suitable finished part mechanical properties, the decision between atmospheres must be based on other considerations which include:

1. Carbon control
2. Size control
3. Soot control
4. Operating considerations
5. Availability
6. Sintered part appearance
7. And of course COST

When all the relevant factors are weighed, the case for the continued growth of nitrogen-based atmospheres is very strong.

AN ADVANCED P/M MATERIAL FOR HIGH
TENSILE, IMPACT AND HARDNESS APPLICATIONS

H. R. McCurdy and W. B. Aufderhaar

IPM Corporation
Columbus, Ohio

Introduction

In 1976, a review of our customer needs and the properties of our P/M materials showed a need for a material with improved tensile strength and good impact resistance. After evaluating several ways of improving the strength and reducing the notch sensitivity of our P/M materials, it was decided to evaluate the copper infiltration of an atomized, low alloy steel powder. The result was a copper infiltrated, low alloy steel called IPM 338IH, which is receiving favorable reception by many of our customers.

This paper describes some of the properties and applications of IPM 338IH, which is a versatile P/M material. It can be heat treated to a high strength level with medium ductility for a P/M material or it can be heat treated to a medium strength level with good ductility. Some of the general benefits of IPM 338IH follow.

1. Good Mechanical Properties - Good tensile strength, hardness and impact strength are obtained through alloy addition and copper infiltration.

2. Uniform Density - P/M parts can be infiltrated to obtain a uniform density distribution of 7.45 to 7.55 g/cc.

3. Good Plating Characteristics - Because IPM 338IH is a copper infiltrated, low alloy steel with no interconnected porosity, parts can be easily plated. The sealed porosity avoids damaging the interior of the part with plating fluids and avoids subsequent bleeding problems which tend to stain the plated surface of some plated P/M parts.

4. Good Brazing Characteristics - Again, because IPM 338IH is a copper

infiltrated, low alloy steel with no interconnected porosity, P/M parts made from IPM 338IH can be easily assembled to other components by brazing procedures. Infiltration seals the surface porosity and the braze metal remains at the joint being brazed rather than infiltrating into the pores of the P/M part as is sometimes the case with noninfiltrated P/M parts.

5. Pressure Tightness - Because IPM 338IH is a copper infiltrated, low alloy steel, it has good pressure tightness and can be used in many low pressure hydraulic applications.

6. Excellent Machinability - The excellent machinability of IPM 338IH in the annealed condition permits secondary machining (lathe turning, drilling and tapping). Close tolerances can be held and machined parts have a good surface finish.

7. Precise Dimensional Control - Because IPM 338IH is a low growth material, precise dimensional control from part to part can be obtained in production parts.

8. Assembly of Multiple Parts - Different sections of a final part, pressed separately, can be assembled in the green condition and bonded into one part by a one step infiltration procedure.

Material

The typical composition of IPM 338IH follows:

Element	W. T. Percent
C	0.7
Ni	0.45
Mo	0.5
Mn	0.3
Fe	Balance
Cu	15 (By infiltration)

The P/M material is prepared by the infiltration of a commercially available low alloy steel powder with a commercially available copper alloy infiltrant.

Properties

Machinability

Machinability of IPM 338IH was determined in a drill life test that was previously described by O. W. Reen(1). The machinability rating was determined by the number of inches of a material removed by drilling in a drill press under specified machining conditions until tool failure occurred. With sintered material, as many holes as feasible for the shape and size of the part were drilled. Drilling continued in identically produced parts until the end point was reached. A description of the test conditions follows.

(1) O. W. Reen, The Machinability of P/M Materials, Modern Developments in Powder Metallurgy, Vol. 10, pp. 431-452.

Press - The drill press was a two horsepower AVEY, Size 2, Style VMA. The press, used only for drilling tests, was equipped with special close tolerance bearings.

Drills - Each lot of test drill bits was produced from a single heat of high speed steel. The specifications for the drill bits were:

a. Material: Type M-1 high speed steel
b. Finish: UCW
c. Style: Cleveland Twist Drill Company list 967, 2 1/8-inch overall length with 1 3/8-inch flute
d. Diameter: 0.2500-inch plus 0.0000 minus 0.0005
e. Flute: Ground. Spacing to be within 0.003 T.I.R.
f. Clearance: Regular
g. Hardness: Rc 64.5 - 66.0
h. Cutting Tips: Straight
i. Web: Central to be held within 0.002 T.I.R.
   At point to be 0.042 plus 0.001 inch
   Increase to be 0.024 to 0.026 per inch
j. Point Angle: 118° included
k. Back Taper: 0.0005 per inch of flute

Coolant - Gulf Grade 43A, a sulfurized, chlorinated mineral oil was applied through three orifices at a fixed rate of 2.5 gallons per minute. A swivel vise with modified jaws held the circular test samples.

Feed and Speed - The feed was constant at 0.005 inch per revolution. The speed was variable and was dependent upon the test material. Usually, a trial speed was selected. If excessive time for tool failure or excessive test material consumption was apparent, a faster speed was tried. Tests were conducted for a minimum of three speeds.

Depth of Drilling - The depth of drilling was governed by the size and shape of the sintered part. The specimens were 1 inch diameter by 1 inch long. This allowed uniform drilling of 3/4 inch deep blind holes.

End Point - The end point was the catastrophic drill bit failure. This occurred suddenly after a wearland of about 0.015 inch had been formed on the cutting edges.

Drill Life Rating - Because of the variability in this type of test which was partially caused by the tools, the test data were analyzed statistically to insure that they met specified probability limits. Statistical limits, based on means and ranges, to give 95 percent confidence levels have been tabulated for quick reference. A sequential sampling plan was used to minimize the number of tests.

Tests were run in groups of three. If the mean and range for the first three tests met the statistical criteria for the confidence limits, testing was stopped. If they did not, a second series of three tests was run and the mean and combined ranges for the six tests were analyzed. If they were within the limits prescribed for six tests, testing was stopped at this stage. In unusual cases, up to nine tests might be required to obtain a valid rating.

Machinability ratings were based on mean drill life expressed in the average total inches drilled (number of holes x 0.75 inch). The number of tests required to obtain a statistical valid mean could be used as a measure of material uniformity. With IPM 338IH, only three tests were required to obtain a statistical valid mean showing good material uniformity.

Interpretation of Data

Two methods may be used to intepret the data. For a given drill life, the material utilizing the highest drill speed would rate highest in machinability. Similarly, for a given drill speed, the material with the greatest number of inches drilled would rate highest in machinability. In either case, the relationship of one sintered P/M material with another is essentially the same.

The machinability curves for IPM 338I in the infiltrated and in the infiltrated and annealed conditions are plotted in Figure 1. The machinability of IPM 338I at a density of 7.5 g/cc is compared with the following materials:

1. MPIF Designation F-0008-P (0.9 percent carbon-sponge iron at a density of 6.3 g/cc)

2. MPIF Designation F-0008-P-Sulfurized (0.7 percent carbon, 0.5 percent sulfur at a density of 6.2 g/cc)

3. MPIF Designation FX-2008-T (Copper infiltrated 0.9 percent carbon-sponge iron at a density of 7.4 g/cc)

The data show that IPM 338I (infiltrated condition) is machinable, but it is more difficult to machine than FX-2008-T and sulfurized sintered steel with 0.5 percent sulfur and 0.7 percent carbon. Also, IPM 338I (infiltrated condition) is easier to machine than F-0008-P, which machines with some difficulty. IPM 338I (annealed condition) machined without difficulty, easier than sulfurized steel and FX-2008-T materials.

An examination of the microstructure of the infiltrated IPM 338I specimens and the annealed IPM 338I specimens explained the differences in machinability ratings. The microstructure of the steel matrix material of IPM 338I (infiltrated condition) revealed a mixed microstructure of martensite, bainite and pearlite. The microstructure of the steel matrix low alloy steel of IPM 338I (annealed condition) was primarily pearlite with only traces of tempered martensite.

Mechanical Properties

In Table I, the typical room temperature mechanical properties are listed for IPM 338IH for two different conditions of heat treatment. The properties are compared with the properties of the infiltrated steel, MPIF Designation FX-1008-T (heat treated condition). The tensile properties of IPM 338IH were determined at room temperature according to ASTM Specification E-8 using TS-4 tensile bars (Figure 2) that were machined from a disc 5 inches

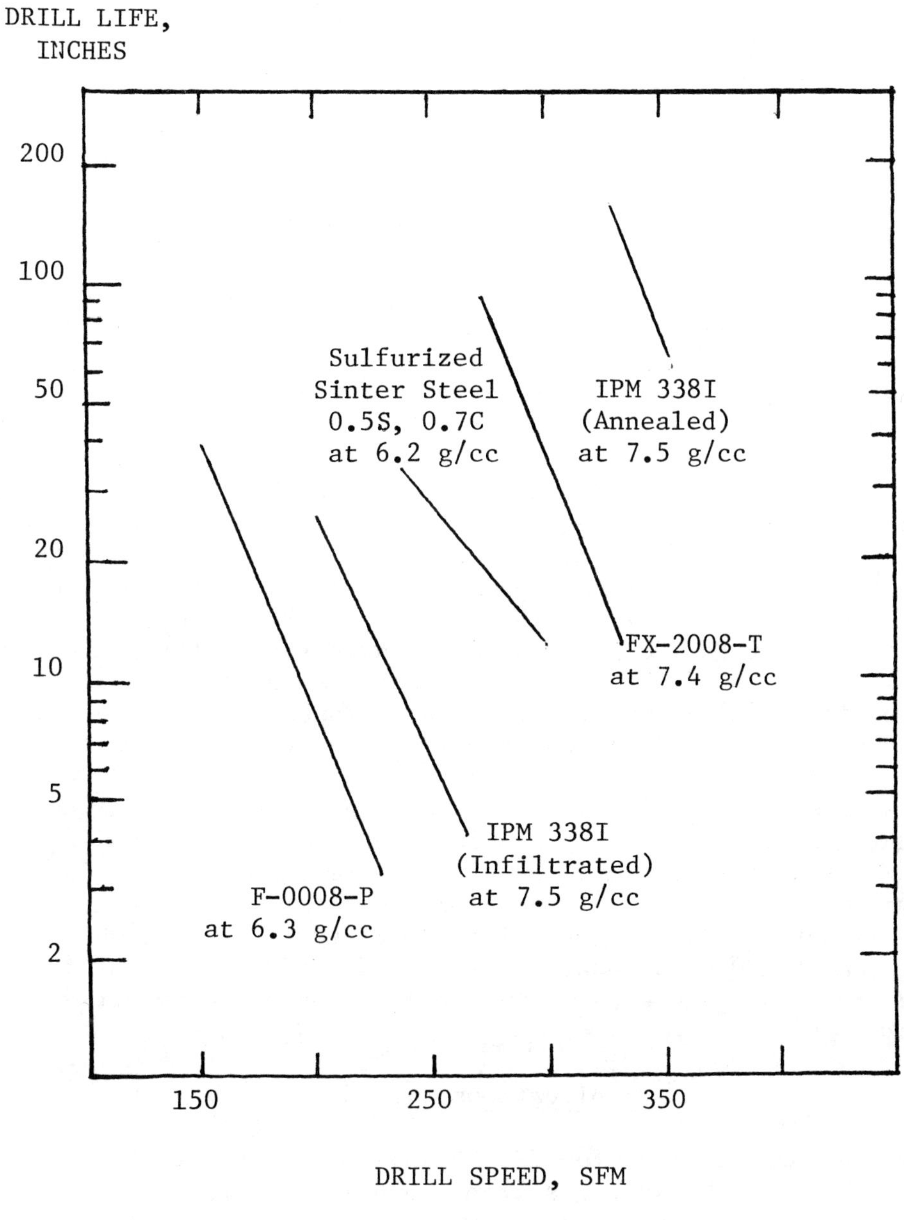

Figure 1. A COMPARISON OF THE MACHINABILITY OF IPM 338I (INFILTRATED AND ANNEALED CONDITIONS) WITH F-0008-P, FX-2008-T AND A SULFURIZED SINTERED STEEL

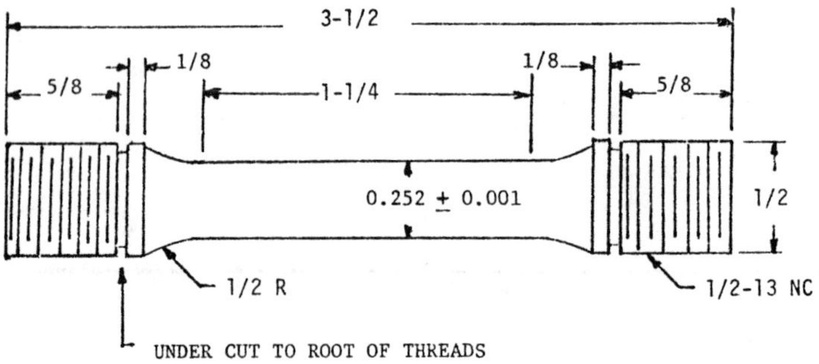

Figure 2. SKETCH OF TS-4 TENSILE BAR

diameter by 5/8 inch long and heat treated. Impact strength properties were determined at room temperature on standard, unnotched Charpy bars 10 x 10 x 75 mm that were pressed to size and infiltrated. The unnotched Charpy properties were determined according to ASTM Specification E-23 for metal powder structural parts.

IPM 338IH is an extremely versatile material as shown in Table I. The ultimate tensile strength of specimens that were heat treated at Condition 1 were 7 to 17 percent higher than those of MPIF Standard for FX-1008-T at an elongation in 1 inch of less than 2 percent. In addition, the unnotched Charpy impact strength of 8 to 13 ft-lbs for IPM 338IH is compared with the standard of 7 ft-lbs for FX-1008-T. The apparent hardness for IPM 338IH (Condition 1) is Rc 35 to 45 with a particle hardness of the steel matrix material equivalent to Rc 55-60. The apparent hardness of FX-1008-T material is about the same as IPM 338IH at heat treating Condition 1 or Rc 40.

In applications where ultimate tensile strength can be sacrificed for impact strength, a second heat treating condition (Condition 2 in Table I) can be applied. At this condition, the ultimate tensile strength drops to 104-112 ksi and the elongation in 1 inch increases to 3-4 percent. The unnotched Charpy impact strength values climb to 21 to 26 ft-lbs, which is triple that of the FX-1008-T standard of 7 ft-lbs. P/M parts heat treated at Condition 2 are finding acceptance in applications where impact strength is important.

Applications

There are many applications for IPM 338IH. Most require a demanding balance of tensile strength, wear resistance and impact strength. In addition, the sealing of the pores by copper infiltration promotes pressure tightness for low pressure hydraulic applications, good plating characteristics and good brazing characteristics. Some users have reported a reduction in noise level of running gears when the gear material was switched from wrought steel to IPM 338IH.

Some specific applications follow:

1. Spacer - A hollow cylindrical spacer (approximately 2.5 inch OD x 1.4 inch ID x 2.3 inch length) is used in a

TABLE I.

COMPARISON OF THE TYPICAL ROOM TEMPERATURE MECHANICAL PROPERTIES OF IPM 338IH AT A DENSITY OF 7.5 g/cc AND IN THE HEAT TREATED CONDITION (1 OR 2) WITH MPIF STANDARD FX-1008-T IN THE HEAT TREATED CONDITION

PROPERTY	IPM 338IH CONDITION		FX-1008-T
	1	2	
Ultimate Tensile Strength, ksi	140 - 152	104 - 112	130
Yield Strength, (0.2 Offset), ksi	125 - 135	83 - 99	105
Elongation (1 inch), percent	0 - 2	3 - 4	0.5
Impact Strength, ft-lbs (Unnotched Charpy)	8 - 13	21 - 26	7
Apparent Hardness (Rockwell)	Rc 35-45	Rb 85-95	Rc 40
Particle Hardness (Rockwell Equivalent)	Rc 55-60	---	---

diesel engine mount. The material is heat treated to an apparent hardness of Rc 32 to 38. The critical properties in this application are:

a. Compressive Strength - The faces of the spacer must maintain a parallelism of 0.005 inch when a bolt is inserted in the ID of the spacer and the bolt is torqued to the engine.

b. Impact Strength - The spacer must withstand the shock loading that the spacer receives in the engine operation.

2. Clutch Collar - A clutch collar (approximately 3.8 inch OD with an ID spline and a 0.8 inch length) is used in a truck transmission. The critical properties are:

   a. Machinability - An OD groove is machined into the part after infiltration.

   b. Particle Hardness - After heat treating, a final particle hardness of the steel matrix equivalent to Rc 55-60 is required for good wear resistance.

3. Worm Shaft Lock - A worm shaft lock (approximately 0.8 inch OD x 0.8 inch long) is used in a slack adjuster. The critical property in this application is torque strength, which is obtained by heat treating to Condition No. 2.

4. Camber Bushing - A flanged bushing (approximately 1.8 inch OD x 1.4 inch long) is used to adjust the camber of the front wheels of a recreational vehicle. In this application, the critical properties are:

   a. Machinability - Good machinability is required for secondary machining of a groove in the flange of the bushing and a reverse taper in the ID of the part.

   b. Torque Strength - The material needs good torque strength to permit camber adjustment without bushing failure.

5. Drive Pinion and Bevel Gear - A drive pinion and its mating bevel gear, which are transmission components in a riding lawn mower, are fabricated from IPM 338IH. The critical property is impact strength. The drive pinion and bevel gear must withstand the impact of gear shifting as the lawn mower is shifted from forward to reverse.

Summary

The room temperature mechanical properties and some applications for IPM 338IH, a copper infiltrated, low alloy steel, were discussed. In the infiltrated and annealed condition, IPM 338IH has excellent machinability. At a density of 7.5 g/cc, it can be heat treated to an ultimate tensile strength of 140,000 to 152,000 psi, a tensile yield strength (0.2 offset) of 125,000 to 135,000 psi and an elongation in 1 inch of 0 to 2 percent. The unnotched Charpy impact strength at this condition is 8 to 13 ft-lbs and the apparent hardness is Rockwell C 35 to 45.

If tensile strength can be sacrificed for ductility, IPM 338IH, at a density of 7.5 g/cc, can be heat treated at a second condition which results in an ultimate tensile strength of 104,000 to 112,000 psi, a yield strength (0.2 offset) of 83,000 to 99,000 psi and an elongation in 1 inch of 3 to 4 percent. The unnotched Charpy impact strength at this condition is 21 to 26 ft-lbs and the apparent hardness is Rockwell B 85 to 95.

IPM 338IH is finding customer acceptance in applications where a balance between tensile strength, wear resistance and impact strength is required.

SURFACE POROSITY IN P/M STEEL FORGINGS

Andrew Crowson
Materials and Manufacturing Technology Division
Fire Control and Small Caliber Weapon Systems Laboratory
U. S. Army Armament Research and Development Command
Dover, New Jersey 07801

Abstract

The effects of forging parameters on surface porosity of P/M steel forgings were determined. Forging at a temperature of 1200°C in a preheated die (200-260°C) gave the optimum conditions for minimizing surface porosity. In this case, the amount of chilling and frictional restraints encountered at the die wall surfaces were minimized. Lower forging temperatures (950, 1040 and 1120°C) and/or die temperatures (less than 200°C) resulted in substantial increases in surface porosity. Preheating of dies beyond 260°C produced only a negligible reduction on the remaining surface porosity. Forging pressures (above 275 MPa) and lateral deformation, on the other hand, had only a minimal effect on final surface porosity. In both cases, the frictional behavior at the die wall surfaces was found to offer increased resistance to pore closure. Porosity content across the die forging's surface followed the general trend: corner edge > end > center. This was attributed to the increasing lack of restraints at these locations. A slight deviation to this trend was observed for forged surfaces with large amounts of lateral flow (above 25%). In these cases, chilling and frictional effects took precedence over the increasing lack of restraint.

Introduction

The powder metallurgy (P/M) forging process involves deforming a conventional sintered P/M preform under sufficient temperature and pressure to obtain full densification. Ideally, one hundred percent densification is achieved throughout the part. In reality, however, density variations are often encountered. Most of these variations are confined to the residual porosity near the external surface. Earlier work by Fischmeister, Aren, and Easterling on open die, as well as closed die, forgings indicates nonuniform material flow during deformation causes these variations[1]. This is a consequence of the nonuniformity of the stress-strain field within the workpiece during the forging stroke. Causes for this nonuniformity have been related to chilling and frictional effects between the die and the workpiece.

This situation has undersirable consequences. The detrimental effect that residual porosity has on mechanical properties has been the subject of concern by several investigators. Kaufman and Mocarski have shown that small amounts of residual porosity in a P/M steel forging can substantially affect the resultant dynamic and static properties[2]. In fact, as much as a 40 percent reduction in yield strength was seen to occur within the first four percent porosity. Likewise, in ductility experiments, brittle fracture became predominant when porosity levels above 2.2 percent were present. Antes and Pietrocini have shown that the impact strength is affected by porosity as well[3,4]. Impact data on a 4630 type (1.8 Ni - 0.5% Mo) P/M steel alloy showed a threefold increase over the density range 7.7 g/cc to 7.86 g/cc. Similar results have been documented by Cook and Bockstiegel[5,6]. The effect of residual porosity on fatigue properties has not been sufficiently investigated to present conclusive data. However, Ferguson, Kuhn, and Lawley have shown that fatigue strength is enhanced by lateral flow; which, in turn, has been related to porosity levels[7].

One of the major features of hot forging of metal powder is that the formed surface is to be the operational surface. Considering surface porosity could range as high as four percent, this can be a serious disadvantage to the process. Besides the delerious effect on the mechanical properties, additional problems during secondary finishing operations can also occur. This can lead to a loss in inherent hardness and wear resistance for a particular part. This is especially critical in sliding applications; such as bearings or gears. Here high Hertz stresses are developed which can cause surface fatigue or spalling. Consequently, elimination or a substantial reduction in the amount of surface porosity present in P/M forgings is necessary to alleviate these shortcomings. Since the forging cycle of the P/M forging route has the greatest influence on surface porosity, the intent of this investigation was to determine how the forging parameters affect the final surface porosity content. Forging parameters studied include forging temperature, forging pressure, lateral flow, and die temperature. Upon completion of this study, the optimum forging conditions necessary to keep the surface porosity at a minimum level will be resolved.

Procedure

The powder used in this investigation was a commercially available 4600 prealloyed steel powder. The chemical composition and powder characteristics for this powder are given in Tables I and II, respectively. The powder has been

TABLE I

CHEMICAL ANALYSIS OF 4600 PREALLOYED POWDER

Element	AISI Specification (%)	4600 Powder (%)
Carbon	--	0.01
Nickel	1.65 - 2.00	1.78
Molybdenum	0.2 - 0.3	0.49
Manganese	0.6 - 0.8	0.32
Copper	--	0.10
Chromium	--	0.03
Phosphorus	0.04 max.	<0.005
Sulfur	0.04 max.	0.05
Silicon	0.02 - 0.03	0.04
Oxygen	--	0.193

TABLE II

CHARACTERISTICS OF 4600 PREALLOYED POWDER

U.S. Screen Size	Sieve Analysis, wt %
+80 (180μ)	1.5
-80+100	9.4
-100+140	16.5
-140+200	23.8
-200+230	6.7
-230+325	16.4
-325 (44μ)	25.7
Apparent Density	3.04 g/cm^3
Flow Rate	26.7 sec

modified (lower manganese, higher molybdenum) to promote sintering conditions. Flake graphite was blended with the powder to obtain a 0.4 percent carbon content. The blended powder was compacted into 8.90 cm X 1.90 cm X 2.54 cm rectangular bars to 80 percent theoretical density. The compaction was conducted using a closed confined die and a Riehle 205,000 kg tensile testing machine. The die assembly was supported by springs which enabled the compaction operation to simulate a double acting press, thus yielding uniform density profiles in the compacts. The compacted preforms were sintered in a 95 hydrogen-5 methane atmosphere for 40 minutes at 1200°C. Subsequently, forging of the sintered preforms was carried out in a preheated closed forging die. Forging pressures were adjusted to achieve full densification.

The surface porosity investigation was carried out in four parts. A summary of the experimental variables used is shown in Table III. Die preheat

TABLE III

SUMMARY OF EXPERIMENTAL VARIABLES

Run No.	Td (°C)	T_f (°C)	P_f (MPa)	Deformation (%)
1	25	1200	275	5
2	120	-	-	-
3	230	-	-	-
4	315	-	-	-
5	230	950	-	-
6	-	1040	-	-
7	-	1120	-	-
8	-	1200	-	-
9	-	-	140	-
10	-	-	275	-
11	-	-	415	-
12	-	-	550	-
13	-	-	275	-
14	-	-	-	15
15	-	-	-	25
16	-	-	-	35
17	-	-	-	45

Td = Die temperature, T_f = Forging temperature, P_f = Forging pressure temperature, forging temperature, forging pressure, and amount of lateral deformation were each varied respectively while maintaining constant conditions on the remaining three variables. A die preheat temperature of 230°C, forging temperature of 1200°C, forging pressure of 275MPa, and lateral deformation of 5 percent were used as the standard forging conditions.

The analysis of the surface porosity content was conducted using quantitive metallography on the as-forged sample surfaces. Porosity measurements were made on rectangular sections 0.2mm depth by 0.6mm length. A statistical survey of the entire sample surface was made. This included the top, bottom, and sides of the resulting forging. The edge , end, and center sections of the surface were also evaluated to determine the surface porosity uniformity (see Figure 1).

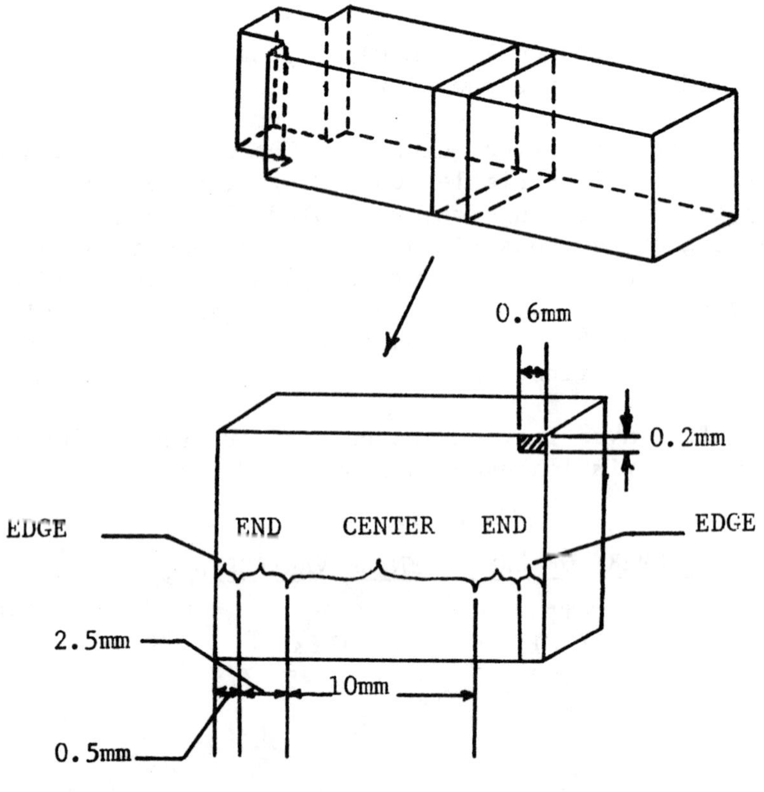

FIGURE 1. REPRESENTATIVE SAMPLE OF SECTION USED IN SURFACE POROSITY MEASUREMENTS (Shaded area represents typical section analyzed for surface porosity)

The corner edge comprised the first 0.5mm of the forging's surface. The end section was the next 2.5mm from either end of the corner edge section. The center section covered the remaining 10.0mm of the forging's surface. Surface porosity measurements up to 1.0mm depth (in 0.2mm increments) were also made to ascertain the extent of the surface porosity. Correlations between the experimental data and the forging parameters were then determined.

Results and Discussion

In evaluating surface porosity content in P/M steel forgings, it was first necessary to determine the depth to which significant porosity levels occurred. Figure 2 depicts the relationship established between porosity and depth.

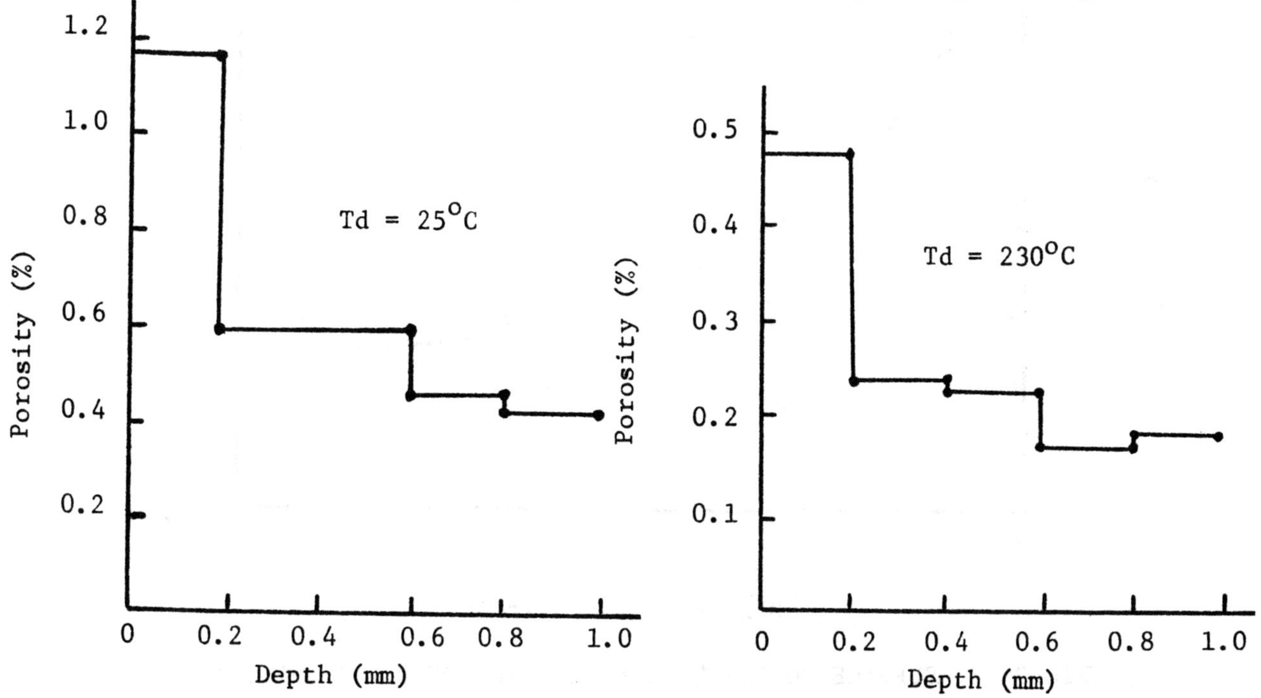

FIGURE 2. POROSITY LEVELS AS A FUNCTION OF DEPTH IN P/M STEEL FORGINGS

The two samples shown here were forged under identical conditions except for die preheat temperature. In both cases, the first 0.2mm depth comprised the highest percentage of porosity. From 0.2 to 1.0mm depth the porosity content decreased substantially and a leveling-off trend was apparent. This was evident throughout the forged surfaces except in the case of the corner edges of some of the forgings. In these cases, the porosity content was found to be significant up to 0.4mm depth. Therefore, for experimental purposes, the first 0.2mm depth was used as the basic criteria for the analysis of surface porosity.

Die Temperature

Heating of P/M forming dies prior to forging of metal powder preforms is common practice. This is done to minimize chilling effects, which can lead to enhanced surface porosity. The effect of die temperature on surface porosity is shown in Figure 3. A substantial reduction in surface porosity is evident as the forging die temperature is increased. Surface porosity levels range from 1.32 percent for dies at room temperature (25°C) down to a low of 0.41 percent for dies heated to 315°C. However, this reduction is significant only up to 230°C. Heating of dies above this temperature results in only a slight additional reduction of surface porosity.

Empirically, this can be described by the exponential equation

$$\sigma = 5.0\ Td^{-0.43} \qquad (1)$$

where
σ = percent surface porosity
Td = die temperature

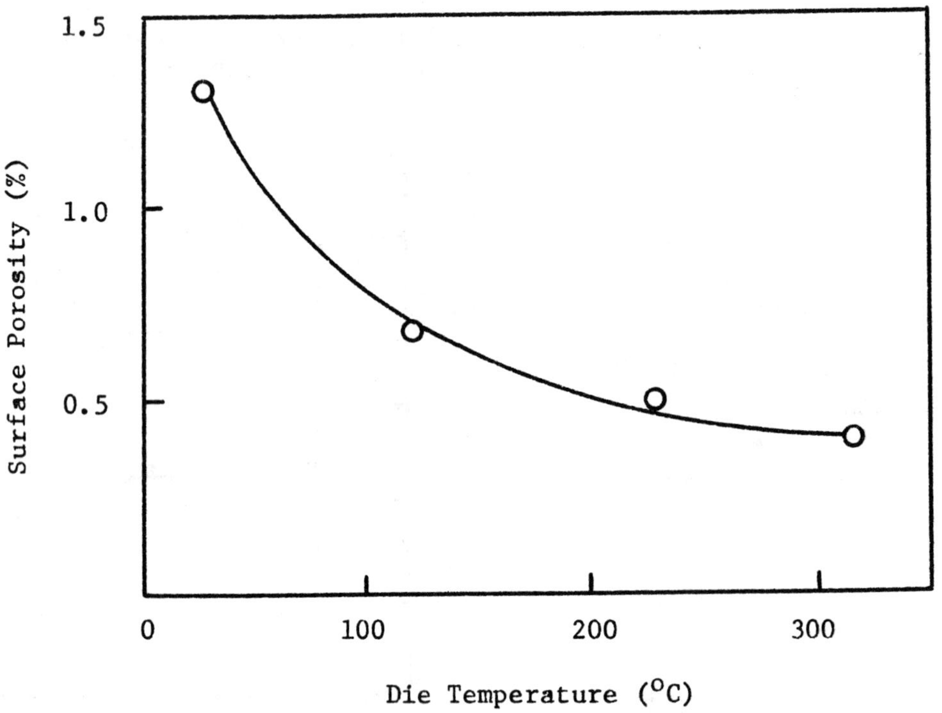

FIGURE 3. SURFACE POROSITY AS A FUNCTION OF DIE TEMPERATURE

In Figure 4, metallographic sections of sample porosity variance as a function of die temperatures are shown. The photographs are representative examples of the center sections of the forging's surface. Again, noticeable improvement

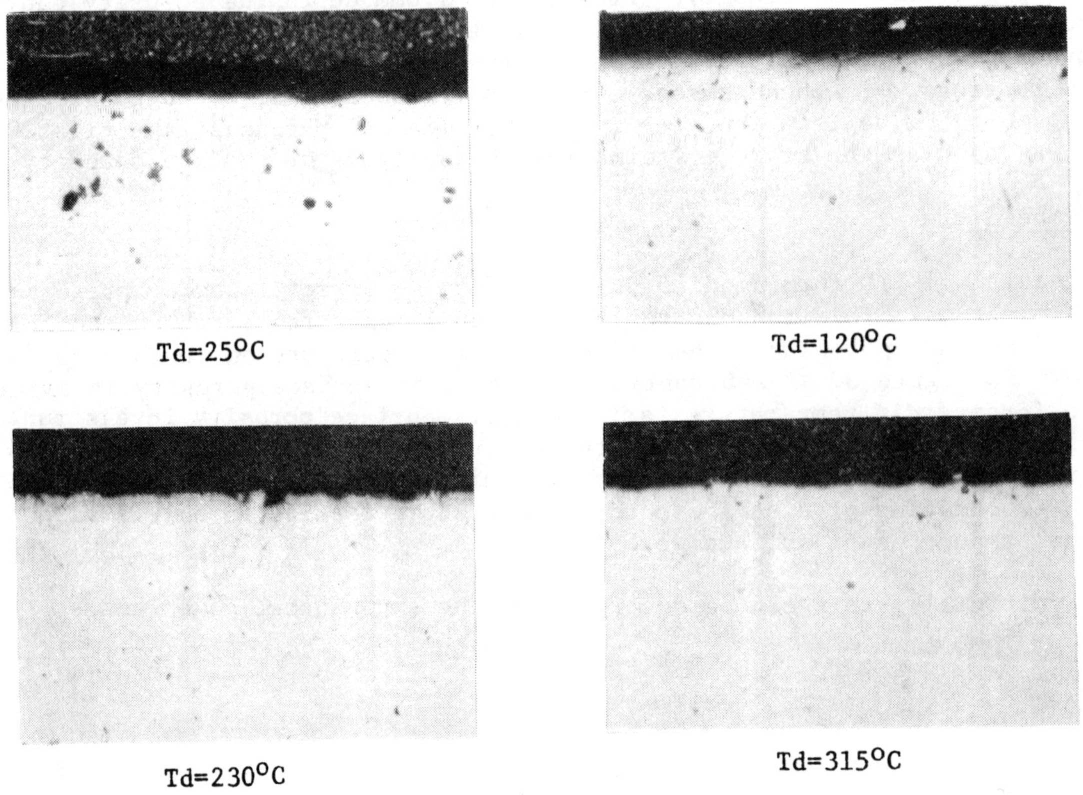

FIGURE 4. PHOTOMICROGRAPHS OF THE EFFECT OF DIE TEMPERATURES (Td) ON SURFACE POROSITY IN P/M STEEL FORGINGS (100X)

in surface porosity levels is typified as the die temperature is increased. Corresponding corner edge section photographs are shown in Figure 5. Here the reduction in surface porosity with increasing die temperature is even more perceptible.

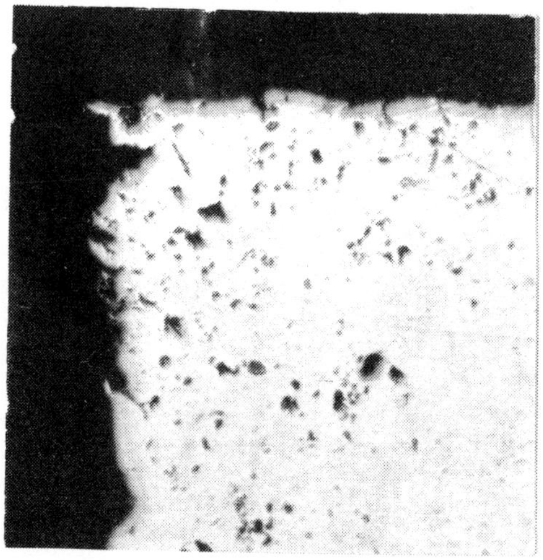

Td=25°C

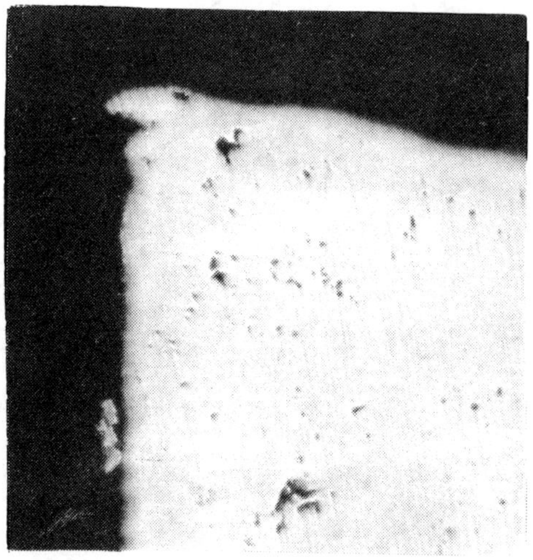

Td=120°C

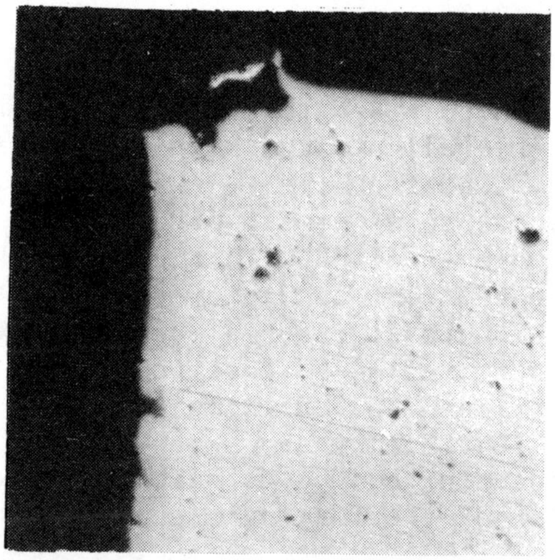

Td=230°C

Td=315°C

FIGURE 5. CORNER EDGE PHOTOMICROGRAPHS OF THE EFFECT OF DIE TEMPERATURE (Td) ON SURFACE POROSITY IN P/M STEEL FORGINGS (100X)

Porosity uniformity across the P/M forging's surface is analyzed in Figure 6. The corner edge, end, and center sections of the sample surface are shown for comparison. Exponential relationships between surface porosity and die temperature is evident in all sections. Also, as expected the surface porosity content is found to increase as the corner edge of the forging is approached. This effect is quite discernible when the data points for a particular sample are plotted, as shown in Figure 7. This is attributed to an increasing lack of restraint at these locations.

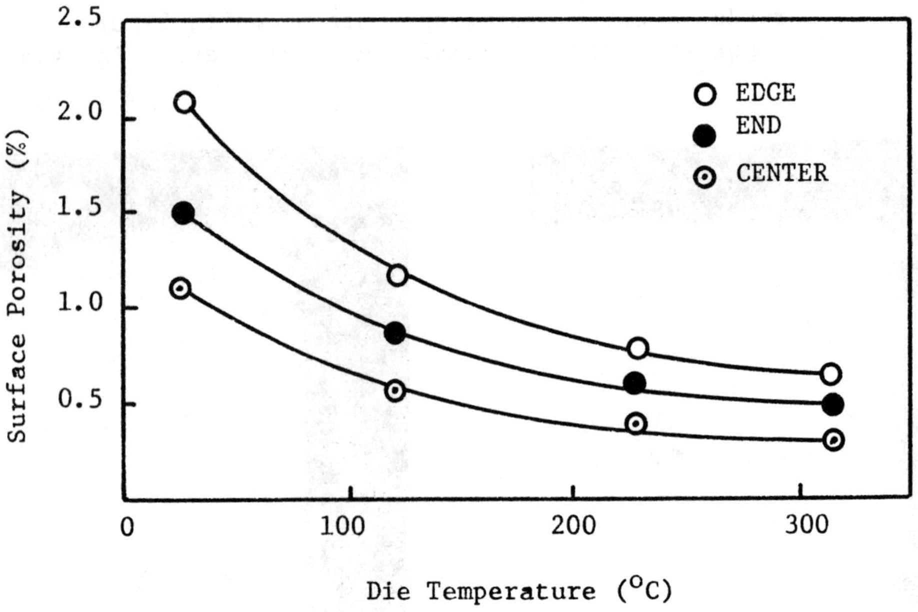

FIGURE 6. SURFACE POROSITY - DIE TEMPERATURE UNIFORMITY PROFILES

Lateral Deformation

The amount of lateral deformation imposed on a P/M forging is known to effect its final density. Figure 8 shows the overall average surface porosity as a function of lateral deformation. It is evident from the figure that deformation had only a minimal effect. A slight increase in sample porosity was noted when large amounts of lateral deformations were involved. In Figure 9, the uniformity of surface porosity under these conditions is depicted. An interesting aspect noted here was that under large amounts of deformations, the center section of the forging's surface actually increased in surface porosity content. This can be attributed to the increased frictional and chilling effects imposed on this section during deformation. Figure 10 visually shows representative sections of the surface porosity content under large amounts of deformations. Again, it is apparent that the center section of the forging's surface contains the highest porosity content.

Forging Temperature

The effect of forging temperature on the overall P/M steel forging's density has been documented in previous investigations[9]. Increasing the forging temperature will correspondingly increase the final density. In Figure 11, a similar effect on surface porosity is apparent. A substantial reduction in surface porosity is noted with increasing forging temperature over the range of 950 to 1200°C. Surface porosity levels range from a high of 2 percent at 950°C down to a low of 0.46 percent at 1200°C. In Figure 12, this effect is shown in photomicrographs. This decrease is to be expected since forging at a higher temperature will reduce the chilling and frictional constraints of the die wall surface.

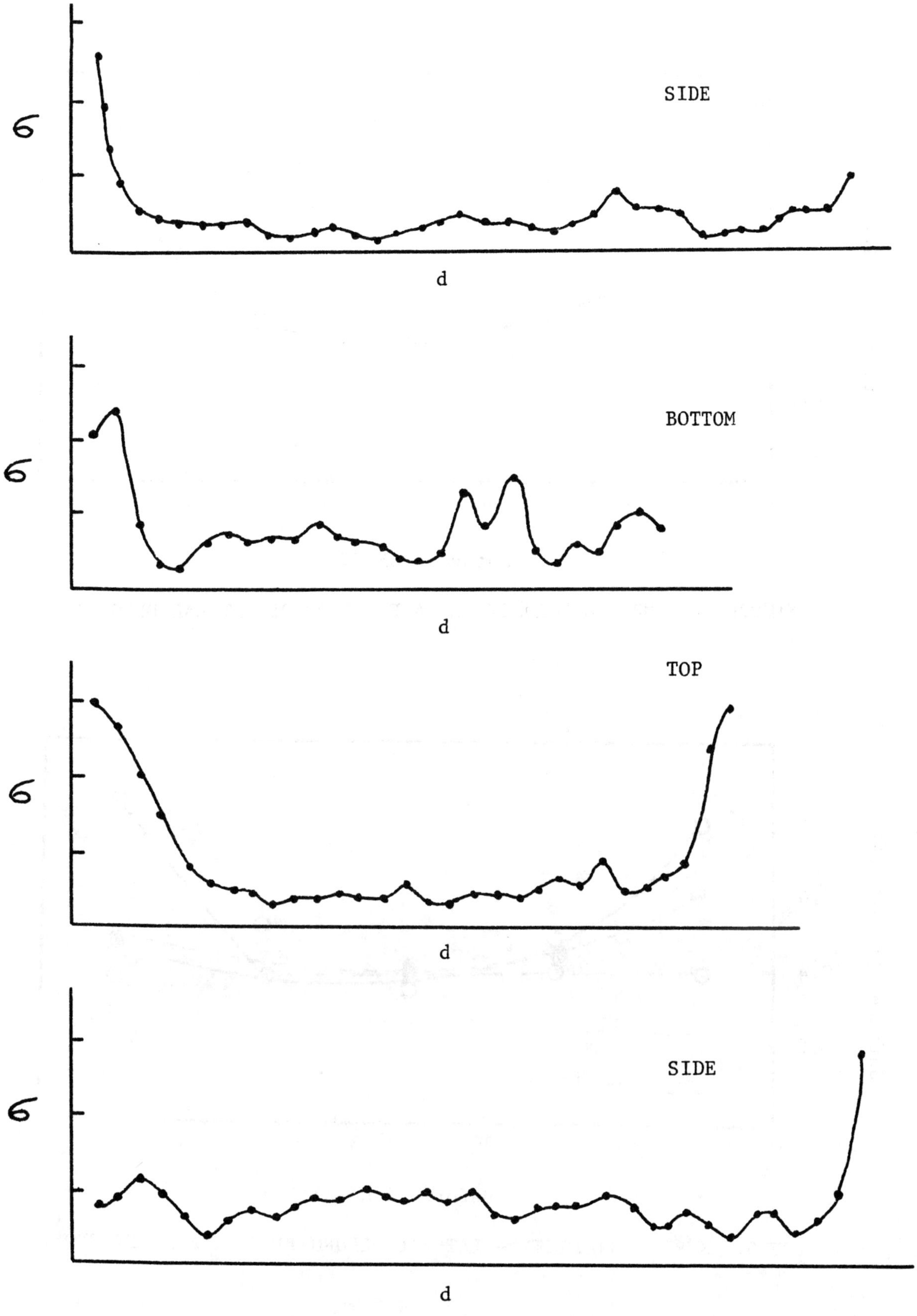

FIGURE 7. TYPICAL POROSITY PROFILES ON THE SURFACE OF P/M STEEL FORGINGS

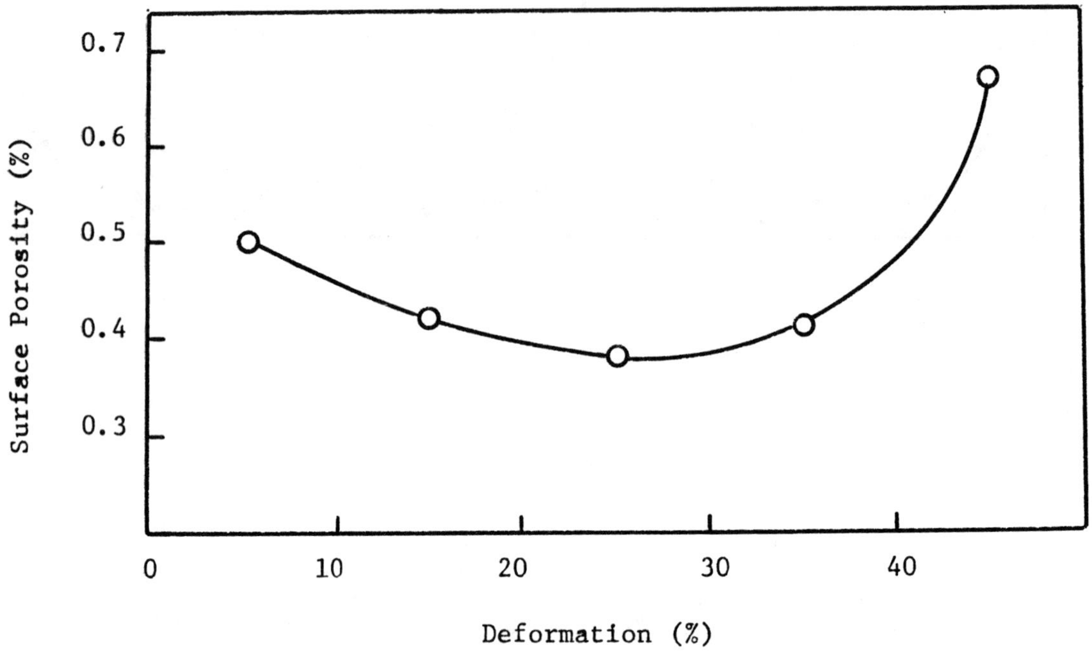

FIGURE 8. SURFACE POROSITY AS A FUNCTION OF LATERAL DEFORMATION

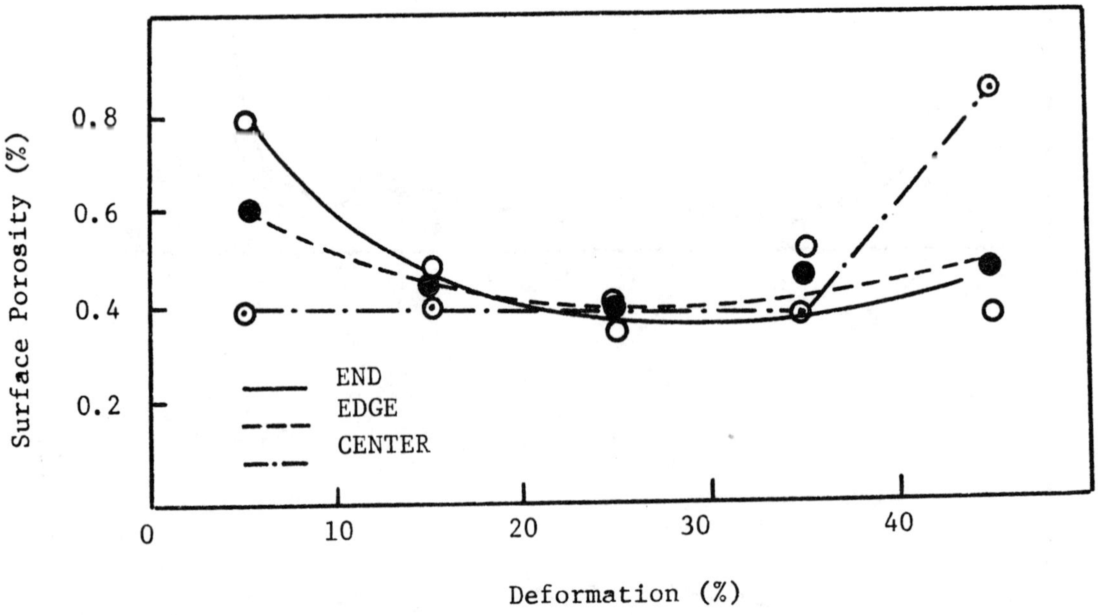

FIGURE 9. SURFACE POROSITY - LATERAL DEFORMATION UNIFORMITY PROFILES

END

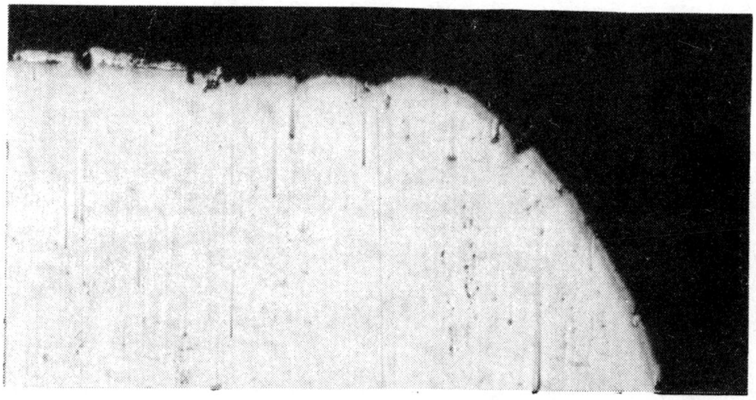

EDGE

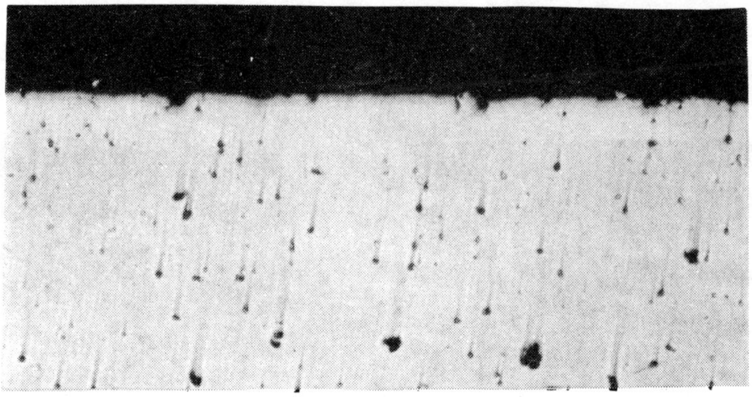

CENTER

FIGURE 10. PHOTOMICROGRAPHS SHOWING THE EFFECT OF LARGE AMOUNTS OF DEFORMATIONS ON THE POROSITY CONTENT AT VARIOUS SECTIONS ACROSS THE P/M STEEL FORGED SURFACE (100X)

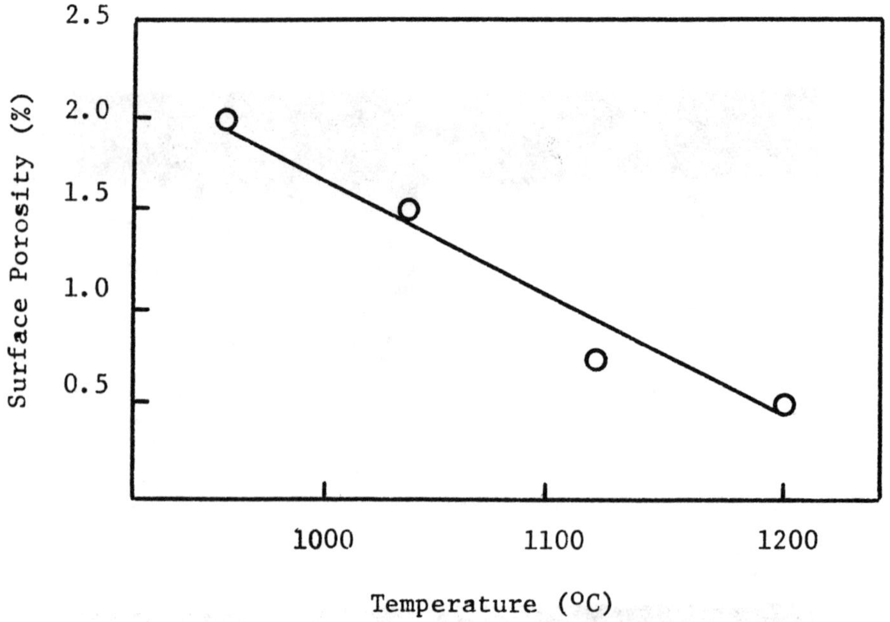

FIGURE 11. SURFACE POROSITY AS A FUNCTION OF FORGING TEMPERATURE

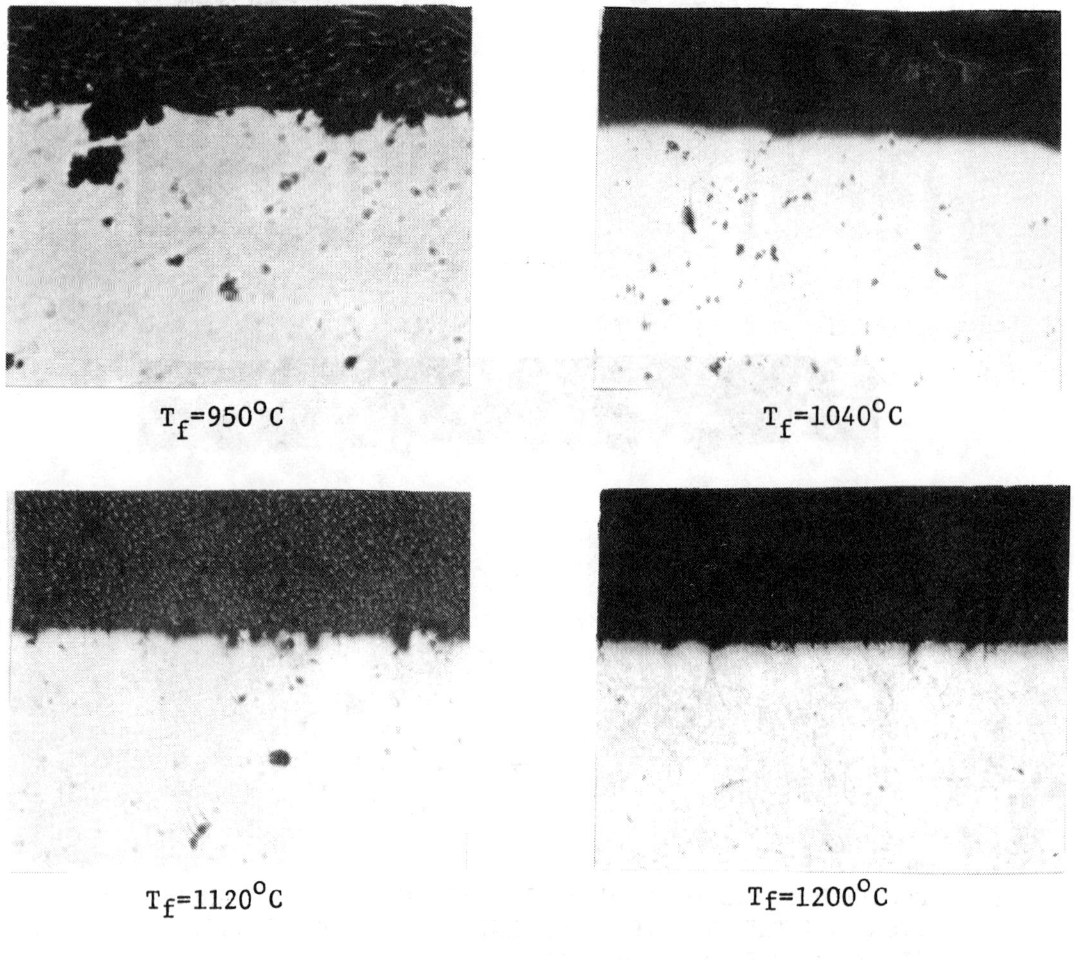

FIGURE 12. PHOTOMICROGRAPHS OF THE EFFECT OF FORGING TEMPERATURE (T_f) ON SURFACE POROSITY IN P/M STEEL FORGINGS (100X)

Quantitatively, this can be described by the empirical equation.

$$\sigma = 7.60 - 6.0 \times 10^{-3} \qquad (2)$$

where

σ = percent surface porosity
T_f = forging temperature

The uniformity of the surface porosity at the various forging temperatures is depicted in Figure 13. Reduction of surface porosity in all sections is apparent throughout the temperature range. Also, as in the case of the die temperature study, the general trend

corner edge $>$ end $>$ center

in surface porosity content is obvious.

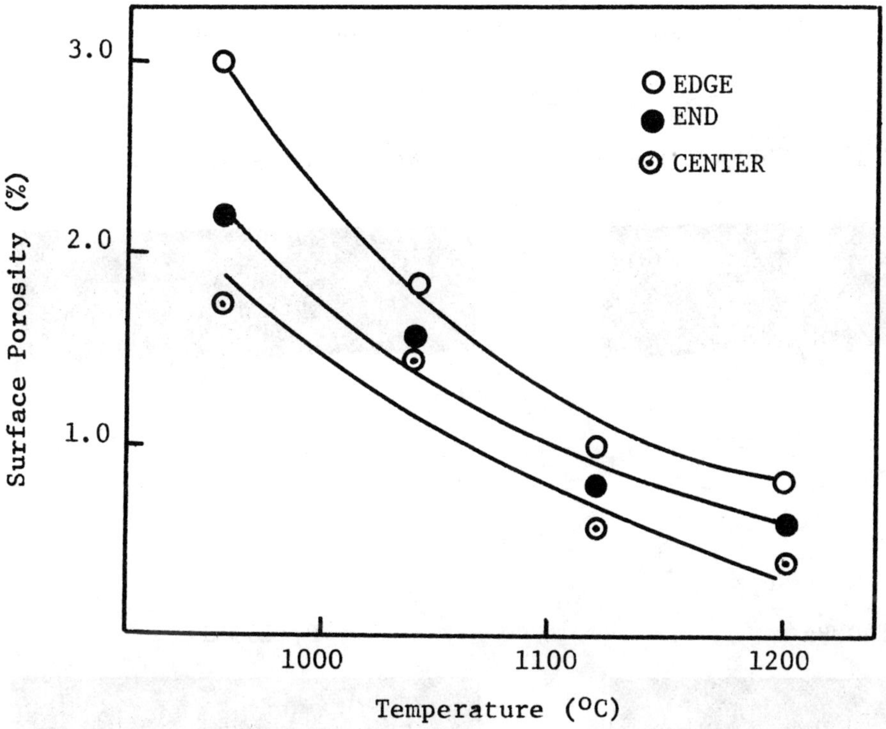

FIGURE 13. SURFACE POROSITY - FORGING TEMPERATURE UNIFORMITY PROFILES

Forging Pressure

It has been ascertained in prior investigations that full theoretical density can be achieved at a forging temperature of 1200°C under a forging pressure of 275 MPa[10]. Figure 14 demonstrates how surface porosity varies with forging pressures. As can be seen, a substantial reduction in surface porosity occurs from 140 MPa to 275 MPa. However, pressures greater than this have only a minimal effect on the remaining surface porosity. In Figure 15, this effect is shown in photomicrographs. Again, it is apparent that above 275 MPa little difference can be deduced concerning the surface porosity content.

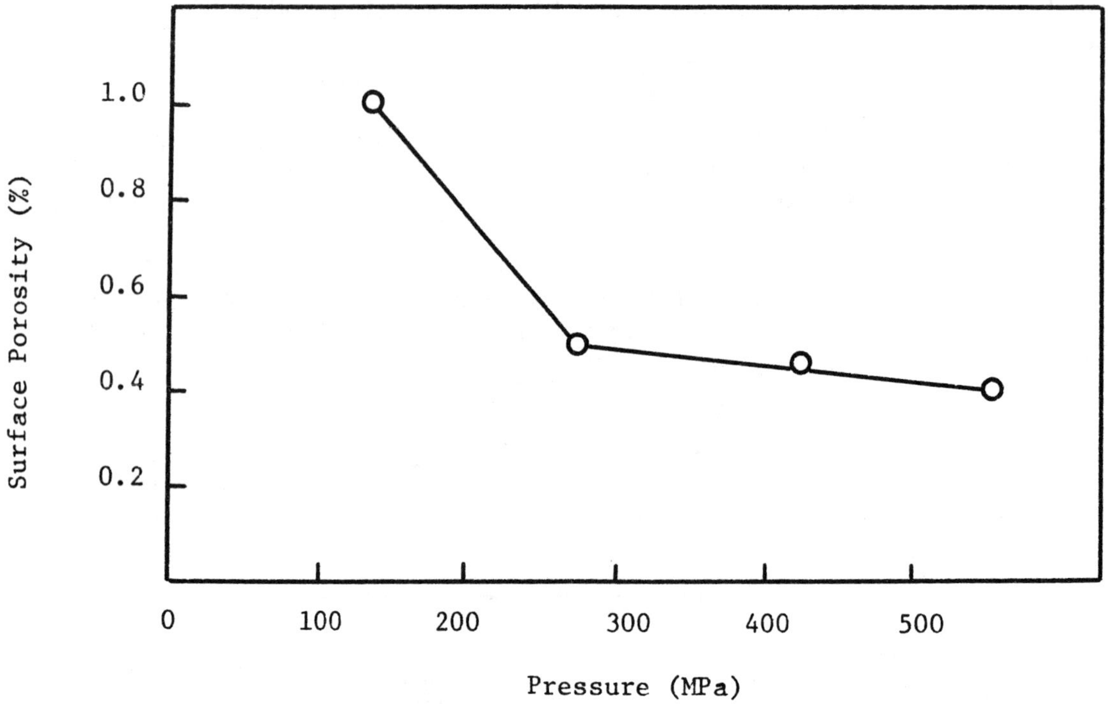

FIGURE 14. SURFACE POROSITY AS A FUNCTION OF FORGING PRESSURE

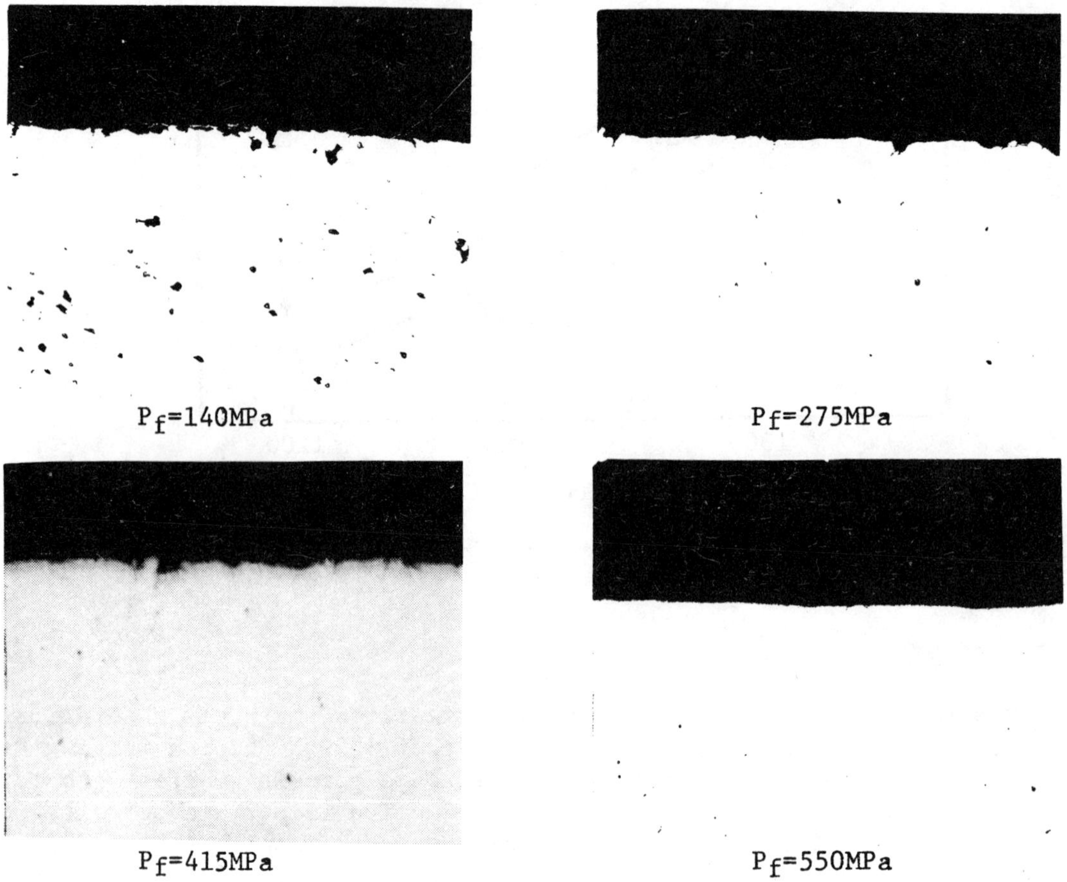

FIGURE 15. PHOTOMICROGRAPHS OF THE EFFECT OF FORGING PRESSURE (P_f) ON SURFACE POROSITY IN P/M STEEL FORGINGS (100X)

Uniformity of surface porosity throughout the various sections of the forging's surface followed the general trend reported in the die temperature and forging temperature studies. Figure 16 exhibits this phenomenon. The main difference noted was that the forging pressures more strongly influenced surface porosity in the end and center sections than in the corner edge section. This shows that additional forging pressure concentrates in reducing the surface porosity in these areas as opposed to the corners where the surface/volume ratio leads to higher frictional forces.

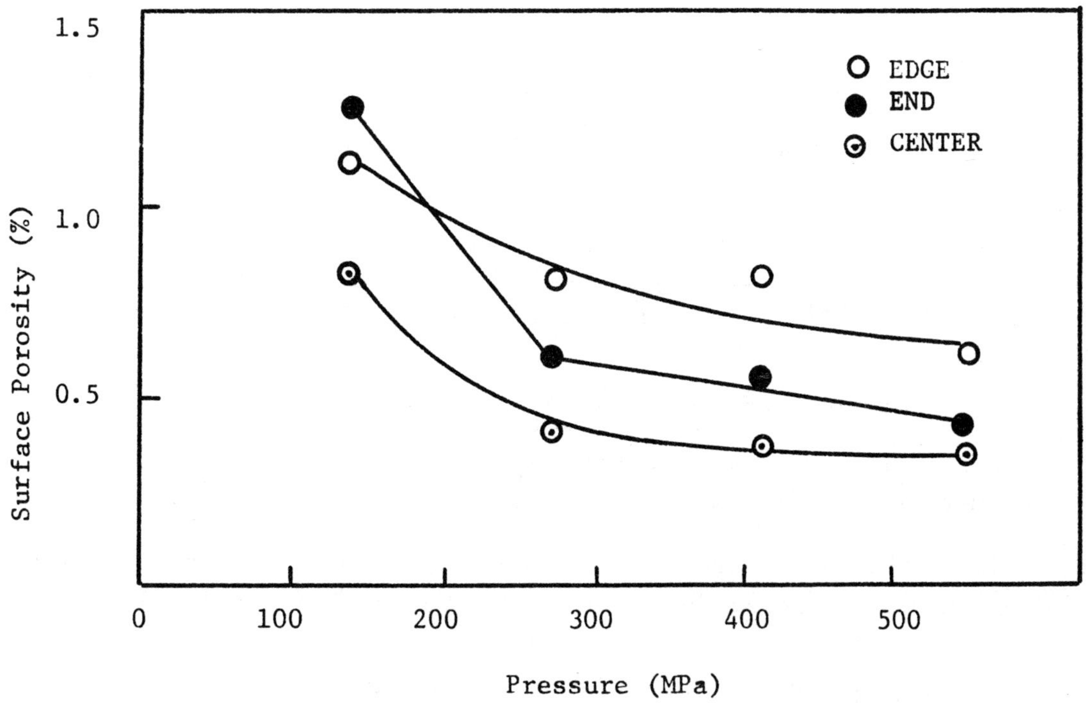

FIGURE 16. SURFACE POROSITY - FORGING PRESSURE UNIFORMITY PROFILES

Conclusions

Residual porosity in P/M steel forgings is confined to a rather shallow surface layer. This is mainly attributed to the chilling and frictional effects of the forging tools on the flow of the work-piece during deformation. Previous investigators have postulated various means for minimizing these effects. Using high die temperatures, insulating the tooling from the work-piece, using hammers instead of presses, and proper preform design include some of the criteria suggested. In summation, however, the general view is that the forging cycle predominately controls the final surface porosity level. This was evident in this investigation. Temperature, whether it be forging or die preheat, was found to have the greatest influence. A substantial decrease in surface porosity is apparent as the temperature is increased. This is expected since an increase in temperature undoubtedly reduced the chilling effects of the punches and die walls during forging. Forging pressure and lateral deformation, on the other hand had only a negligible effect on surface porosity. Once a forging pressure of 275MPa is obtained, higher pressures (up to 550MPa) resulted in less than an additional one-tenth percent decrease in surface porosity. Likewise, increased lateral flow (from 5 to 25%) resulted in reduced surface porosity to less than an additional one-tenth percent. Therefore, use of very high pressures and/or high rates of deformation to reduce surface porosity is unjustified. The forging parameters which were found to give the optimum surface porosity, as well as minimizing the die wear are:

1. Forging temperature = 1200°C
2. Forging pressure = 275 MPa
3. Die wall temperature = 230°C
4. Lateral deformation = 5%

Use of these conditions will limit surface porosity in any region of the forging's surface (0.2mm depth) to less than 0.5 percent.

References

1. H. F. Fischmeister, B. Aren, and K. E. Easterling, "Deformation and Densification of Porous Preforms in Hot Forging", Powder Metallurgy, Vol. 14, No. 27, 1971, pg. 144-163.

2. S. M. Kaufman and S. Mocarski, "The Effect of Small Amounts of Residual Porosity on the Mechanical Properties of P/M Forgings", The International Journal of Powder Metallurgy and Powder Technology, Vol. 7, No. 3, 1971, pg. 19-32.

3. H. Antes, "Forming Metal Powders", SME, MF72-504.

4. T. W. Pietrocini, "Hot Formed P/M - Applications", Modern Developments in Powder Metallurgy, Plenum Press, NY, 1974, Vol. 7, pg. 395-410.

5. J. P. Cook, "Manganese Steels for P/M Hot Forming", The International Journal of Powder Metallurgy and Powder Technology, Vol. 10, No. 1, 1974, pg. 15-19.

6. G. Bockstiegel and C. A. Blande, "The Influence of Slag Inclusions and Pores on Impact Strength and Fatigue Strength of Powder Forged Iron and Steel", Powder Metallurgy International, Vol. 8, No. 4, 1976, pg. 155-160.

7. B. L. Ferguson, H. A. Kuhn, and A. Lawley, "Fatigue of Iron-Base P/M Forgings", Proceedings: Fifth International Metallurgy Conference, 1976.

8. H. A. Kuhn and C. L. Downey, "P/M Preform Design for Hot Forging", Forging of Powder Metallurgy Preforms, Plenum Press, NY, 1973, Vol. 6, pg. 45-56.

9. Crowson, A. and Anderson, F. E. "Properties of Powder Metallurgy Steel Forgings" presented at 1977 National Powder Metallurgy Conference, in press.

10. Lally, F. T., Toth, I. J., and DiBenedetto, J., "Forged Metal Powder Products", SWERR-TR-72-51, August 1972.

11. R. B. Bargainnier and J. S. Hirschhom, "Forging Studies of Ni-Mo P/M Steel", Forging of Powder Metallurgy Preforms, Plenum Press, NY, 1973, Vol. 6, pg. 229-266.

DEVELOPMENT AND APPLICATIONS OF HEAT-TREATED P/M STEELS CONTAINING MANGANESE, CHROMIUM, VANADIUM, AND MOLYBDENUM

G. Schlieper[1], F.Thümmler[1], G.Zapf[2]

1) University and Nuclear Research Centre Karlsruhe Germany

2) Sintermetallwerk Krebsöge GmbH, Germany

ABSTRACT

The master alloys, MCM and MVM, containing high oxygen affinity elements Mn, Cr and V and their influence as alloying additives on the properties of sintered steels have been investigated. These master alloys allow the production of high strength heat-treatable sintered steels at low alloying cost. Double-pressed sintered steels containing 2% MCM and 0,45% C_{tot} have a maximum UTS of 1400 N/mm^2 and a yield strength of 950 N/mm^2 after heat treatment. With the forging technique, a maximum UTS of 1950 N/mm^2 and a yield strength of 1550 N/mm^2 is reached with 2% MCM and 0,6% C_{tot}. MVM alloyed steels are superior in strength to MCM alloys in the as-sintered state, but their response to heat treatment is less pronounced. Sinterforged MCM as well as MVM steels are equal in strength to the cast and wrought heat-treatable steel SAE 1045. Finally, some examples for the application of MCM alloys in the automotive industry are presented.

INTRODUCTION

The application of elements with high oxygen affinity in powder metallurgy is only possible by use of master alloys in order to protect them from oxidation /1/. In recent years, two new master alloys have been developed which contain transition metals in the form of carbidic compounds /2/. They are named MCM and MVM after their main alloying components Mn, Cr, Mo and Mn, V, Mo, respectively. Preceding investigations revealed that these master alloys allow the production of high strength heat-treatable sintered steels at low alloying cost /3,4,5/.

In this paper, the phase composition of MCM and MVM master alloys is presented. Further, the hardenability of double-pressed sintered steels produced with these master alloys has been investigated. The mechanical properties of double-pressed as well as hot-forged steels were studied in the as-sintered and the heat-treated condition and are compared with the properties of wrought steels. The plot of ultimate tensile strength, yield strength, hardness, and elongation versus annealing temperature can be used as a guide to enable the producer to choose the alloy and heat treatment best suited for his application. The occurring microstructures are also discussed. Finally, some applications of MCM and MVM alloys in industrial production are presented.

COMPOSITION OF MCM AND MVM

The two master alloys contain about 20 - 25% each of Fe, Mn, Mo, and Cr or V. The amount of carbon in MCM is 7%, in MVM it is 5%. MCM has a relatively low oxygen content of 0.18%, but in MVM the amount of oxygen is higher. When stored for some time without protection from the air, the oxygen content may increase up 1%. As a result, we find a considerable amount of oxide in the MVM master alloy. The oxygen content in MVM is relatively high. However, the oxygen does not exist in the form of slags, but in the form of finely distributed oxides. Since generally only 1-2% of the master alloy is used for the production of sintered steel, the amount of oxygen introducted by the master alloy reduces to a maximum of 200 ppm.

The phase composition of the master alloys was established by metallographic means, X-ray analysis, and electron beam microanalysis. The chemical and phase compositions of the master alloys are given in Fig. 1.

MCM						MVM					
Mn	Cr	Mo	Fe	C	O	Mn	V	Mo	Fe	C	O
25	23	22	22	7	o.18	25.5	23	25.5	2o	5	o.2-1.o
8o % $(Cr,Mn,Fe,Mo)_7C_3$ 1o % $(Fe,Mo,Cr,Mn)_6C$ 5 % $(Fe,Mn,Mo)_3C$ 5 % α-(Fe,Mn,Cr,Mo)						25 % (V,Mo)C 25 % $(V,Mo)_2C$ 2o-25 % α-(Mn,Fe,Mo) 2o-25 % γ-(Fe,Mn,Mo) MnO (Percentage depending on oxygen content)					

Fig. 1: Chemical and Phase Composition of MCM and MVM Master Alloys

MCM contains three complex carbides, M_7C_3, M_6C, and M_3C, which make up about 95% of the master alloy. Only a small amount of a metallic phase is present which is probably a ferritic solid solution. Thus the alloying elements are well protected against oxidation. As most of the master alloy is melting between 1150°C and 1190°C (2100°F and 2175°F), which is well below the sintering temperature of 1280°C (2336°F), the sintering process is activated by a liquid phase.

The phase composition of MVM is less satisfactory, since five different phases are present. The two carbide phases (V,Mo)C and $(V,Mo)_2C$ make up only 50% of the master alloy, the two metallic phases are α-Mn and γ-Fe solid solutions. As a result, we find that manganese is well homogenized after sintering, since the metallic phases are melting during the heating-up process. On the other hand, vanadium goes very slowly into solution, because (V,Mo)C is stable at sintering temperature and has a very small solubility in γ-iron /6/. Molybdenum, which is contained in all MVM phases, exhibits an average behaviour.

HARDENABILITY

The positive effect of MCM and MVM on hardenability has already been established for the single-sintering technique /3/. In order to obtain hardenability curves for the double-sintering technique, further tests were carried out, the results of which are presented here. Jominy standard test bars were isostatically compacted at 6000 bar, presintered at 900°C for 30 min, and compacted again at the same pressure. Then they were sintered at 1280°C under cracked ammonia in a walking beam furnace. Densities of the compacts ranged between 7.2 and 7.3 g/cm^3.

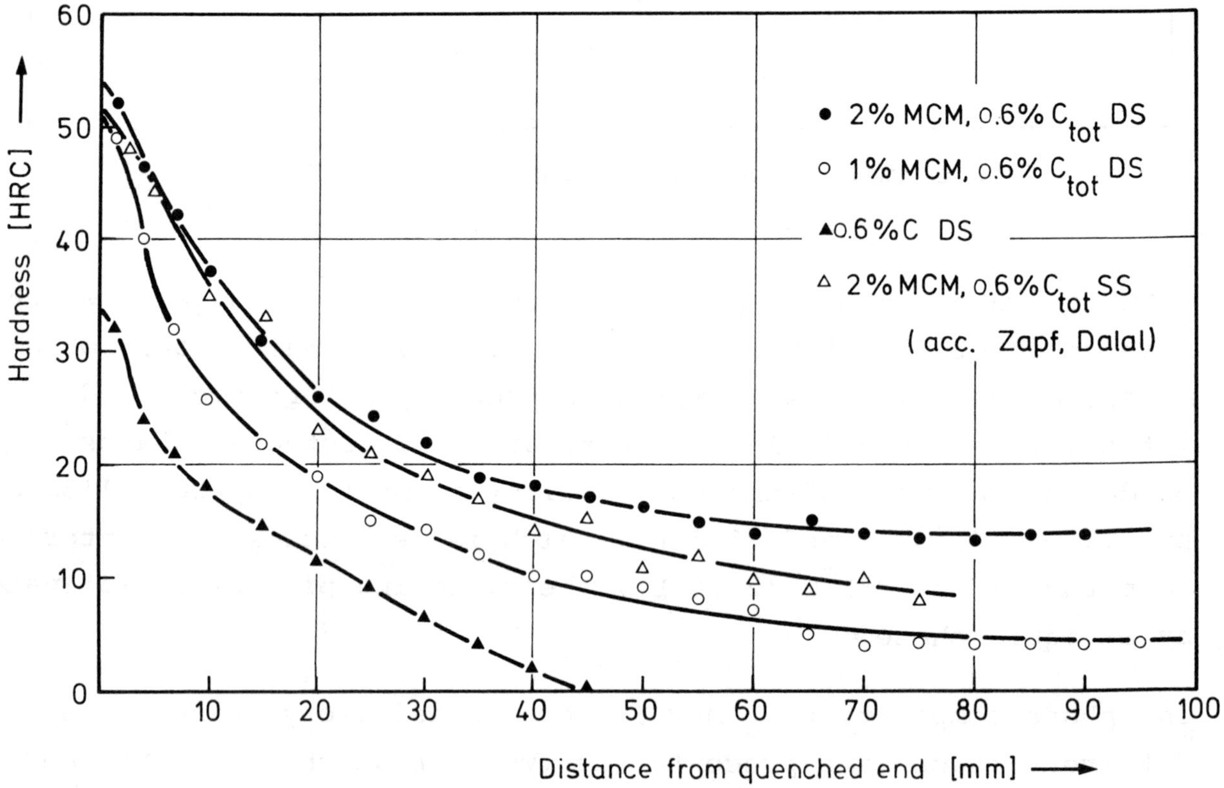

Fig. 2: Hardenability of MCM Alloys

Fig. 2 shows hardenability curves of double-sintered alloys (DS) containing 0, 1, and 2% MCM and a total amount of 0.6% C. With increasing MCM content we find a considerable increase in hardenability. In comparison with the single-sintered alloy (SS) there is an increase in hardness due to the higher density, as well.

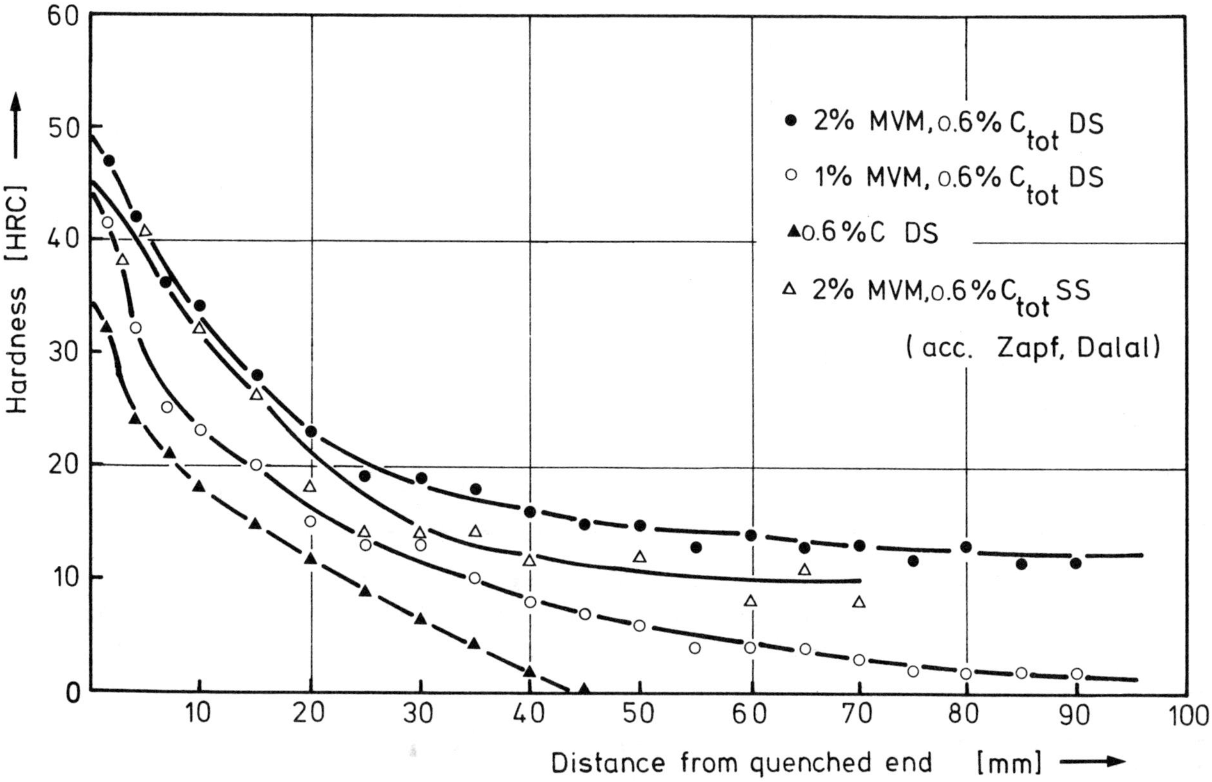

Fig. 3: Hardenability of MVM Alloys

The hardenability of MVM alloys is distinctly inferior to that of MCM alloys (Fig. 3). As can be seen from the microstructures, this is caused by the poor homogeneity of these alloys.

MECHANICAL PROPERTIES

The mechanical properties of the double-pressed alloys were determined using MPA standard test bars. A highly compressible iron powder was used to obtain densities of about 7.3 g/cm^3. With the forging technique rectangular bars were manufactured from which test samples with a circular cross-section were machined /7,8,9/.

Without heat treatment, the ultimate tensile strength (σ_B) of double-pressed alloys reaches values up to 730 N/mm^2 with a master alloy content of as low as 2%, corresponding to 0.4% each of Mn, V, and Mo. The maximum yield strength ($\sigma_{0.1}$) is 540 N/mm^2, the elongation (δ) almost 3% (Fig. 4). In the as-sintered condition, the strength of MVM alloys is considerably higher than that of MCM

Alloy	σ_B [N/mm^2]	$\sigma_{0.1}$ [N/mm^2]	δ [%]	HRB	E [N/mm^2]
0.6 % C DS	451	254	8.8	70	197 000
2 % MCM, 0.6 % C_{tot} DS	679	432	5.3	87	184 000
2 % MVM, 0.6 % C_{tot} DS	730	542	2.8	89	180 000
0.6 % C SF	643	384	19.6	85	199 000
2 % MCM, 0.6 % C_{tot} SF	961	590	12.0	98	211 000
2 % MVM, 0.6 % C_{tot} SF	1053	768	10.5	24 HRC	208 000
SAE 1045	717	423	25.3	90	208 000

Fig. 4: Mechanical Properties of Double-sintered (DS) and Sinterforged (SF) Steels, No Heat Treatment

alloys containing the same amount of additives.

The forged alloys combine strengths up to 1050 N/mm^2 and an elongation of over 10%. The strength of the normalized conventional steel SAE 1045 is substantially lower than that of the forged alloys, but the elongation is much higher.

Heat treatment comprised of austenitization at 850°C for 20 min under cracked ammonia and hardening in oil that had been preheated to about 60°C. The samples were then annealed in air for 2 hours at various temperatures up to 600°C. The mechanical properties of heat-treated alloys are presented in Fig 5 and 6.

The diagrams reveal a maximum of ultimate tensile strength (UTS) and yield strength at annealing temperatures between 200°C and 300°C. At lower annealing temperatures the strength decreases due to the brittleness of the martensite which prevents that local stress maxima can be compensated by microplastic deformation. Thus cracks are formed that reduce the cross-section of the sample and lead to an early fracture. At higher annealing temperatures the strength decreases due to the decomposition of the martensite. At the same time toughness increases.

Fig. 5: Mechanical Properties of Heat-treated Sintered Steels, Double-pressing Technique

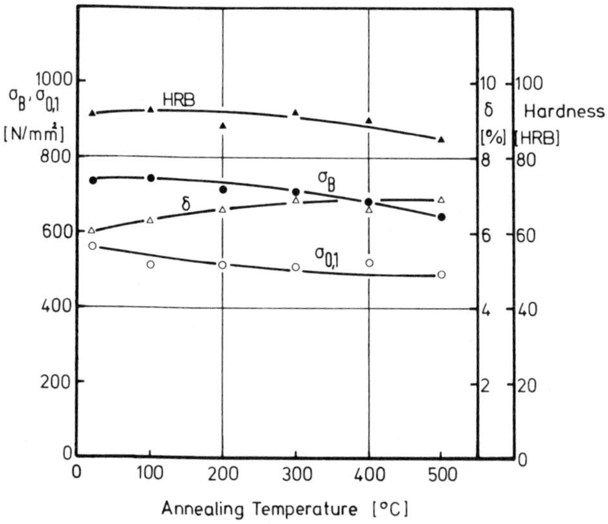

a) 0.45% C DS

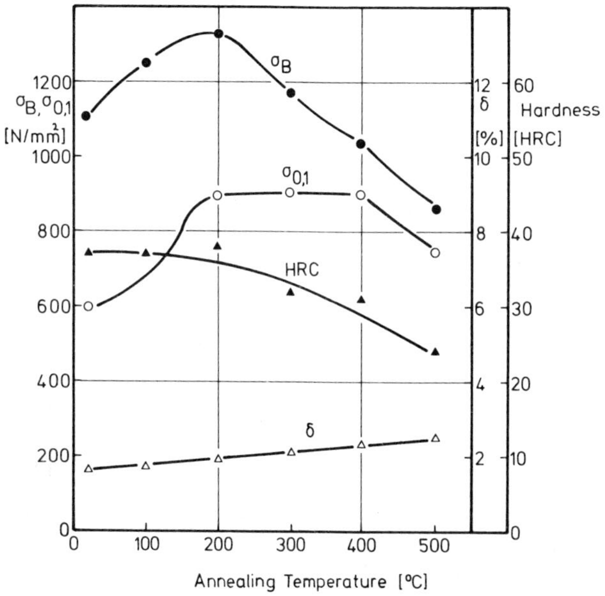

b) 1% MCM, 0.45% C_{tot} DS

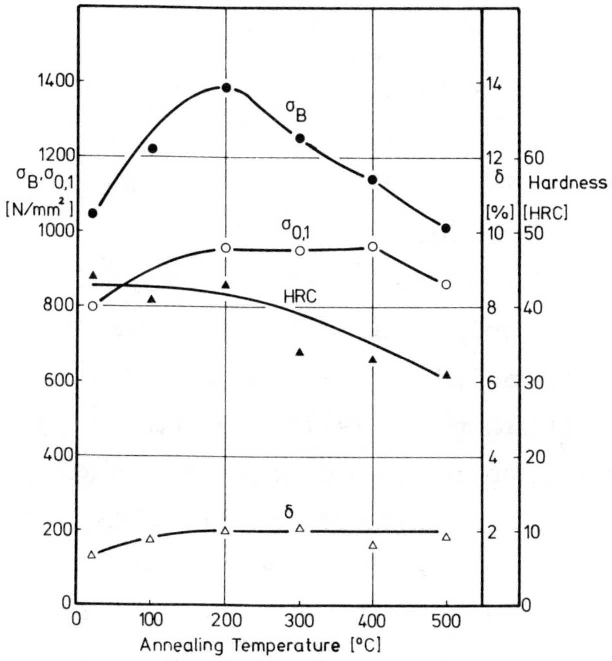

c) 2% MCM, 0.45% C_{tot} DS

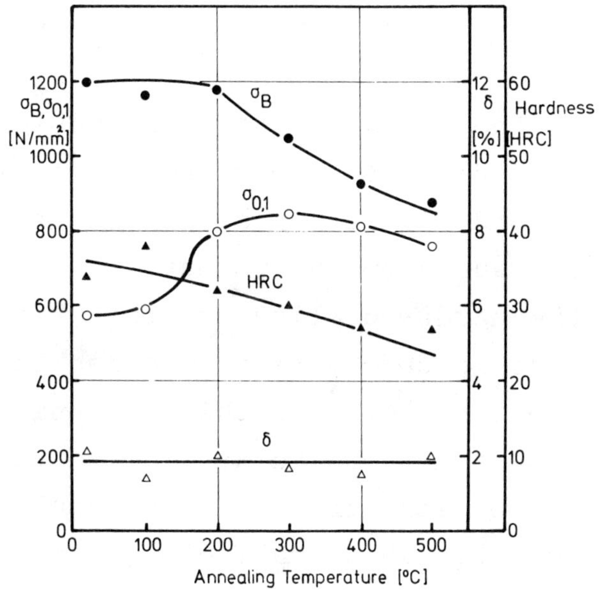

d) 2% MVM, 0.45% C_{tot} DS

Fig. 6: Mechanical Properties of Heat-treated Sintered Steels, Forging Technique

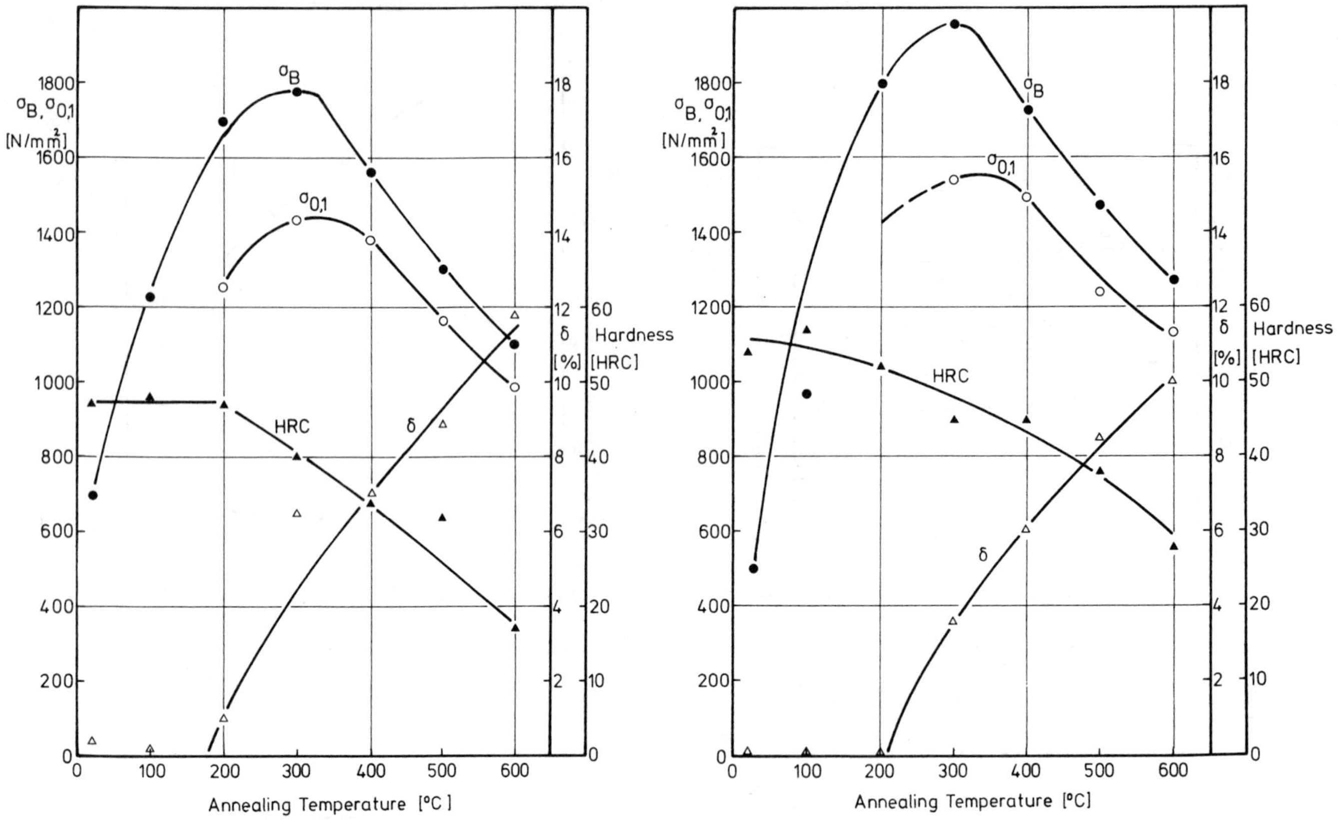

a) 1% MCM, 0.6% C_{tot} SF   b) 2% MCM, 0.6% C_{tot} SF

The elongation of the double-pressed steels is almost independent of the heat treatment. In most cases it lies below or around 2%. At annealing temperatures below 200°C this elongation is caused by the formation of cracks during the deformation. Even the hardened alloys contain region of bainite that allow some plastic deformation, whereas the brittle martensite breaks. Only above 400°C, when the martensite is completely decomposed, the elongation can be attributed to a plastic deformation of the whole microstructure. It should be added that the margin of error is large when small elongations are measured. Therefore, elongations below 2% do not give great evidence on the ductility of the material. Because of their small elongations, heat-treated double-sintered MCM steels are only suitable for static and not for dynamic loading applications.

The double-sintered plain carbon steels reveal only a poor response to heat treatment which is expressed by the constant values of tensile and yield strength (Fig. 5a). The addition of only 1% MCM leads to a sharp increase in strength, and with 2% MCM even higher strengths are achieved. Maximum UTS values of 1400 N/mm^2 and a yield strength of 950 N/mm^2 are reached with the double-sintering technique (Fig. 5b,c). Sinterforged MCM alloys have maximum UTS values of 1950 N/mm^2 and a yield strength of 1550 N/mm^2 (Fig. 6a,b).

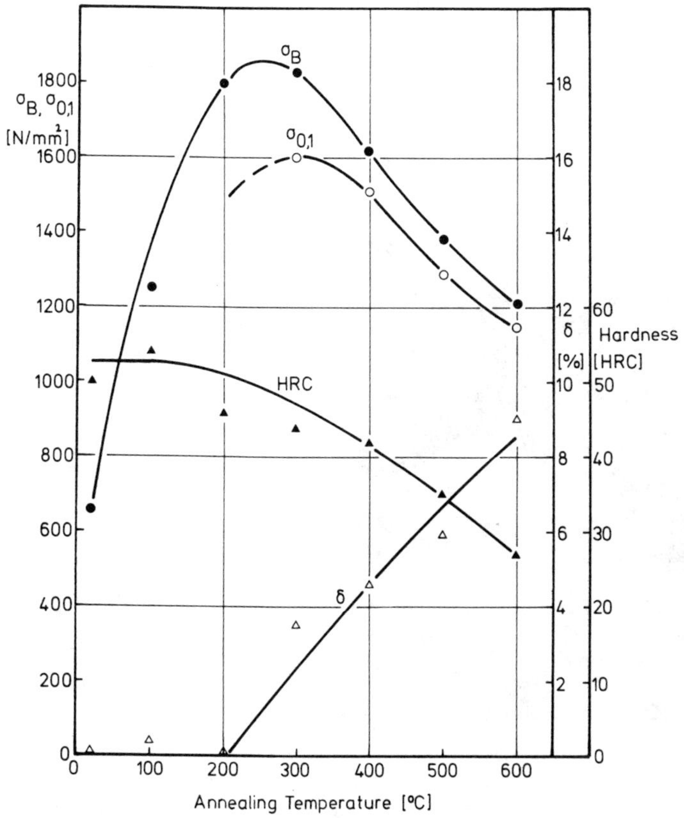

c) 2% MVM, 0.6% C_{tot} SF

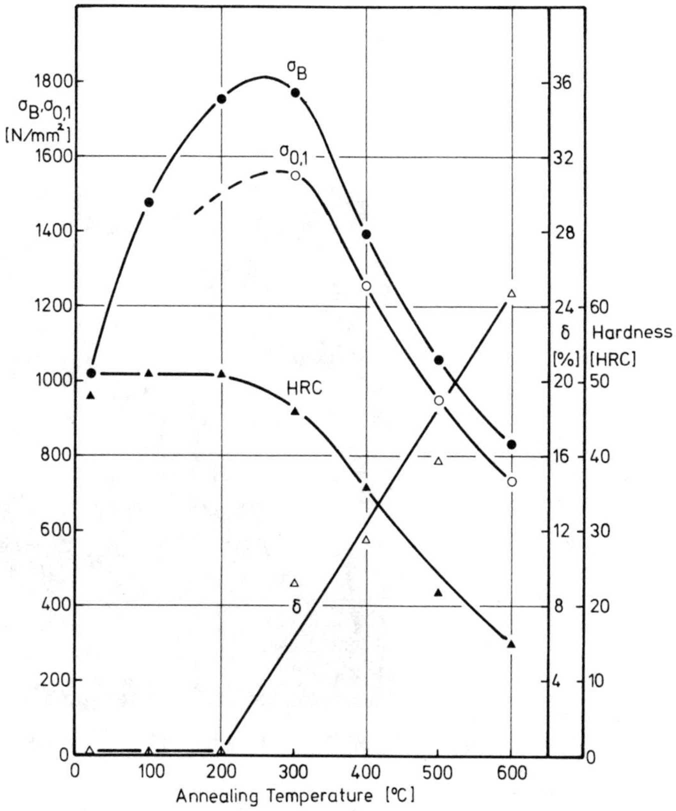

d) SAE 1045 conventionally wrought Steel

The properties of heat-treated MVM alloys are not as good as those of the MCM alloys which is in agreement with their inferior hardenability. The maximum UTS and yield strength of the double-sintered steels are 1200 N/mm^2 and 850 N/mm^2, respectively (Fig. 5d). The sinterforged MVM alloys have a maximum UTS of over 1800 N/mm^2 and the maximum yield strength of 1600 N/mm^2 is even higher than that of MCM alloys (Fig. 6c).

UTS, yield strength, and hardness of sinterforged MCM and MVM alloys are comparable with the properties of the heat-treated conventional steel SAE 1045 (Fig. 6d). The elongation however is substantially smaller, is reaches only one half of the values of conventional steel.

MICROSTRUCTURES

Fig. 7: Microstructure of Double-pressed Sintered Steels, No Heat Treatment

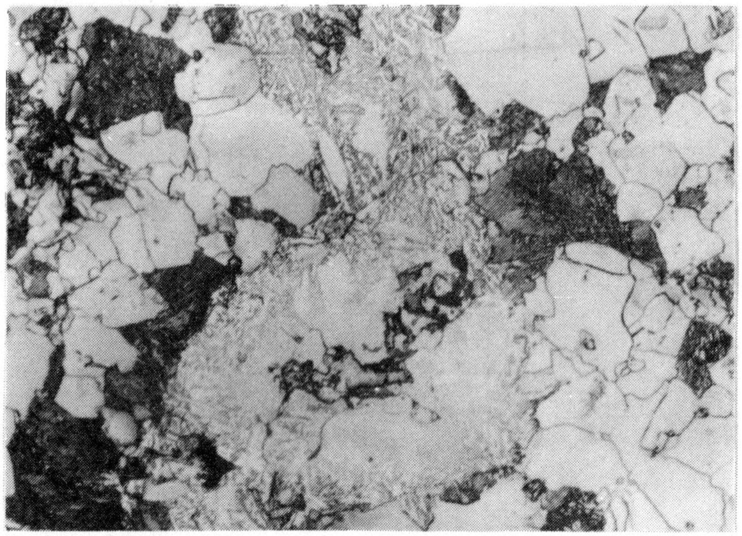

a) 2% MVM, 0.45% C_{tot} DS 500 x

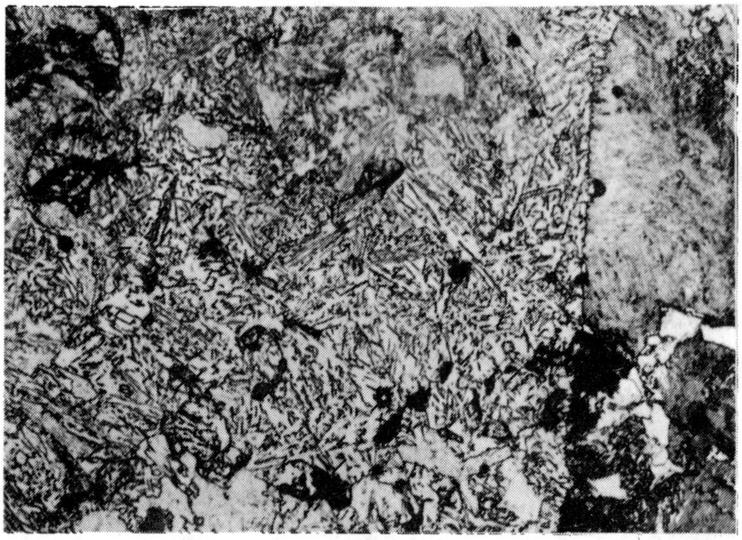

b) 2% MCM, 0.6% C_{tot} DS 500 x

Fig. 8: Microstructure of Sinterforged Steels, No Heat Treatment

a) 2% MVM, 0.6% C_{tot} 500 x

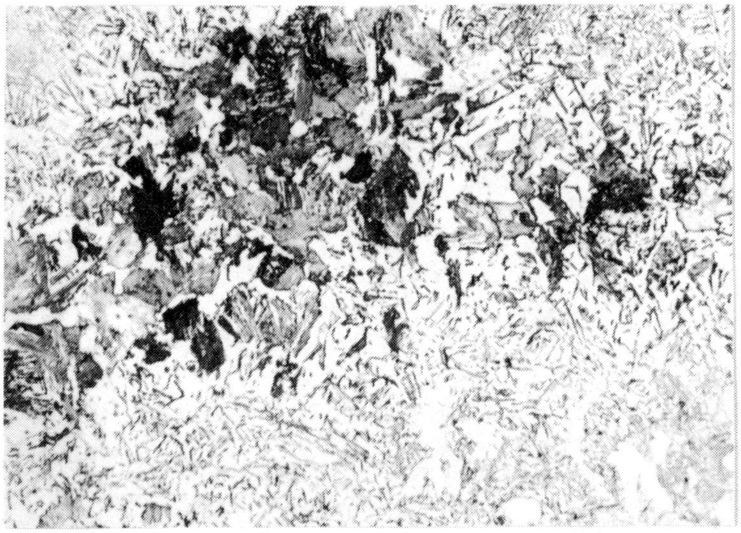

b) 2% MCM, 0.6% C_{tot} 500 x

In the as-sintered condition we find bainite, pearlite, and ferrite in the microstructure. Double-sintered MVM alloys contain isolated regions of bainite amounting up to 25% of the microstructure. Homogenization is incomplete, there are inclusions of undissolved (V,Mo)C and of MnO present (Fig. 7). MCM alloys are better homogenized and contain up to 70% of bainite. The superior strength of the MVM alloys in spite of their insufficient homogeneity is obviously caused by the fact that a relatively coarse bainite can have lower strength than a finely laminated pearlite, even though the bainite contains more alloying elements. The good homogenization of Mn in MVM alloys may exert an additional strengthening effect.

The grain size of the ferritic-pearlitic microstructure of double-pressed steels containing 0.45% C is 8 on the ASTM scale, with 0.6% C it is 9. Forged steels have much smaller grain sizes of 10 to 11 (Fig. 8). The bainite content of forged steels is somewhat higher than that of double-pressed steels. It runs up to 40% in MVM alloys and 75% in MCM alloys.

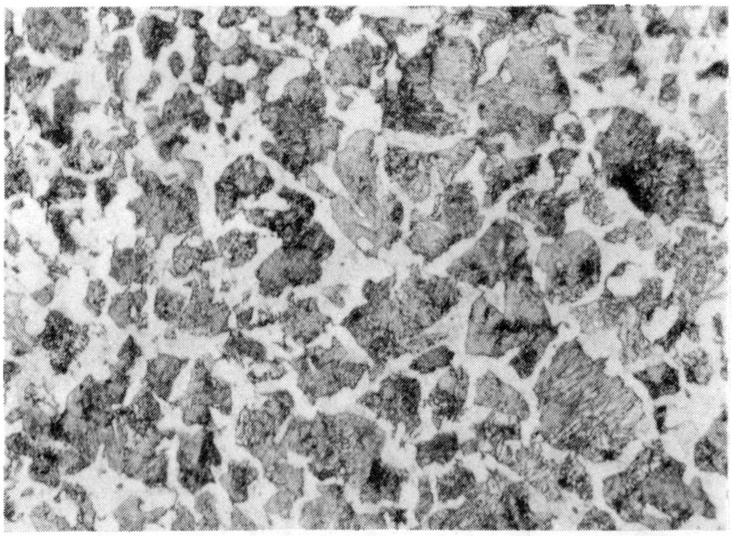

Fig. 9: Microstructure of Conventional Steel SAE 1045 No Heat Treatment 500 x

The microstructure of the conventional unalloyed steel SAE 1045 is completely ferritic-pearlitic without any bainite content. The grain size is lo on the ASTM scale (Fig. 9).

After oil hardening, MVM alloys contain up to 75% of martensite and the balance is bainite. Again inclusions of undissolved carbide and oxide are visible (Fig. 10 a). These inclusions retain some of the alloying elements which are therefore not available for improving hardenability and strength.

Fig. 10: Microstructure of Hardened Sinterforged Steels

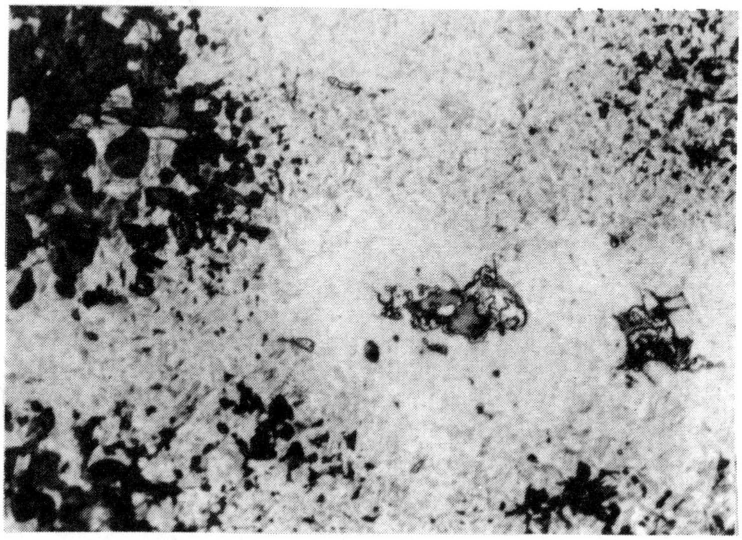

a) 2% MVM, 0,6% C_{tot}SF 500 x

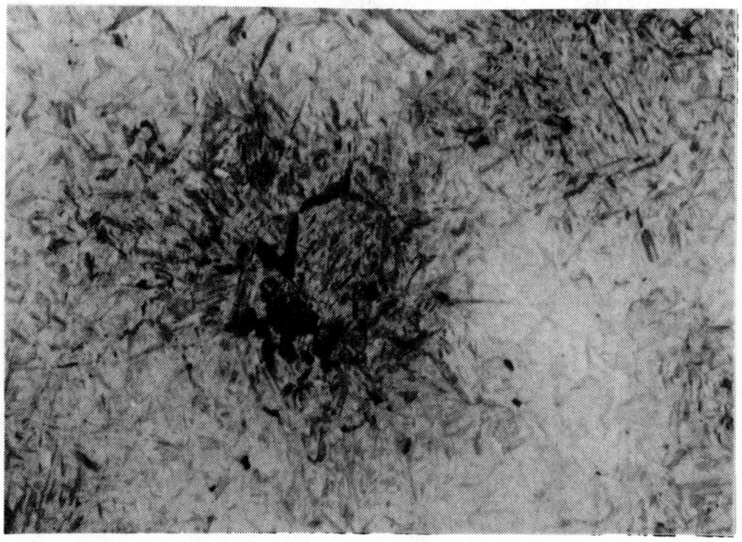

b) 2% MCM, 0,6% C_{tot} SF 500 x

By reason of their better homogeneity, MCM alloys contain up to 95% of martensite in the hardened condition (Fig.10 b). However, by X-ray analysis it was established that the martensite is cubic rather than tetragonal. The cooling rate in oil is not high enough to produce tetragonal martensite.

Fig. 11: Microstructure of Heat-treated Sinter-forged Steels

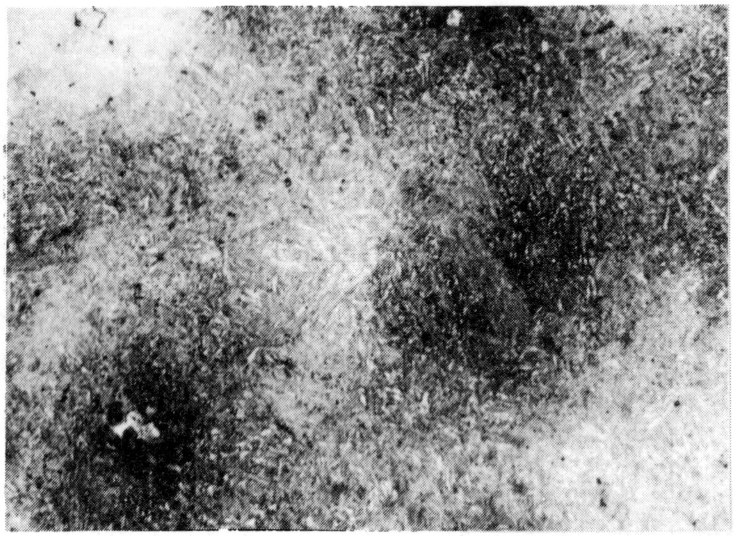

a) 2% MCM, 0.6% C_{tot}, annealed $300^{O}C/2h$ 500 x

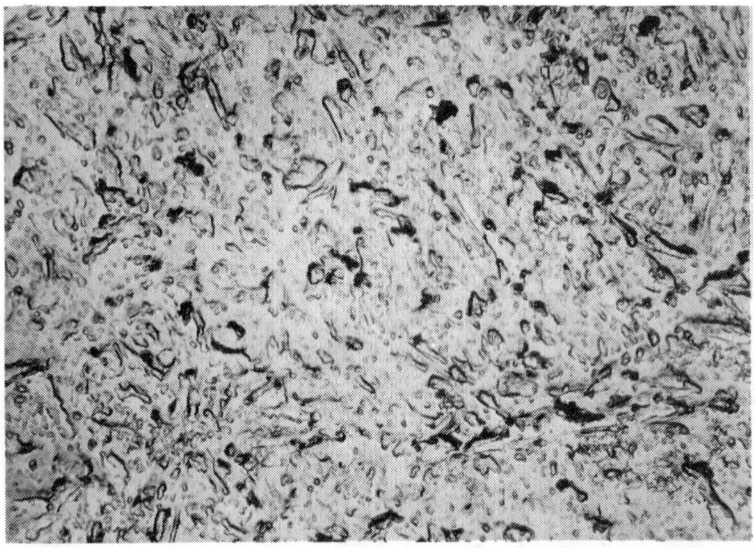

b) Annealed Martensite, TEM 12 000 x

On annealing, the martensite is decomposed and finely distributed cementite particles are precipitated in the microstructure (Fig. 11). The higher the annealing temperature, the larger the growth of the cementite particles so that hardness and strength decrease whereas toughness increases.

APPLICATIONS

MCM and MVM alloyed steels can be used for the production of sintered parts with high static as well as dynamic strength properties. Today, applications are mainly in the field of high pressure hydraulics and in the automotive industry. Parts for do-it-yourself tools are being developed, but they are still at an early stage of production.

Fig. 12: Plug Universal Joint Slip Yoke made from Single-sintered MVM Alloy

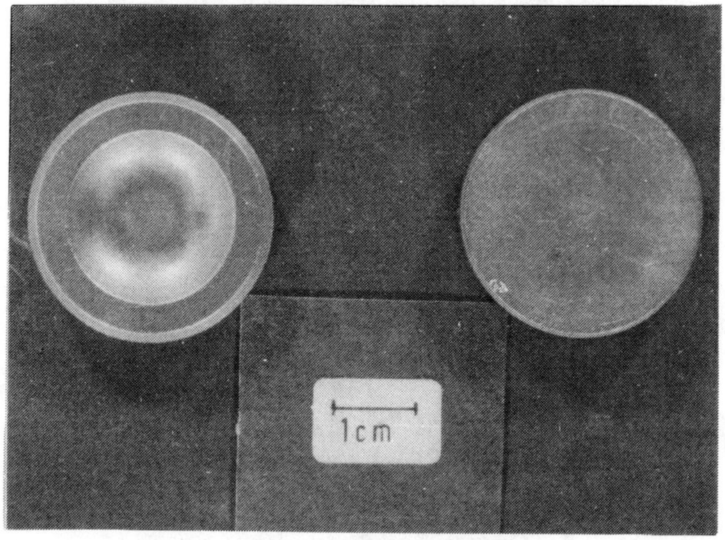

Fig. 12 shows a plug slip yoke for the cardan shaft which is built into Ford automobiles produced in Germany. It is made from an MVM alloy by single-sintering and sizing to obtain a high quality of the spherical surface. No heat treatment is needed for this part. Further examples of single-sintered MCM and MVM alloyed parts produced for the automotive industry are given in Fig. 13. An example for the use of heat-treated sinterforged steel is the brake support for the hand-brake system in a Mercedes-Benz passenger car. Fig.14 shows the part and its location in an anchor plate. After forging it is sintered for a second time to obtain complete homogenization. Heat treatment consists of austenization at 900°C for 30 min, hardening in oil, and annealing for 1 hour at 400°C. The importance of this application lies in the fact that for the first time a safety part in a passenger car could be produced by powder metallurgy methods.

Fig. 13: Parts made from Single-sintered MCM and MVM Alloys

Fig. 14: Brake Support made from Heat-treated Sinterforged MVM Alloy

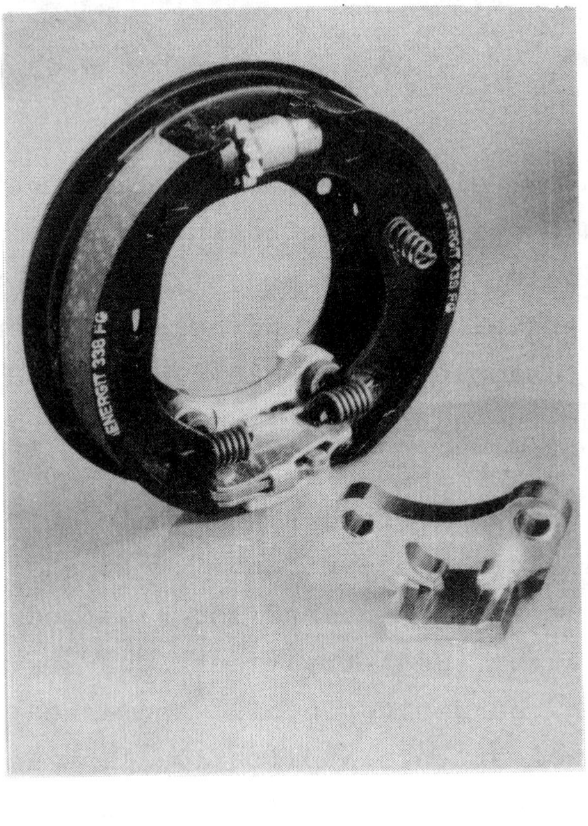

REFERENCES

/1/ P.Lindskog, G.Skoglund, Powder Metallurgy, Third European Powder Metallurgy Symposium 1971, Conf. Suppl. Part I, p. 375-396.

/2/ H.-J. Retelsdorf, R.M. Fichte, G. Hoffmann. K. Dalal, Metall, Vol. 29 (1975), No.10

/3/ G. Zapf, K. Dalal, Modern Developments in Powder Metallurgy, Proceedings of the 1976 P/M Conference Chicago

/4/ G. Zapf, G. Hoffmann, K. Dalal, Powder Metallurgy, Vol.18 (1975), No. 35, p. 214-236.

/5/ G. Zapf, G. Hoffmann. K. Dalal, Arch. Eisenhüttenwesen, Vol. 46 (1975), No.5, p. 347-352.

/6/ E. Houdremont, Handbuch der Sonderstahlkunde, 1956

/7/ G. Zapf, G. Hoffmann, K. Dalal, Zeitschr. f. Werkstofftechnik (1975), No.11 and 12

/8/ G. Zapf, G. Hoffmann, K. Dalal, Zeitschr. f. Werkstofftechnik (1976), No.7

/9/ G. Hoffmann, K. Dalal, Modern Developments in Powder Metallurgy, Proceedings of the 1976 P/M Conference Chicago

THE FATIGUE RESISTANCE OF HIGH-STRENGTH 7XXX P/M ALUMINUM ALLOY EXTRUSIONS

R. E. Sanders, Jr.
Alloy Technology Division

W. L. Otto, Jr.
Powder & Forgings Division

ALCOA LABORATORIES
ALCOA CENTER, PA 15069

Abstract

Research at Alcoa Laboratories has led to the production of high-strength 7XXX P/M aluminum alloys with outstanding secondary properties of fracture toughness and stress-corrosion cracking resistance. The materials are processed by cold compaction of fine (approximately 15 μm A.P.D.) prealloyed atomized powders, vacuum preheating and hot pressing to form fully dense billets. Subsequent forging, rolling or extrusion completes the fabrication of the wrought product. The utilization of these alloys in aerospace applications necessitates a complete evaluation and understanding of their fatigue properties. This investigation was performed under the sponsorship of the Air Force Materials Laboratory to study the effects of fabrication practice, alloy chemistry and aging treatment (temper) on the microstructure and fatigue resistance of P/M alloys. Notched ($K_t = 3$) S/N and fatigue crack growth (FCG) tests were performed to assess the resistance of these materials to the initiation and propagation of fatigue cracks. In notched fatigue tests, P/M alloys exhibited a considerably higher fatigue strength at 10^7 cycles than conventional 7XXX products. Fatigue crack growth testing showed that the P/M alloys had equivalent or slightly lower FCG resistance compared to I/M alloy 7050.

Introduction

The importance of precipitation-strengthened 7XXX aluminum alloys to the aircraft industry has stimulated many investigations aimed at the improvement of properties in the Al-Zn-Mg-Cu system. While yield strengths of more than 800 MPa are obtainable in 7XXX alloys (1), designers are invariably restricted to lower strength levels by so-called "secondary" properties of fracture toughness, corrosion resistance, and fatigue (2). The consideration of these properties in design generally limits useful yield strength levels in these alloys to 500-600 MPa.

The development of powder metallurgy (P/M) techniques at Alcoa Laboratories has resulted in the production of 7XXX alloys with superior combinations of mechanical properties compared to alloys produced by conventional ingot metallurgy (I/M) techniques. At equivalent levels of stress-corrosion cracking resistance, P/M alloys have been shown to exhibit equivalent fracture toughness, with higher strength and exfoliation resistance than the high-performance I/M alloys, 7050 and 7475 (3,4). These products were fabricated using a process that involves vacuum preheating and hot pressing cold compacts of prealloyed, atomized powders with subsequent hot forging, extrusion, or rolling to final product form. An attempt to scale-up this process from small laboratory-produced hot compacted billets to 1415 Kg plant-produced compacts has been partially successful thus far and seems to offer the potential for fabrication of many large scale parts.

The potential use of such P/M materials in aerospace applications has made the evaluation of their fatigue properties extremely important. A complete characterization of resistance to both fatigue crack initiation and crack growth is required before full commercial acceptance of 7XXX P/M alloys can be realized. An investigation has been made under the sponsorship of the Air Force Materials Laboratory (5) to determine interrelationships among composition, fabricating practice, microstructure, and fatigue in high-strength 7XXX P/M alloys. This paper will describe the results of notched fatigue and fatigue crack growth testing of a series of alloys fabricated as extrusions for this program. A similar evaluation of an I/M control material, alloy 7050, will also be presented. A limited discussion will be made of the results with regard to the microstructural features and mechanisms controlling fatigue in these alloys.

Experimental Procedures

Materials and Fabrication

Five lots of powder were prepared for use in the experimental program. The first three powder lots (A1, A2, and A3) were prepared by air atomizing alloy melts with a target base composition of 6.5 wt.% Zn, 2.5 wt.% Mg, and 1.5 wt.% Cu and nominal Co additions of 0.0, 0.4, or 0.8 wt.%. Average melt analyses for these powders are shown in Table I. The two remaining powder lots (B and C) were produced with varying amounts of Zn and Cu at a constant (0.4 wt.%) Co content. Particle size distributions for samples of each powder lot were determined by a U.S. standard screen analysis, and APD measurements from a Fisher Sub-Sieve Sizer.

Portions of each powder lot were cold isostatically pressed at 207 MPa in a wet bag system to form 66 Kg cylindrical compacts, 18.7 cm diameter by 109.2 cm

TABLE I. MELT ANALYSIS

	Composition - wt.%					
Alloy	Si	Fe	Cu	Mg	Zn	Co
A1	0.06	0.06	1.43/1.51	2.40/2.48	6.19/6.36	0.0
A2	0.06/0.07	0.06	1.47/1.51	2.36/2.47	6.40/6.52	0.36/0.40
A3	0.05/0.08	0.07	1.48/1.57	2.50/2.58	6.59/6.60	0.83/0.87
B	0.04	0.05	1.58	2.61	8.44	0.40
C	0.05	0.06	2.36	2.59	6.68	0.41

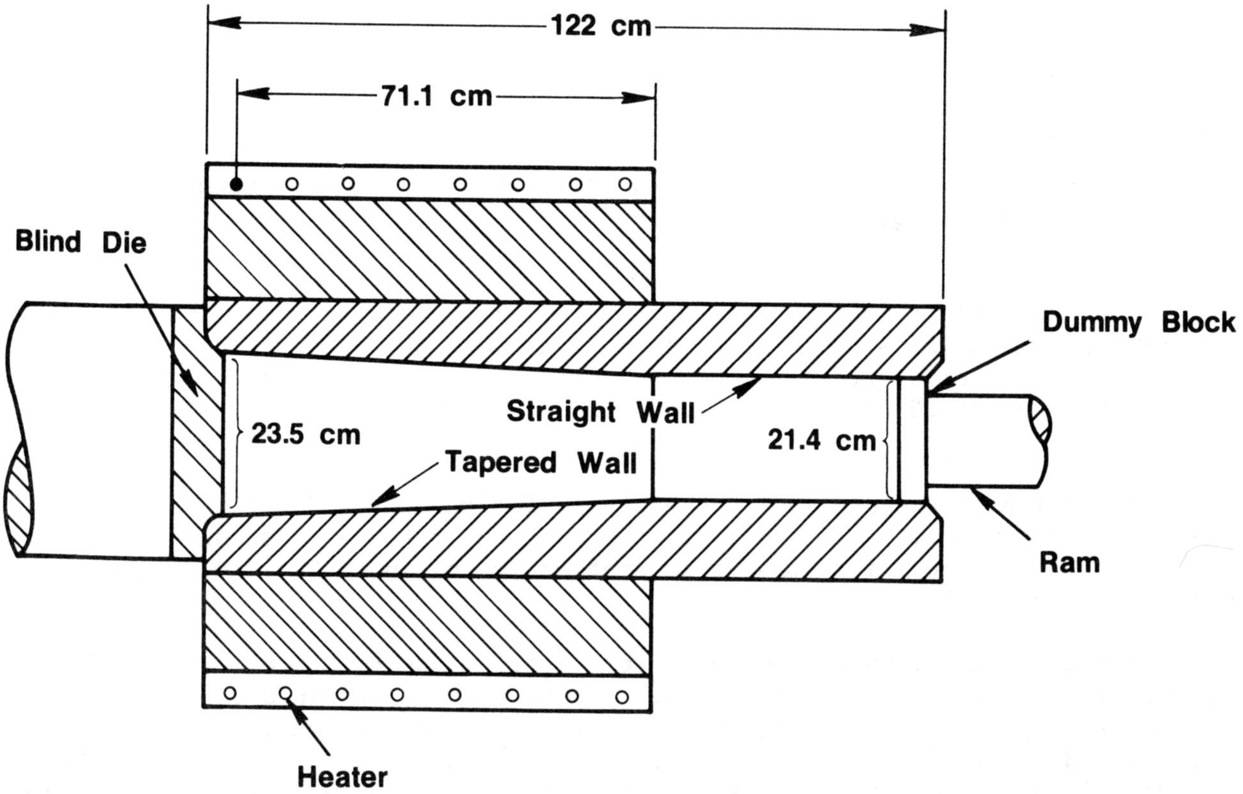

Figure 1. Schematic diagram of 21.4 cm diameter hot compacting cylinder used to fabricate billets for the experimental program.

long. Cold pressed density was ∿75% of theoretical.

The cold compacts were encapsulated in a welded 3003 aluminum alloy can and preheated at 521°C using a dynamic vacuum level of ∿10 μ of Hg. Total preheat cycle time ranged from 7.0 to 8.0 hr with 1 hr at 521±2.8°C. At completion of the preheat cycle, the compacts were removed from the furnace and the evacuation lines were sealed. The compacts were then hot pressed in a closed end compaction cylinder (Figure 1) at a nominal axial pressure of 621 MPa which was maintained for one minute. The hot pressed compacts were then air cooled to room temperature and scalped to 18.4 cm diameter.

Three hot pressed billets of each composition were extruded to a 1.3 cm x 16.5 cm cross-section bar using a 23.5 cm diameter extrusion cylinder. The combination of extrusion temperature (445°C) and average strain rate (4.8×10^{-2} sec^{-1}) corresponded to an average temperature-compensated strain rate of 2.5×10^{-10} sec^{-1}. Extrusion ram loads and displacements were continuously recorded during each extrusion cycle. These data were used to select portions of each extruded bar produced by steady-state conditions for evaluation.

Portions of each extrusion were solution heat treated 2 hr at 488°C and quenched in cold running water. Each extrusion was stretched 2.0% permanent set in the longitudinal direction after quenching for stress relieval. Each of the products was subsequently aged four days at room temperature followed by artificial aging at 121°C for 24 hrs. Portions of the products were given additional aging treatments at 163°C to bring them to desired levels of strength and electrical conductivity.

A 7050-T76511 extrusion produced by I/M techniques was obtained for evaluation as a control material in the program. The nominal chemical composition of alloy 7050 is listed in Table I.

Testing and Evaluation

Uniaxial tensile properties of the P/M extrusions were determined using 0.91 cm diameter longitudinal specimens. Testing was done in accordance with ASTM E8-69.

Notched fatigue testing was performed on longitudinal round (0.64 cm dia) tapered seat specimens. The specimen design had a 60° notch with a notch tip radius of 0.033 cm, corresponding to a stress concentration factor (K_t) of 3.0. Tests were run at 25 Hz in laboratory air with R (minimum stress/maximum stress) = 0.0.

A modified staircase sequence was used for the notched fatigue tests in order to determine the average fatigue strengths of the extrusions at 10^7 cycles. This method, illustrated schematically in Figure 2, is described in detail elsewhere (6).

Fatigue crack growth (FCG) data were obtained using center crack (CCT) panels 10.2 cm wide by 0.64 cm thick. Specimens were machined so that the stress was applied in the longitudinal (extrusion) direction and the cracks propagated across the width (L-T). Tests were conducted under constant amplitude loading in high humidity (>90% R.H.) or salt fog at R=0.33. Crack length was measured either visually or through the use of crack propagation gages spaced 0.25 cm apart. Crack lengths were recorded at intervals between 1.02 cm and 3.30 cm. The combination of crack length and applied stress yielded FCG data over a range of cyclic stress intensity, ΔK, from 2.0-10.0 $MPa\sqrt{m}$.

Microstructural evaluation of the experimental materials was carried out using standard optical and transmission electron microscopic (TEM) techniques. An etchant of HNO_3, HF, and CrO_3 in H_2O was used for optical examination of the P/M extrusions. A 25% HNO_3/75% H_2O solution at 70°C was used for etching of the 7050 alloy. Scanning electron microscopy (SEM) was used for examination of the as-atomized powders and fatigue specimen fracture surfaces.

Results and Discussion

Microstructure

Particle size distributions and APD measurements for each powder lot are shown in Table II. Metallographic examination by SEM of the powders, Figure 3, revealed that the individual powder particles were irregular in shape. The fine particle size and irregular shape of these powders are extremely beneficial to processing of these alloys. In subsequent compaction steps, these particles may be more closely packed than larger, more rounded particles, allowing for significant increases in cold compact densities. The cold compacts fabricated in this program were approximately 78% of theoretical density. The hot-pressed billets were 100% dense after vacuum hot compaction.

Figure 4 shows the extremely fine dendrite size obtained during solidification of the powders. The dendrite arm spacings ($\sim 0.5\ \mu m$) generally observed in the atomized powders of this program are indicative of cooling rates in the 10^5-10^6 °C/sec range (7). Consolidation and extrusion of the P/M alloys

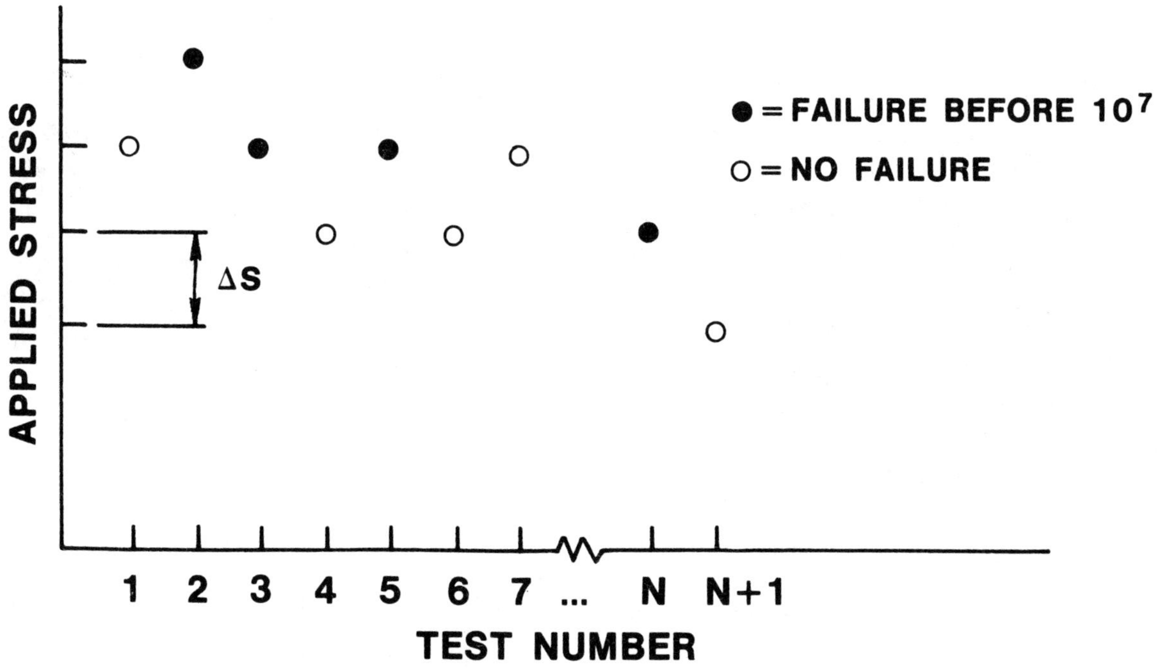

SEQUENTIAL TESTING

ARBITRARY STRESS LEVEL CHOSEN (ΔS)

SUM UP STRESS LEVELS FOR TESTS 2 THROUGH N+1

DIVIDE BY NUMBER OF TESTS (N)

$$\text{AVERAGE FATIGUE STRENGTH } (10^7 \text{ CYCLES}) = \frac{\sum_{i=2}^{N+1} S_i}{N}$$

Figure 2. Modified staircase testing method used to obtain notched fatigue strengths for P/M extrusions.

TABLE II. POWDER SIZE AND DISTRIBUTION

Alloy	U.S. Standard Screen Analysis (wt.%) -50 +100	-100 +200	-200 +325	-325	Average Particle Diameter, μm
A1	trace	1.8	6.2	91.8	12.10
A2	0.2	1.2	10.6	87.6	14.63
A3	trace	1.0	4.4	94.4	11.97
B	trace	3.4	6.2	90.3	13.97
C	trace	2.8	5.6	92.4	13.63

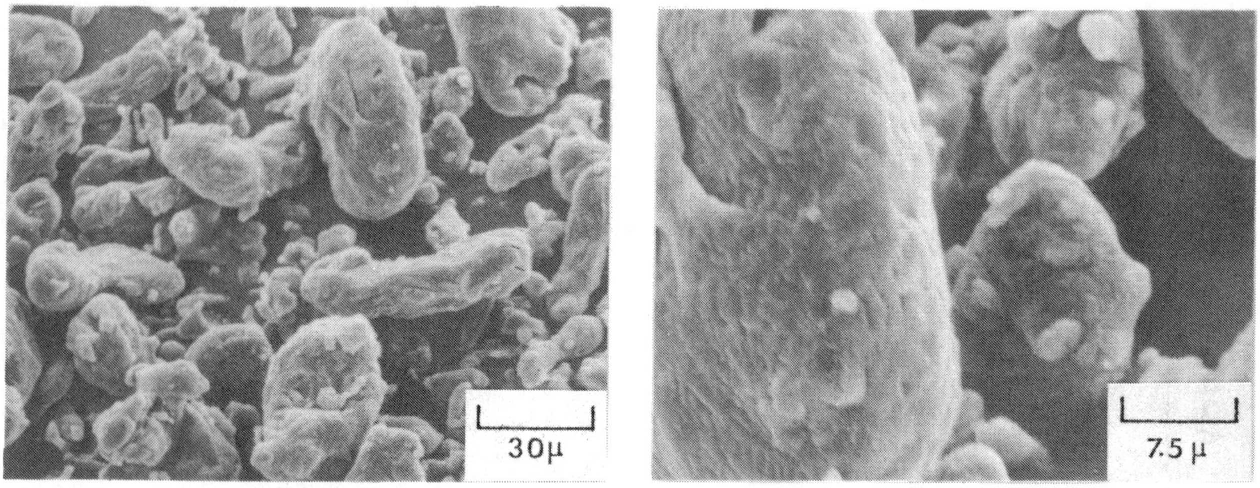

(a) (b)

Figure 3. Scanning electron micrographs of as-atomized alloy A2 (CT91) powder: (a) 500X and (b) 2000X.

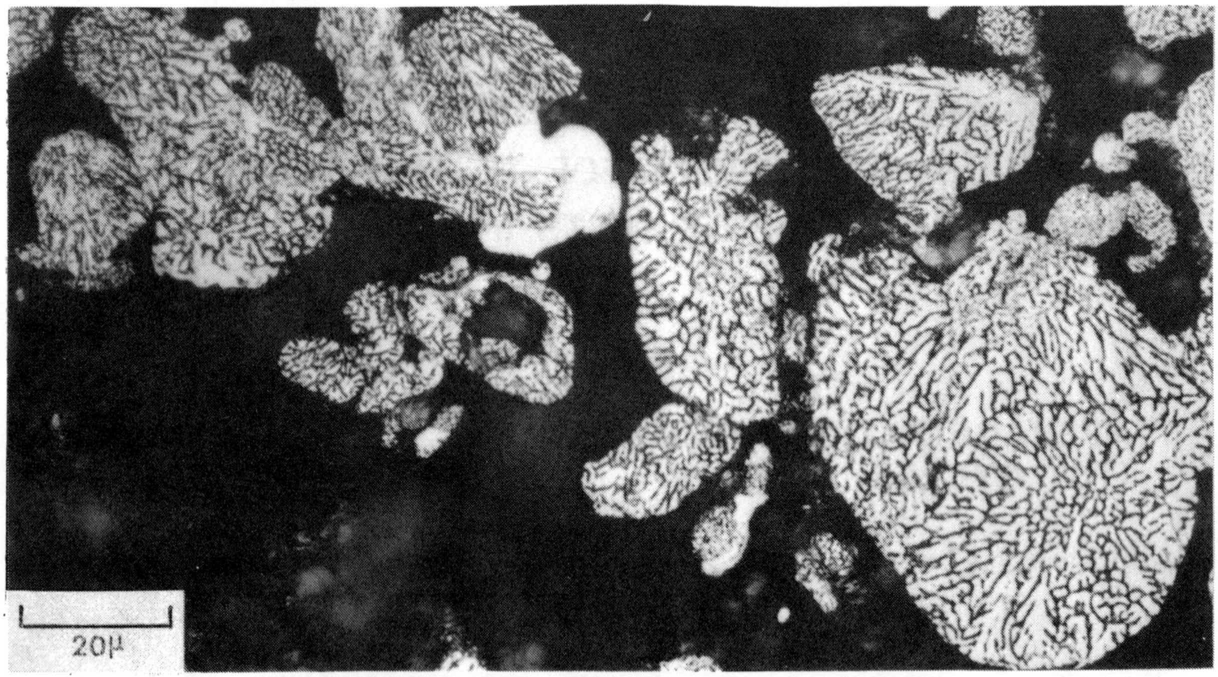

Figure 4. Optical micrograph of polished and etched atomized alloy A2 (CT91) powder particles. 1000X

resulted in the production of extremely fine microstructures such as that shown in Figure 5 (CT91). The fine features (1-5 μm) evident in Figure 5(b) are quite probably recovered subgrains rather than grains in the classical sense. X-ray and optical metallography have shown that such P/M extrusions are almost completely uncrecrystallized and exhibit a (110)[11$\bar{2}$] texture similar to that often found in similarly fabricated wrought I/M alloys.

The optical microstructure of the CT91 alloy may be compared with that of a 7050-T76511 extrusion shown in Figure 6. The I/M material has a partially recrystallized structure and contains coarse (3-5 μm) constituent particles (Mg_2Si, Al_2CuMg, and Al_7Cu_2Fe). These particles, which are known to be somewhat

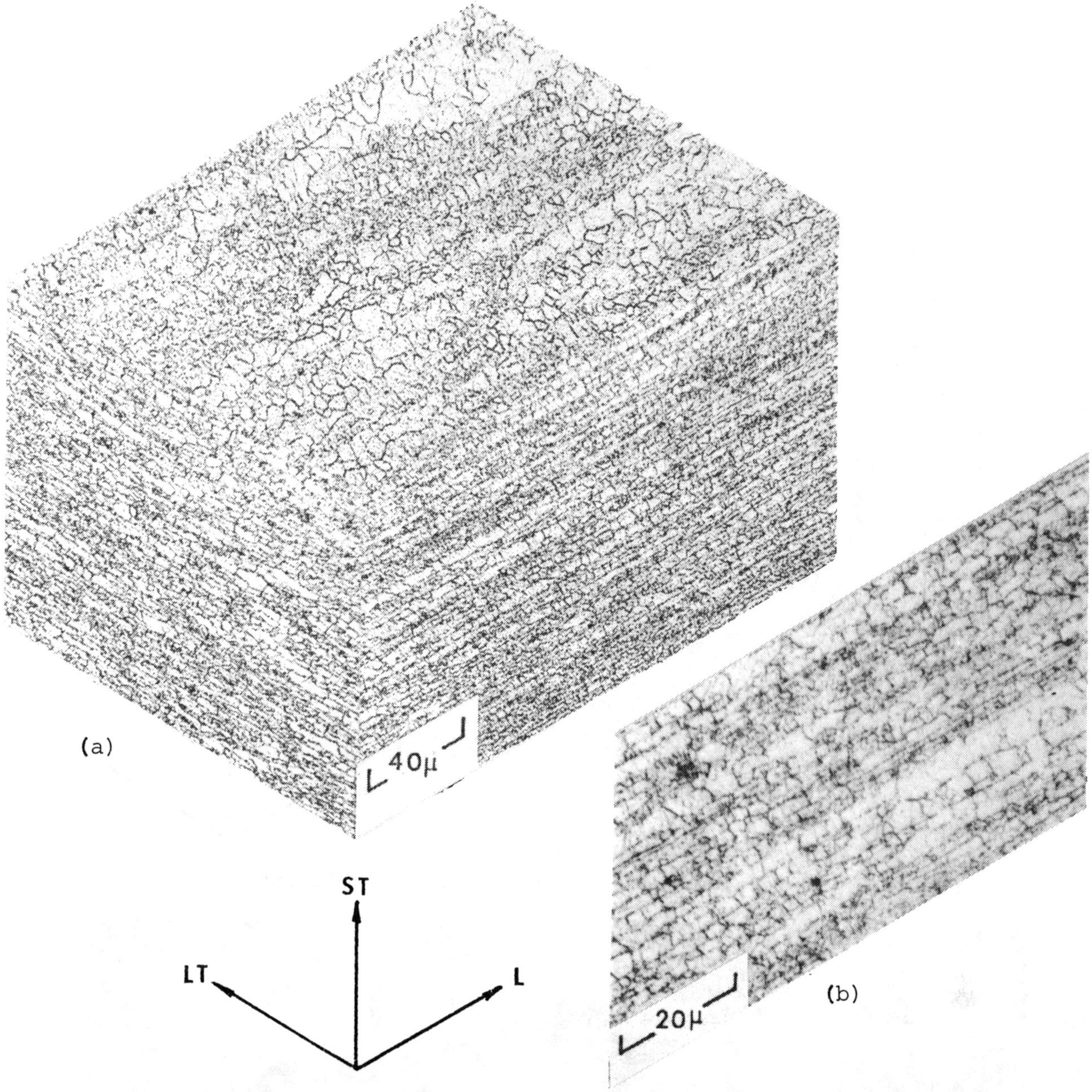

Figure 5. Optical microstructure of CT91-T7X1 extrusion.

detrimental to fracture properties in high-strength 7XXX alloys (8), are not typically seen in the P/M extrusions (Figure 5). The formation of large constituents such as Al_7Cu_2Fe, which are precipitated and coarsened during casting of I/M alloys, is suppressed by the extremely rapid solidification of the powders. Control of both composition and processing of the P/M alloys serves to minimize or eliminate the occurrence of partially soluble constituents such as Al_2CuMg (S phase) often observed in 7XXX I/M alloys.

A TEM examination of the CT91-T7X1 extrusion shows the distribution and type of second-phase particles found in high-strength 7XXX P/M products. Figure 7 shows the presence of large, spherical (0.1-0.5 μm) Co_2Al_9 particles, fine oxide particles, and strengthening precipitates. The Co_2Al_9 phase forms during powder solidification and is quite probably largely unaffected by the subsequent processing of the P/M alloy. These particles act as dispersoids to inhibit grain coarsening during thermal and mechanical processing. The

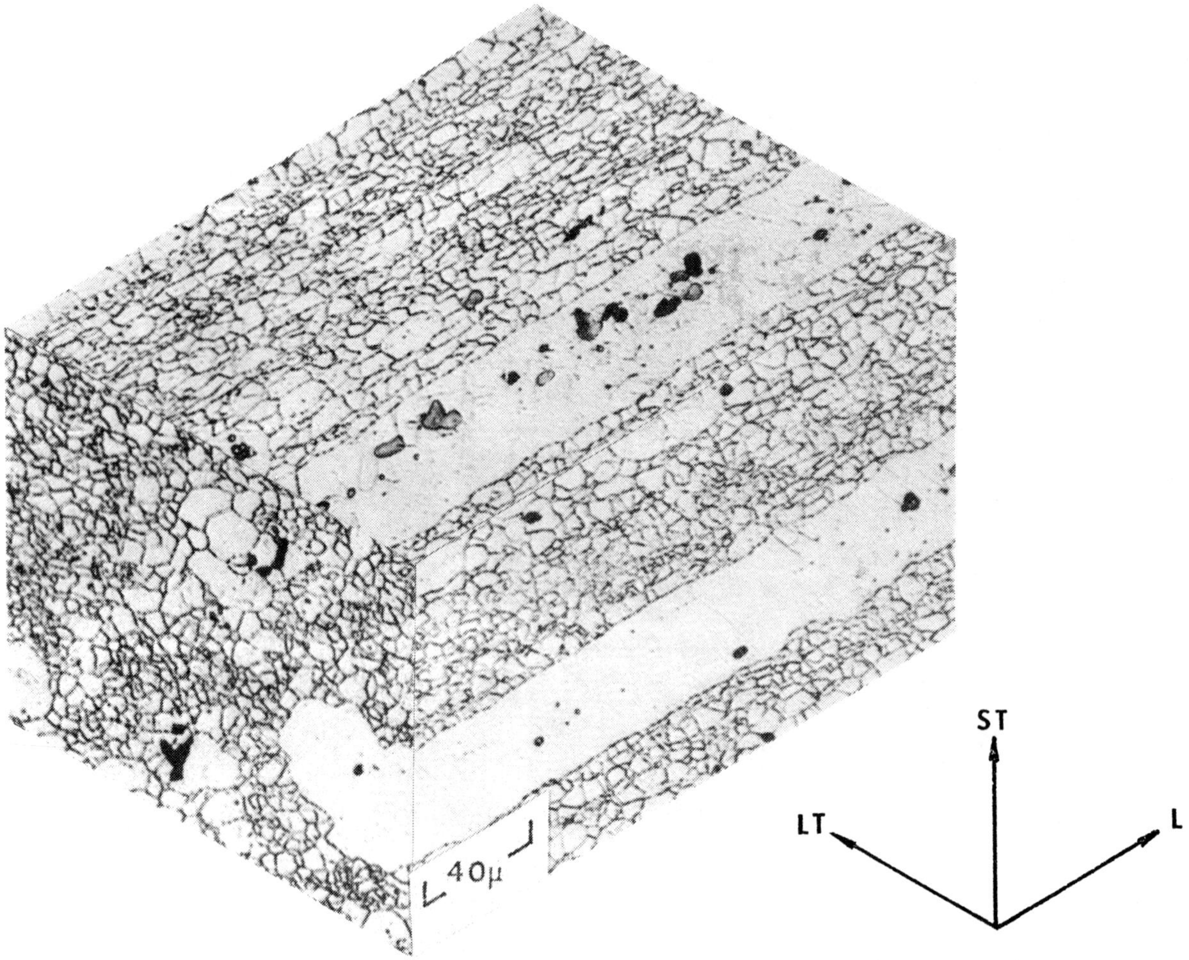

Figure 6. Optical microstructure of 7050-T76511 extrusion.

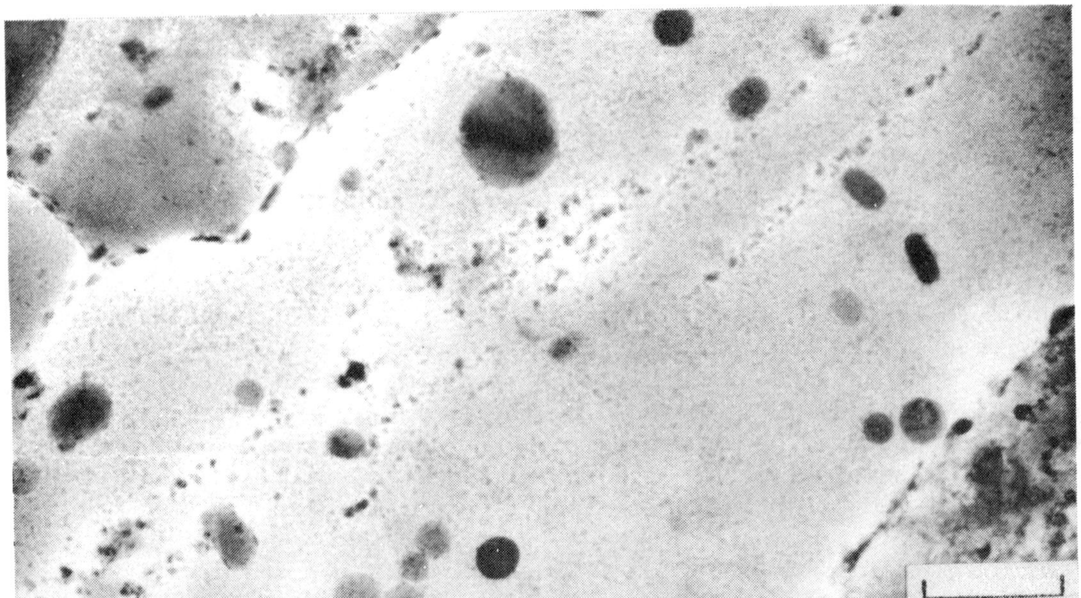

Figure 7. Transmission electron micrograph of alloy A2 (CT91) extrusion in the T7X1 temper.

oxides seen in Figure 7 result from the presence of a thin (50-100A) oxide film originally formed on the surfaces of the individual powder particles. The film is broken down during fabrication of the P/M extrusions, with the result that oxide particles remain strung out in the matrix. The role of oxide particles on strengthening and fracture in these P/M alloys is unclear at present. The strengthening precipitates (based on M phase, $Mg(Cu,Zn,Al)_2$) are barely visible as a fine dispersion throughout the aluminum matrix.

Tensile Properties

Uniaxial tensile properties for the experimental materials are listed in Table III. P/M extrusions in the T7X1 temper showed higher strength (50-60 MPa) with equivalent or slightly lower ductility than the 7050 material. In the T7X2 temper, the P/M alloys had somewhat lower strength than 7050 with better ductility. The tensile data in Table III support the results of previous work which have shown that P/M 7XXX alloys can be produced with superior combinations of strength and ductility compared to I/M alloys (4).

Notched Fatigue Performance

Table IV shows the values of average notched fatigue strength at 10^7 cycles obtained by staircase testing of alloys A1, A2, and A3.* A typical notched fatigue strength previously reported for 7050-T76511 is also shown in Table IV. Comparison of the data reveals that differences in Co content between the "A" alloys had no significant effect on resistance to notched fatigue. Furthermore, all P/M alloys exhibited approximately 40% higher fatigue strength than that reported for the 7050 extrusion.

The superior notched fatigue resistance observed for the P/M extrusions tested is a consequence of the increased strength and homogeneity of deformation of the P/M alloys. A notched specimen undergoing stress-controlled cycling experiences a zone of plasticity at the notch tip (9). The accumulation

TABLE III. TENSILE PROPERTIES OF EXPERIMENTAL ALLOYS

Alloy	Temper	Yield Strength, MPa	Tensile Strength, MPa	Elongation, %	Reduction in Area, %
A1	T7X1	553	594	11.4	26
A2	T7X1	560	598	11.9	29
	T7X2	463	527	13.4	40
A3	T7X1	560	603	10.7	25
B	T7X1	566	605	11.0	26
	T7X2	474	537	14.5	44
C	T7X1	578	632	12.0	21
	T7X2	487	551	14.2	41
7050 (I/M)	T76511	514	576	11.5	--

*In order to obtain larger numbers of data points, three sets of materials with slightly different extrusion conditions were used. These differences in fabrication practice were previously shown to have no appreciable effects on structure or properties of the P/M alloys. All materials were tested in the T7X1 tempers only.

TABLE IV. FATIGUE STRENGTHS (10^7 CYCLES)
P/M EXTRUSIONS

Alloy	% Cobalt	Notched Fatigue Strength, MPa	Number of Specimens
A1	0.0	138	18
A2 (CT91)	0.4	138	17
A3	0.8	144	18
7050 (typical)	--	98	--

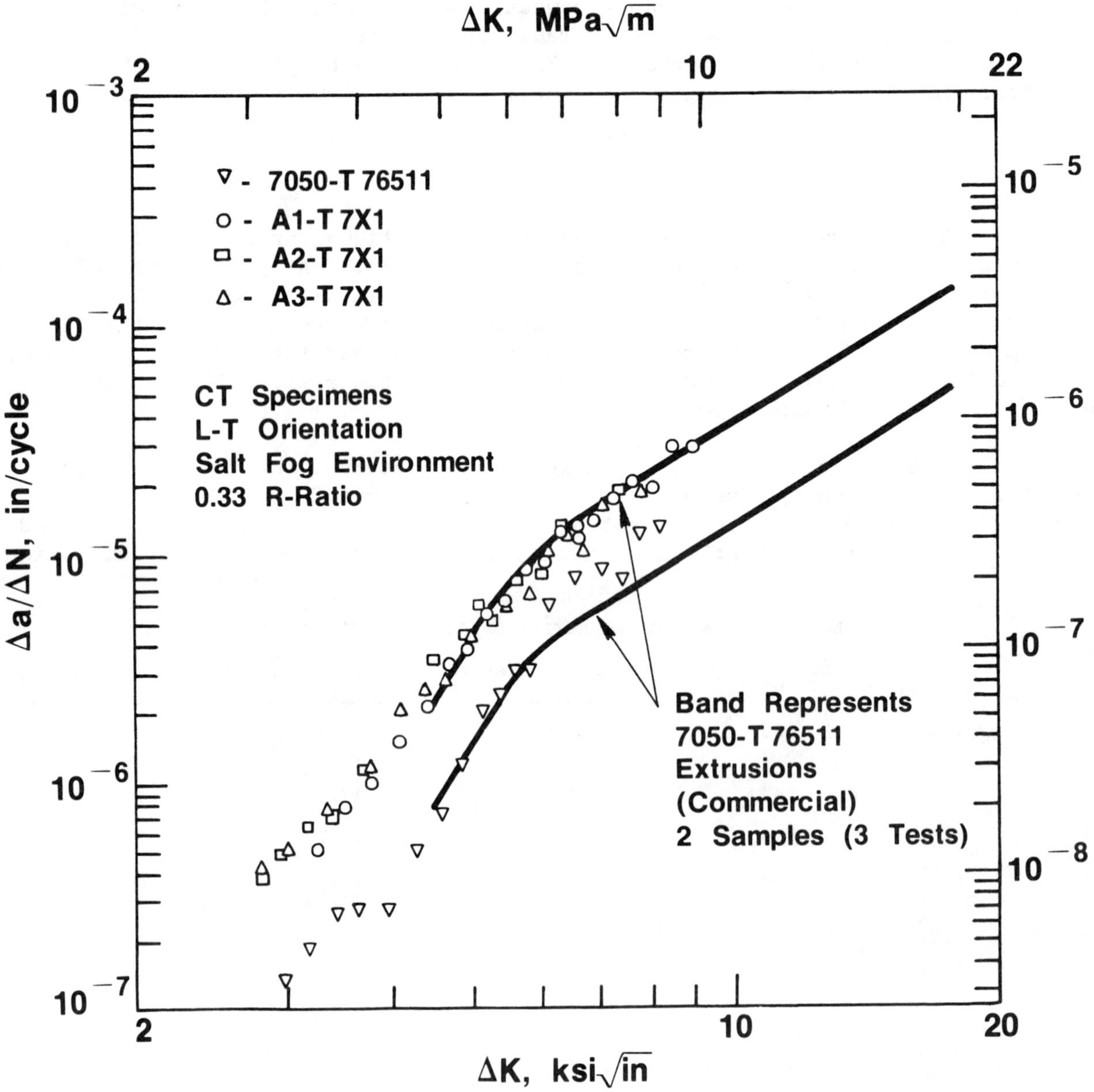

Figure 8. Fatigue crack growth data for alloys A1, A2, and A3 in salt fog atmosphere.

of plastic damage in this zone occurs during fatigue testing until crack initiation takes place. The higher-strength P/M alloys undergo somewhat smaller amounts of plastic strain during each cycle with a resultant increase in total fatigue life before crack initiation. Homogeneous deformation in the P/M materials, promoted by the fine microstructure and the absence of large constituent particles to act as sites for localized stress concentrations, allows for a greater accumulation of plastic strain before cyclic ductility is exhausted and crack initiation occurs.

Fatigue Crack Growth

The results of FCG tests in both salt fog and high humidity air (R.H.>90%) are shown in Figures 8 through 10. The data are plotted as log da/dN (FCG rate) versus log ΔK (cyclic stress intensity factor). Figure 8 shows that varying Co content (alloys A1, A2, and A3) had no significant effects on salt fog FCG resistance of the P/M extrusions in the T7X1 temper. The P/M alloys showed slightly faster crack growth rates than the 7050-T76511 material tested

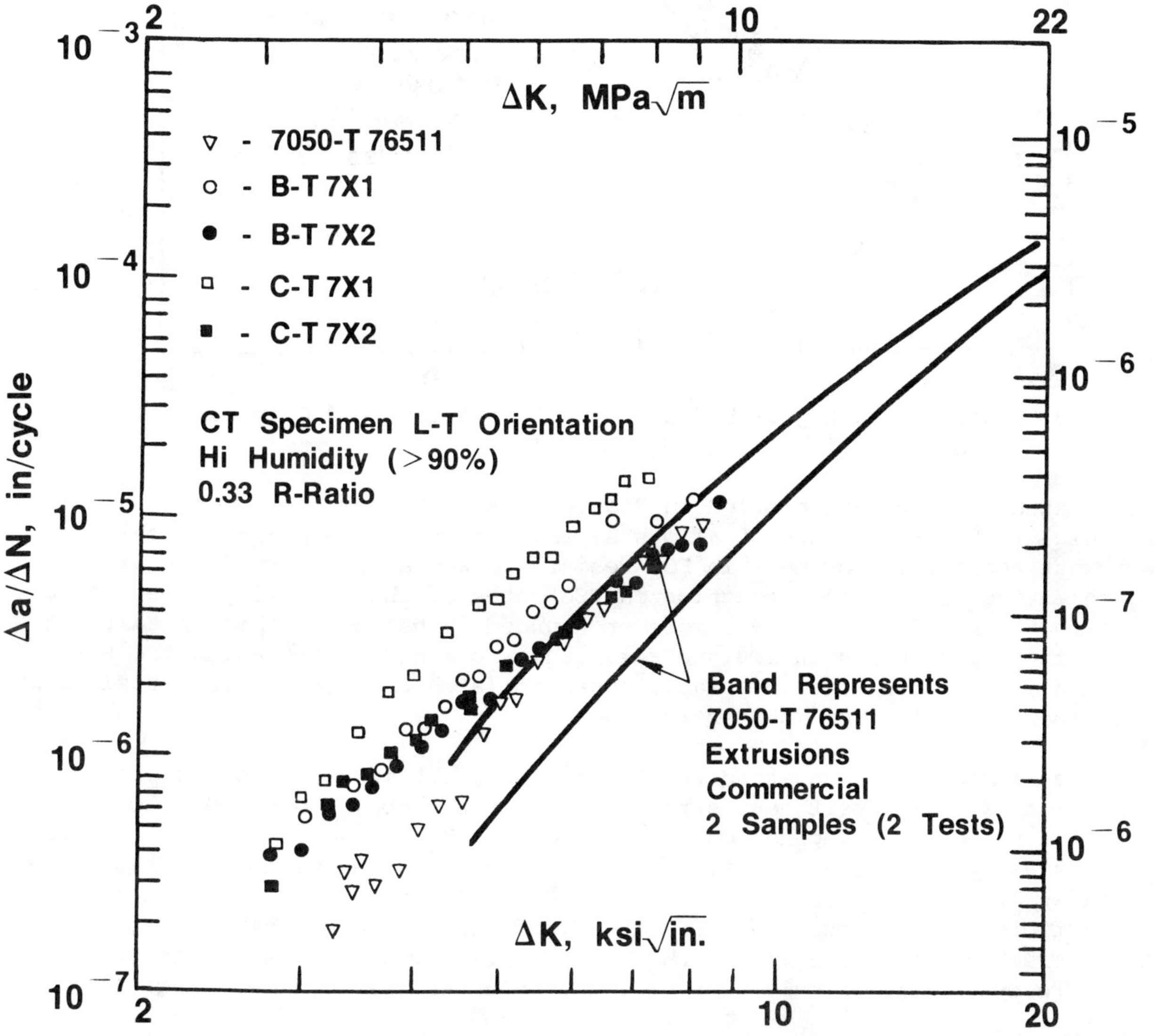

Figure 9. Fatigue crack growth data for Alloys B2 and C in humid air atmosphere.

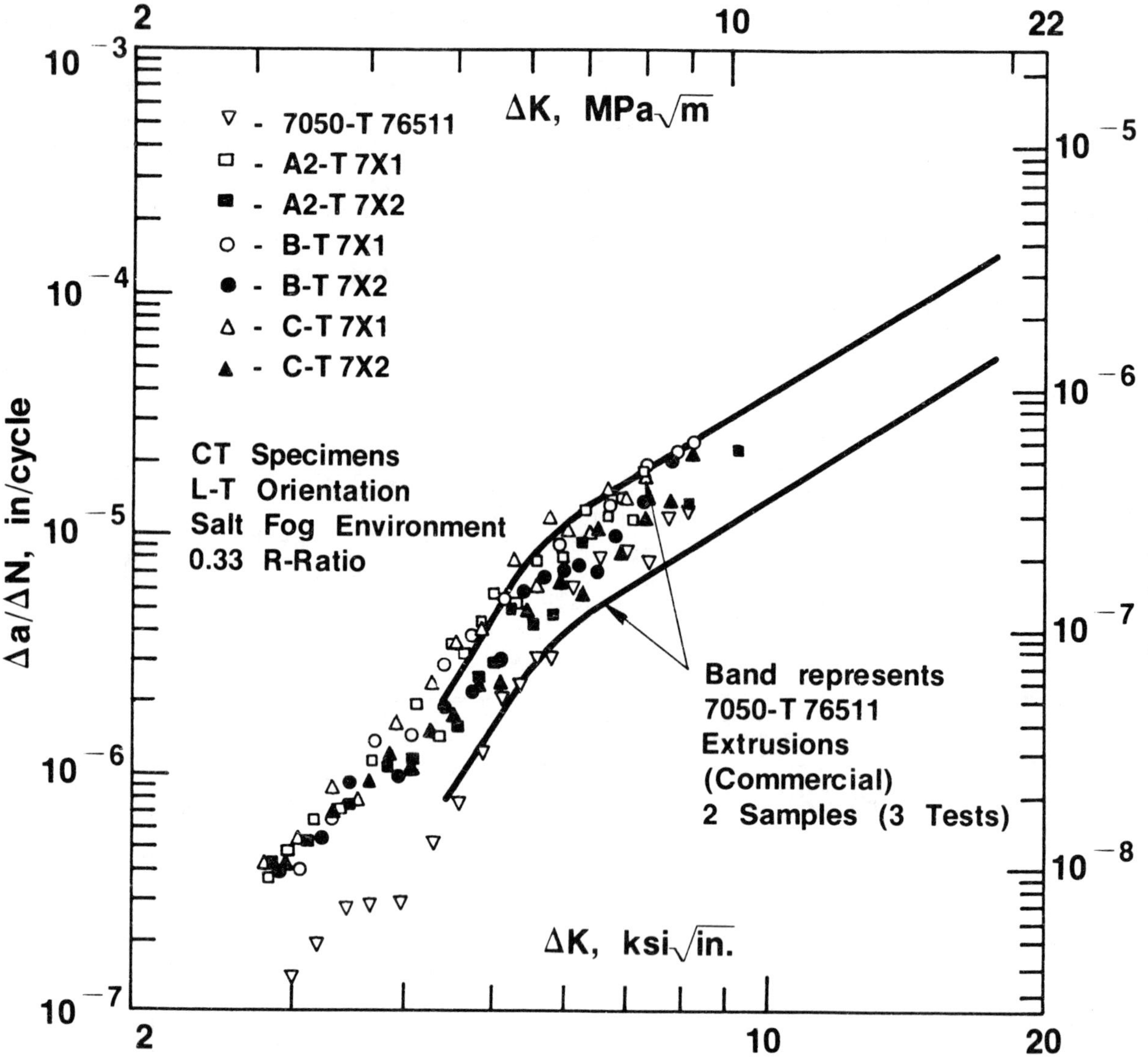

Figure 10. Fatigue crack growth data for alloys A2, B, and C in salt fog atmosphere.

in this study.* The data plotted in Figures 9 and 10 show that overaging was beneficial to FCG resistance of alloys A2 and C in both humid air and salt fog. A somewhat smaller improvement in FCG resistance was also observed for alloy B in both environments. The reduction in FCG rates of the P/M alloys obtained by overaging to the T7X2 temper was more pronounced in humid air than in salt fog. Despite the improvements in fatigue resistance obtained by overaging the P/M extrusions, FCG rates were invariably equal to (at higher ΔK levels) or slightly faster than FCG rates for 7050 extrusions.

In an attempt to understand the FCG results, the monotonic plastic zone size (r_p) relative to pertinent structural features should be taken into account (12). The value of r_p depends on yield strength, K_{max}, and the state of stress (plane strain/plane stress conditions) experienced by the specimen. When r_p is significantly greater than a given microstructural feature, e.g., grain size, interparticle spacing, that feature should have relatively little effect on fatigue crack growth. The microstructure essentially behaves as a continuum, obeying relationships calculated from linear elastic fracture mechanics (12).

*Also shown in Figures 8 through 10 are scatterbands indicative of FCG rates of 7050-T7X511 extruded materials. The widths of these bands are primarily due to the variety of extruded product forms and tempers tested (10,11).

TABLE V. CALCULATED PLANE STRAIN MONOTONIC PLASTIC ZONE SIZES FOR EXPERIMENTAL MATERIALS

ΔK, MPa$\sqrt{m}$	ΔK, (ksi$\sqrt{in.}$)	K_{max}, MPa$\sqrt{m}$	K_{max}, (ksi$\sqrt{in.}$)	R_p (μm), 565 MPa (82 ksi)	R_p (μm), 517 MPa (75 ksi)	R_p (μm), 483 MPa (70 ksi)
2.2	2.0	3.3	3.0	4.8	5.7	6.5
3.3	3.0	5.0	4.5	10.7	12.7	14.7
4.4	4.0	6.6	6.0	19.0	22.8	26.1
5.5	5.0	8.3	7.5	30.0	35.5	40.9
6.6	6.0	9.9	9.0	43.0	51.2	58.8
7.7	7.0	11.6	10.5	58.2	69.6	80.0
8.8	8.0	13.2	12.0	76.0	91.0	104.5
9.9	9.0	14.9	13.5	97.0	115.2	132.0
11.0	10.0	16.5	15.0	119.0	142.2	163.3

565 MPa = approximate yield stress for P/M-T7X1 extrusions.
517 MPa = yield stress for I/M 7050-T76511 extrusions.
483 MPa = approximate yield stress for P/M-T7X2 extrusions.

Table V contains values of r_p for ΔK levels of interest in the present investigation. Comparison of these values to the microstructural features in Figures 5 and 6 indicates that, above 5 MPa$\sqrt{m}$, both I/M and P/M structures are significantly finer than r_p. Both materials act as continuums and exhibit similar FCG rates, regardless of microstructure. A more extensive investigation has also shown that 7XXX I/M alloys are relatively insensitive to microstructural features at intermediate (5-15 MPa$\sqrt{m}$) ΔK levels (13). At lower ΔK levels, where r_p values are smaller, microstructural effects on FCG are possibly more important, particularly in I/M materials. Larger grains (14) or constituent particles (15) may serve to divert cracks from the preferred crack path normal to the stress axis, resulting in the improved low ΔK FCG performance shown in Figures 8-10.

The SEM observations of fatigue fracture surfaces support the above analysis of microstructural effects on FCG resistance. Relatively flat, uniform fracture features were observed in all P/M specimens examined (Figure 11). The 7050 specimens, on the other hand, had extremely rough fracture features as shown in Figure 12. The path of a fatigue crack in a 7050 specimen was very tortuous and irregular compared to that in a P/M specimen. Microstructural features in the I/M material apparently caused localized deviations in the crack path,

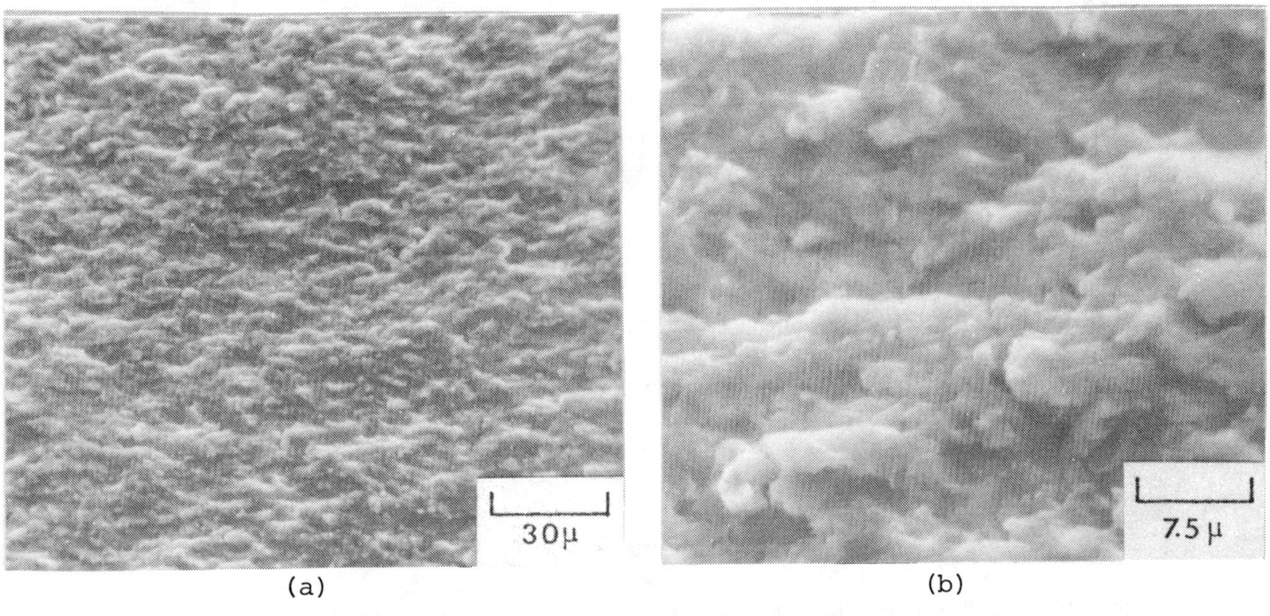

(a) (b)

Figure 11. Fractographs of P/M alloy extrusions tested in salt fog atmosphere, ΔK=7 MPa$\sqrt{m}$: (a) alloy A2, T7X2 temper; (b) alloy B, T7X2 temper.

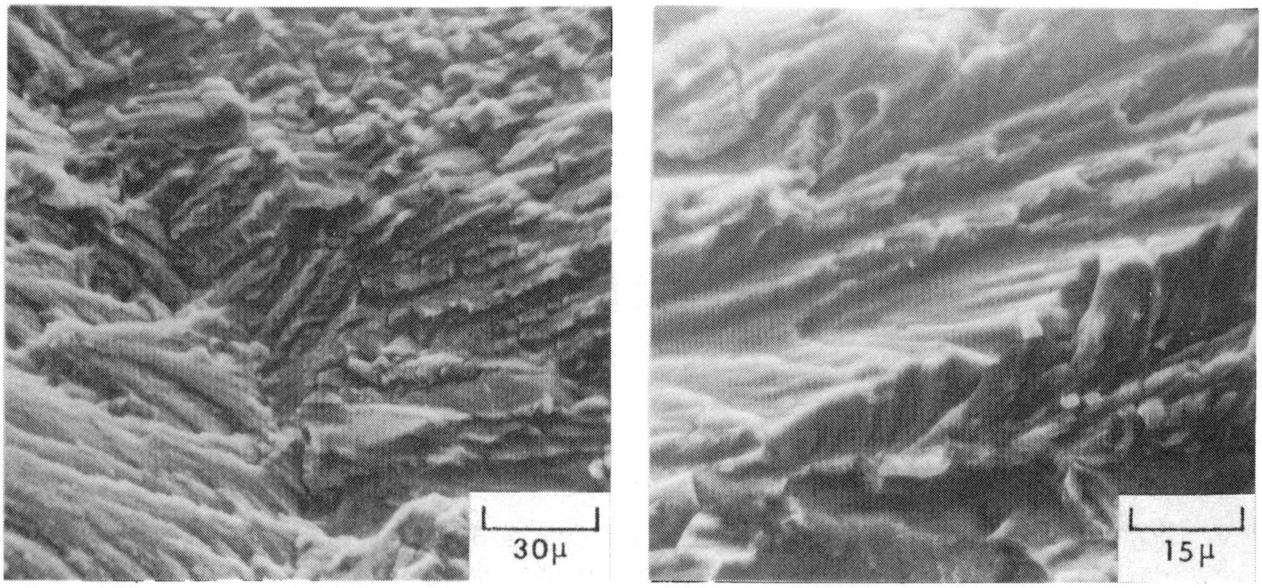

(a) (b)

Figure 12. Fractographs of 7050-T76511 extrusion tested in salt fog atmosphere, $\Delta K=7$ MPa$\sqrt{m}$.

resulting in somewhat better FCG resistance at low ΔK levels. The presence of second-phase particles and a nonhomogeneous grain structure (Figure 6) in the 7050 extrusion probably contributed to this effect. The extremely homogeneous P/M microstructure, however, offered nothing to divert the crack from its macroscopic growth direction and led to somewhat higher measured FCG rates. The increases in FCG resistance of the overaged (T7X2) P/M alloys are probably due, in large part, to an increased resistance to degradation by the aggressive environments used in the present study.

Conclusions

1. Extrusions of 7XXX P/M alloys exhibit improved resistance to fatigue crack initiation in notched ($K_t=3$) tests. The notched fatigue strengths at 10^7 cycles of these alloys are approximately 40% higher than that of I/M 7050-T76511.

2. Fatigue crack growth rates of 7XXX-T7X1 P/M extrusions are slightly higher than those of 7050-T76511 at ΔK levels from 2-10 MPa$\sqrt{m}$ in both salt fog and humid air.

3. Overaging the P/M extrusions to a T7X2 temper increases FCG resistance to approximately that of 7050-T76511 at ΔK levels above 4 MPa$\sqrt{m}$. The effect is more pronounced in humid air than salt fog..

4. The FCG resistance of P/M alloys is primarily controlled by the uniform grain structure, which offers no major microstructural features to divert the fatigue crack from its macroscopic growth direction.

References

1. A. P. Haarr, "Development of Aluminum-Base Alloys, Section III, Contract DA-36-034-ORD-3559RD, AD487764, 1966 May 31.
2. J. A. Nock, Jr., and H. Y. Hunsicker, Journal of Metals, 1963 March, pp 216-224.

3. J. P. Lyle and W. S. Cebulak, "Powder Metallurgy Approach for Control in High Strength Aluminum P/M Products," presented at TMS-AIME Spring Meeting, Philadelphia, PA, 1973 June.

4. W. L. Otto, Jr., "Metallurgical Factors Controlling Structure in High Strength Aluminum P/M Products," AFML TR-76-60, 1976 May.

5. AFML Contract No. F33615-77-C-5174, "Fatigue Resistant Aluminum P/M Alloy Development," Alcoa Laboratories (Contract begun 1977 September).

6. American Society for Testing and Materials, A Tentative Guide for Fatigue Testing and the Statistical Analysis of Fatigue Data, ASTM STP 91-A, Philadelphia, PA, 1958.

7. H. Matyja, B. C. Giessen, and N. J. Grant, Journal of the Institute of Metals, Vol 96, 1968, pp 30-32.

8. T. H. Sanders, Jr., and J. T. Staley, "Review of Fatigue and Fracture Research on High Strength Aluminum Alloys," presented at 1978 ASM Materials Science Seminar, St. Louis, MO, 1978 October 14-15.

9. D. T. Raske, Journal of Testing and Evaluation, Vol 1, 1973, pp 394-404.

10. J. T. Staley, J. E. Jacoby, R. E. Davies, G. E. Nordmark, J. D. Walsh, and F. R. Rudolph, "Aluminum Alloy 7050 Extrusions," AFML TR-76-129, 1977 March.

11. R. E. Davies, G. E. Nordmark, and J. D. Walsh, "Design Mechanical Properties, Fracture Toughness, Fatigue Properties, Exfoliation and Stress-Corrosion Resistance of 7050 Sheet, Plate, Hand Forgings, Die Forgings and Extrusions," Final Report, Contract N00019-72-C-0512, 1975 July.

12. J. Lankford, D. L. Davidson, and T. S. Cook, "Fatigue Crack Tip Plasticity," Cyclic Stress-Strain and Plastic Deformation Aspects of Fatigue Crack Growth, ASTM STP 637, American Society for Testing and Materials, 1977, pp 36-55.

13. W. G. Truckner, J. T. Staley, R. J. Bucci, and A. B. Thakker, "Effects of Microstructure on Fatigue Crack Growth of High-Strength Aluminum Alloys," AFML-TR-76-169, 1976 August.

14. M. E. Fine and R. O. Ritchie, "Fatigue Crack Initiation and Slow Crack Growth," presented at 1978 ASM Materials Science Seminar, Fatigue and Microstructure, St. Louis, MO, 1978 October 14-15.

15. T. H. Sanders, Jr., R. R. Sawtell, J. T. Staley, R. J. Bucci, and A. B. Thakker, "Effect of Microstructure on Fatigue Crack Growth of 7XXX Aluminum Alloys Under Constant Amplitude and Spectrum Loading," Final Report, NADC Contract No. N00019-76-C-0482, 1978 April.

Acknowledgments

The work described in this paper was performed under Contract No. F33615-77-C-5174 for the Air Force Materials Laboratory, Wright-Patterson Air Force Base, Ohio. This investigation was initiated under the direction of Mr. R. R. Sawtell and Mr. J. T. Staley of the Alloy Technology Division, Alcoa Laboratories. The authors are particularly indebted to Mr. W. M. Griffith, who served as contract monitor for AFML, for his assistance during the course of the investigation.

The authors would also like to thank the many workers at Alcoa Laboratories whose efforts contributed to the completion of this work. Discussions with Mr. D. A. Mauney, Dr. R. J. Bucci, and Dr. T. H. Sanders were especially helpful in the preparation of the manuscript.

Fatigue Properties of a Powder Forged

High Strength Aluminum Alloy

H. W. Antes, A. Lawley, and M. J. Koczak

Department of Materials Engineering
Drexel University
Philadelphia, Pa. 19104

ABSTRACT

Atomized aluminum alloy powder was isostatically compacted, sintered, hot-repressed and then warm-forged in plane strain to provide full density material. Visioplasticity was used to establish the exact local deformation in various positions of the forgings. After heat treatment, longitudinal and short-transverse fatigue specimens were excised from regions of the forgings at true local strains (ε) of 0, 0.3, 0.6 and 0.7. Tension-tension axial fatigue tests (R = +0.1) were run to determine the portion of the S-N curves between 60 ksi (414 MPa) and 70 ksi (483 MPa). Weibull analysis was used to establish the median life S-N curves. Fatigue life increased in the longitudinal orientation as the local strain increased as the result of plane strain forging (e.g. @ 60 ksi: $N_{\varepsilon=0} = 45 \times 10^3$, $N_{\varepsilon=0.3} = 65 \times 10^3$, $N_{\varepsilon=0.7} = 300 \times 10^3$ cycles). The fatigue life in the short-transverse direction was relatively unaffected by deformation but was always less than that for the longitudinal orientation. Consequently, fatigue strength anisotropy increased with increasing deformation as a result of enhancement of the longitudinal properties. In constrast, the anisotropy that is observed in conventional ingot metallurgy material is the result of a degradation of the short-transverse properties. Standard metallographic techniques and SEM were used to study the internal structure and fatigue fracture surfaces. The results indicate that the observed fatigue behavior may be explained in terms of fragmented oxide particles, uniformity of aging, localized solute enriched regions and a fine-scale dynamic recovery on the order of 1 to 3μm in size. Work supported by the Air Force Office of Scientific Research under grant #77-3247.

INTRODUCTION

Over the past several years there has been significant interest in processing wrought forms from powder preforms or billets made from metal powders rather than from a cast ingot. It has been found that the amount and mode of deformation has a significant effect on the properties of ferrous powder formed materials (1-9).

Properties and performance of powder metal (P/M) hot formed components have exceeded those of counter-parts made from conventional ingot metallurgy (I/M) processes (4). The superior performance of P/M formed components has been attributed to the greater degree of homogeneity achievable with powder products. However, the superior properties are achieved only when there is sufficient lateral flow during deformation processing to eliminate porosity, fragment powder particle surface oxides, and generate adequate nascent surfaces to develop high integrity metallurgical bonds.

Recent work on wrought high strength aluminum alloys that were made by powder metal processing has indicated that the superior properties realized in P/M processed ferrous materials can be achieved in aluminum alloys too (10-13). It was found that lateral flow has a beneficial effect on the fatigue properties of P/M formed aluminum alloys of equivalent composition to I/M 2014 and 6061 (14). Similar results were also found for aluminum P/M formed material that was equivalent to the high strength I/M 7075 alloy. When comparisons were made in notched axial fatigue testing an enhancement of endurance limit of up to 40% was realized for the P/M formed material (10,15).

In all the work cited above on documenting the superior properties of P/M processed wrought forms there is general agreement on the need for lateral flow in achieving these property levels. Howerver, there were no studies where the precise amount of local deformation or flow was determined. Therefore, it was the purpose of this investigation to determine the effect of microscopic lateral flow on the structure and resulting axial fatigue properties of a warm forged high strength aluminum alloy powder material, when tested in air and a salt-fog environment.

EXPERIMENTAL PROCEDURE

Material Preparation

1. Aluminum Alloy Powder:

The aluminum alloy powder used in this study was made by air atomizing liquid metal. The chemical composition of the powder together with the specifications of two conventional ingot metallurgy (I/M) commercial aluminum alloys are shown in Table I. It can be seen from this table that alloy 7475 is a higher purity form of alloy 7075 (i.e. lower Si and Fe) while the powder is even higher purity (with respect to Si and Fe impurities) than 7475. However, the powder has cobalt rather than chromium as an alloying addition.

The atomized powder had an average particle diameter (APD) of approximately 13 microns. Typical powder size distributions are shown in Tables II and III.

2. Billet Fabrication:

Billet fabrication from powder consisted of: 1-cold compaction, 2-canning and 3-hot compaction. An excess of 145 pounds (65.8 kg) of powder was cold compacted isostatically at a pressure of 30 ksi (206.9 MPa) into a cylinder approximately 8 in. (20.3 cm) diameter x 43 in. (109.2 cm) long. A wet bag process was used and produced a compact density approximately 70% of theoretical. The compact was machined to a weight of 145 pounds (65.8 kg) and encapsulated in a welded 3003 aluminum alloy can 8 1/8 in. (20.6 cm) OD x 1/4 in. (0.64 cm) wall thickness. The canned billet was evacuated to less than 35 microns Hg, preheated to 970°F (521°C) under vacuum and sealed to maintain the vacuum for hot compacting.

Hot compacting was accomplished in a preheated trap-die using a compacting pressure of 90 ksi (620.6 MPa) with a dwell time of 10 minutes under pressure. Hot compacting provided a scalped billet approximately 29 in. (74 cm) long with a diameter of 7 15/16 in. (20.2 cm). Density measurements for the hot compacted material indicate that 99.9+% of the theoretical density was achieved (11).

3. Forging Preform Location:

Preforms were cut from the hot-compacted billet to provide plane strain forged upset ratios of 0, 1.5:1 and 2:1. These upset ratios provide average macroscopic true strains (ε) of 0, 0.4 and 0.7 (where $\varepsilon = -\ln[h_f/h_o]$). The dimensions after forging were 2 in. x 5 in. x 3 in. high (5.1 cm x 12.7 cm x 7.6 cm high). The preform sizes and upset ratios are given in Table IV. In order to make the preforms, the hot compacted billet was cut into 3 in., 4 1/2 in., and 6 in. high cylinders (7.6 cm, 11.4 cm and 15.2 cm high). The height of all preforms was parallel to the axis of symmetry of the billet, which was also the direction of axial compaction for the hot compacted powder billet. All of the six preform surfaces were milled to provide perpendicular faces with the dimensions accurate to within 0.005 in. (0.013 cm).

4. Plane Strain Forging:

a) Visioplasticity Studies. Frictionless compression produces a homogeneous deformation, i.e. all elements are deformed an identical amount and the localized strain is the same as the macro or overall strain. However, if friction is involved local deformation will vary from element to element or a nonhomogeneous deformation will exist.

All metal working operations involve friction and, therefore, nonhomogeneous deformation. One method for determining the local deformation involves the use of grid markings on sectioned billets or preforms. Measurement of the grid spacing after deformation permits calculating the local strain or deformation. This approach is known as visioplasticity.

The plane strain forging upset ratios of 0, 1.5/1 and 2/1 used in this study provided overall true upset strains (ε) of 0, 0.4 and 0.7 ($\varepsilon = -\ln[z_f/z_o]$). The actual preform sizes and finished forging size are shown in Figure 1. All forging was accomplished by an upsetting or reduction of the "z" dimension and an extension in the "y" direction. A confined die forging process was used that maintained the "x" dimension constant at 2 in. (5.1 cm). Therefore, $\varepsilon_z = -\varepsilon_y$ and $\varepsilon_x = 0$, a condition of plane strain.

In order to determine the local deformation, laminated preforms were built-up of two 1/2 in. (1.27 cm) thick slabs and one 1 in. (2.54 cm) thick slab as shown in Figure 2. Grids, 0.394 in. (1 cm) square, were scribed on both faces of the "B"slab (at the A-B and B-C interfaces). A tool maker's height gage with a carbide point was used to scribe the grid on to the surfaces. Inspection with a toolmaker's microscope indicated the grids were accurate to ±0.002 in. (±0.005 cm).

The material used for the laminated preforms was 7075 aluminum alloy plate. Accurate positioning for forging was accomplished by dropping the laminated preforms into the die cavity between transite end spacers. The end spacers were removed before forging.

In order to prevent the preform segments from welding and sticking together during forging, each segment was sprayed with a graphite emulsion from an aerosol dispenser. No sticking of segments was experienced when this technique was used.

All forging was done in a 500 ton (4.45 MN) hydraulic press with a ram speed of 100 in./min. (254 cm/min.). A trap die was used with simultaneous motion of both top and bottom punches during the deformation. Die temperature was maintained at 500°F (250°C) while billets were preheated to 550°F (228°C) for the forging operation. The forging die and punches were sprayed with aerosols of graphite and fine copper particles for lubrication.

Incremental deformation tests were made in order to be assured that plane strain conditions were maintained throughout the forging process. Both the 4 1/2 in. (11.43 cm) high and the 6 in. (15.24 cm) high preforms were upset to heights of 3 1/2, 3 1/8 and 3 in. (8.9, 7.9 and 7.6 cm). The sides of all segments remained planar and parallel for all incremental upsets, see Figure 3.

After forging the segments were separated and the grid spacings were measured at a magnification of 2.5X so that the true local strains could be calculated to an accurace of $\varepsilon = 0.005$. True strains were reported to a value of $\varepsilon = 0.01$. The deformed grid patterns are shown in Figure 4. Calculated local true strains (deformations) are given in Figure 5 for the center of each of the 1/2 in. x 1/2 in. (1.27 cm x 1.27 cm) square areas indicated. Only one quadrant is shown for each upset since deformation was symmetrical. The significant differences in the amount of local deformations in the various regions of the forging are shown for the 1.5:1 upset by the three dimensional plot in Figure 6.

b) Forging Powder Preforms. The identical procedure was used for forging both the segmented preforms and the powder preforms. In addition, the die temperature, preform size, preheat temperature, lubrication conditions and the use of transite location spacers were all the same in both cases. Therefore, it was assumed that the local strain distribution in the powder forged material was the same as that determined for the segmented billet forgings.

5. Heat Treatment:

Forgings were solution treated in an air circulating furnace at a temperature of 910 ±2°F (488 ±1°C) for two hours at temperature and then quenched into ice water. The aging treatment that followed consisted of three stages:

1. Five (5) days at room temperature 70°F. (21°C)

2. Twenty-four (24) hours at 250°F. (121°C)
3. Four (4) hours at 325°F. (163°C)

After heat treatment, the hardness was uniform on the y-z face at a value of 92 $\pm 1R_b$ and was the same for all three upset ratios.

6. Specimen Location:

The dimensions of the axial fatigue specimens are shown in Figure 7. Local deformation strains of ε = 0, 0.3, 0.6 and 0.7 were selected for study. Specimen orientation (with reference to Figure 1) was positioned as follows:

Longitudinal - specimen axis parallel to "y" direction
Short Transverse - specimen axis parallel to "z" direction

The center of the gage length or reduced section was located in the center of the zone shown in Table V. It should be noted that a local true strain deformation of 0.6 was tested from both the 1.5:1 and the 2:1 upset material.

7. Specimen Preparation:

The fatigue specimens were turned on a lathe and polished through a series of abrasive papers (250, 325, 400 and 600 grit). Mechanically polished specimens were electrolytically polished with a solution of 7 parts by volume of glacial acetic acid and 2 parts perchloric acid. The solution was maintained at 0°C with an ice water bath. The cathode was a pure aluminum ring, 2 in. (5.08 cm) ID x 3 in. (7.62 cm) OD x 1/2 in. (1.27 cm) high. The specimen (anode) was placed in the center of the cathode ring with the center of the gage length of the specimen at the midheight of the cathode ring. After polishing, specimens were washed immediately in distilled water and dried with a blast of compressed air. All electrolytically polished specimens were examined with a binocular microscope at 20X to insure that no scratches remained and that there was no pitting on the surface of the specimen owing to the electrolytic polishing.

Minimum diametrical measurements were made at two points 90° apart. An average of the two readings was used to calculate the minimum cross sectional area.

Axial Fatigue Tests in Air and Salt Fog

The axial fatigue tests were made at R = +0.1 ($R = \sigma_{min}/\sigma_{max}$) on a Sonntag testing machine for both air and 3 1/2% NaCl solution salt-fog environments. This is a constant load type testing machine with an automatic adjustment devise for maintaining constant preload. This insures no stress relaxation, owing to deformation of the specimen, during the test.

Maximum stresses for testing were selected at 60, 65 and 70 ksi (414, 448 and 483 MPa). Three to five specimens were tested at each stress level with the lower number of specimens being used at the highest stress. Median value fatigue life was determined from Weibull statistics (16). All cyclic loading was at 30 Hertz.

The apparatus used for making the tests in the salt-fog environment is shown in Figures 8 and 9. In Figure 8 the testing machine is "A", the salt solution reservoir is "B" with the filtered air line shown as "C". In Figure 9a, the plexiglas fogging nozzle is "D", the specimen is "E", one half of the lower shoulder grip is "F" and the rubber chamber in the folded down position is "G". The seal between the rubber chamber and the fogging nozzle was accomplished by burning a pin hole in the wall of the rubber chamber and then stretching this hole over the small spherical protuberance "d" on the figgong nozzle. The seals at the top and bottom of the rubber chamber were accomplished by simply stretching the rubber over flanges and clamping with hose clamps as shown in Figure 9b. In this figure the incoming air line is "H", 3 1/2% NaCl solution line is "I" and the return or drain line is "J". The construction and operation of the salt-fog nozzle was in compliance with ASTM specifications on salt-fog testing.

In the salt-fog tests, the fog was emitted into the specimen chamber approximately 3 to 5 seconds before the cyclic loading began. The salt-fog was continuous during the entire test. Upon fracture the salt-fog was turned off immediately, specimen removed from the apparatus, fracture surfaces flushed with water for 30 seconds, dried with an air blast and then the specimen was stored in a desiccator.

RESULTS

Fatigue Properties in Air

The median fatigue life as a function of maximum stress is plotted in Figure 10. It can be seen from this figure that for $\varepsilon = 0$ there is very little difference in the curves for the longitudinal and short-transverse orientations; indicating reasonably isotropic behavior. However, as ε is increased the span between the two orientations increases; indicating anisotropy increases with increasing local deformation. It can be seen that there is a significant enhancement in the longitudinal fatigue properties as the local deformation (ε) is increased. There is some shifting of the short transverse curve but the net change is small in comparison to the change for the longitudinal orientation.

No distinction could be made between the fatigue properties of the $\varepsilon = 0.6$ specimens from the 1.5:1 and the 2:1 upsets. Therefore, properties seem to be independent of the macro deformation and depend only on the local deformation.

Fatigue Properites in Salt-Fog

The data for median life in salt fog are plotted in Figure 11. The fatigue life is very short in a salt-fog environment. There is very little change in the longitudinal properties as the local deformation is increased. However, there is a decrease in the short-transverse properties (especially at the higher stress levels) with increasing local deformation (ε). Once again the anisotropy is seen to increase with increasing local deformation. But, in salt-fog the anisotropy increase is caused by a degradation of the short-transverse property.

In summary, fatigue properties in air are enhanced in the longitudinal orientation by increasing local deformation. There is a very serious degradation of the fatigue properties when tested in a 3 1/2% NaCl solution salt-fog. The degrading effect of the salt-fog overshadows the beneficial effect that increasing local deformation had when the material was tested in air. Although there was no effect of local deformation on the longitudinal properties in salt-fog, the short transverse properties in salt-fog decreased with increasing deformation.

Metallographic Studies

There are three types of second phase particles in high strength 7XXX series aluminum alloys (17).

1. Constituent Particles - large brittle (∿1 to 30μm) insoluble phases of Fe, Si, Al plus solute elements (e.g. Al_7Cu_2Fe, Mg_2Si and in the case of powder alloys significant oxides) and some partially soluble phases (e.g. S[Al_2CuMg])

2. Smaller Dispersoids - (0.2 to 0.5μm); these are precipitates that form at temperatures above ∿780°F (425°C), (e.g. $Al_{12}Mg_2Cr$, $ZrAl_3$ Co_2Al_9)

3. Hardening Precipitates - fine (0.002 to 0.01μm) containing Zn, Mg and Cu that form as the result of quenching and aging (e.g. η': Mg(Zn_2Al,Cu)]

The general sequench that produces η is:

Supersaturated S.S. → GP zones → η' → η

The transition phase (η') nucleates directly on dislocations and on subgrain boundaries. The equilibrium phase (η) nucleates on grain boundaries.

In order to illustrate the morphology of the grain structure and the distribution of phases at various stages of processing a series of three dimensional photomicrographs are shown in Figures 12 through 14. The orientation in these figures is the same as that shown in Figure 1. It should be noted that the fracture plane for longitudinal specimens is essentially the x-z plane (perpendicular to the y axis) while for the short transverse specimens it is the x-y plane (perpendicular to the z axis).

The structure of the hot compacted material is shown in Figure 12. There is gross precipitation in this structure owing to the slow cooling after hot compacting. Darker etching constituents at the grain boundaries are oxides along with possibly some "S" phase (Al_2CuMg) and probably minor amounts of Al_7Cu_2Fe plus Mg_2Si (owing to the low Fe and Si levels of the alloy). Furthermore, η($MgZn_2$) will appear as platelets along the grain boundaries. Within the grains the precipitate is predominantly η($MgZn_2$) with the dispersoids of Co_2Al_9 and internal oxides. It will be shown later that the Co_2Al_9 phase tends to be associated with triple points of the subgrain boundary structure. Although it is not vivid, careful scrutiny will reveal an equi-axed grain structure. The grains in this material were originally the individual powder particles.

The structure after solution treating and aging is shown in Figure 13. It can be seen in this figure that there are subgrain boundaries within the prior powder particles, which are now the grains. The grains are reasonably equiaxed with a very slight indication of elongation or flattening as a result of the axial compression in the "z" direction during hot compaction. The visible second phases present are predominantly at the grain boundaries and are probably oxides, η($MgZn_2Al_2Cu$) and possibly some S (Al_2CuMg) that did not dissolve during solution treatment.

The structure that results when the hot compacted material is forged and then solution treated and aged is shown in Figure 14. The deformation of the grains is clearly evident. Furthermore, the subgrains are much smaller after forging. The inclined angle of the elongated grains is the result of shearing during the forging operation. Once again the visible second phases present are predominantly at the grain boundaries and are the same as described for Figure 13. It should be noted that the structure of powder forged material is significantly finer than commercially processed (CP) 7XXX alloys (18).

Detailed metallographic studies were made of the material tested in air. The microstructures of hot compacted ($\varepsilon = 0$) specimens in the longitudinal and short-transverse orientations are shown in Figures 15 and 16. The equivalent microstructures for $\varepsilon = 0.7$ material are shown in Figures 17 and 18. The fracture mode for the $\varepsilon = 0$ material was mixed transgranular and intergranular for both the longitudinal and short-transverse orientations, but with a greater proportion of intergranular fracture in the short-transverse orientation. The material that received a local deformation of $\varepsilon = 0.7$ had a fracture mode that was transgranular for the longitudinal orientation but mixed transgranular and intergranular for the short-transverse orientation.

Another important observation can be made by comparing the polarized light microstructures in Figures 15, 16, 17 and 18. It can be seen that precipitation and second phase material is more concentrated in the grain boundary regions of the unworked ($\varepsilon = 0$) specimen, while in the $\varepsilon = 0.7$ specimen the precipitates and second phases are more homogeneously dispersed throughout the structure.

A similar metallographic study was made of the S-N specimens tested in the salt-fog environment. The modes of fracture in salt-fog were similar to those observed in air. However, extensive secondary cracking was observed in the salt-fog tested specimens for all conditions of local deformation and specimen orientation except for $\varepsilon = 0.7$ longitudinal specimens. Crack initiation takes place at grain boundaries. Crack propagation is a mixed mode in the short-transverse orientation, but appears to be mostly intergranular. In the $\varepsilon = 0.7$ short-transverse specimen, the crack appears to avoid the grains that exhibit little substructure (lighter grains). However, in the $\varepsilon = 0.7$ longitudinal specimen the primary crack cuts right through the grains that exhibited little substructure while on its transgranular path.

SEM Fractography

Scanning electron microscopy (SEM) studies were made from 20X to 10,000X. Most of the fractures exhibited a single fatigue initiation site for the specimens tested in air. However, there was a noticeable number of fractures with multiple initiation sites (2 to 3 sites maximum). The fraction of specimens with multiple initiation sites (MIS) varied with σ_{max} in the following manner:

σ_{max} = 70 ksi (483 MPa) 25% with MIS

σ_{max} = 65 ksi (448 MPa) 5% with MIS

σ_{max} = 60 ksi (414 MPa) 1% with MIS

Fatigue fracture initiation was very different in salt-fog by comparison to that in air. Virtually every specimen had multiple initiation sites almost completely around the outer edge of the fracture. A comparison between air and salt-fog fracture surfaces can be seen in Figure 19.

The extent of Stage I is illustrated in Figure 20 as "A". This figure illustrates the fact that the extent of Stage I increases as the maximum stress decreases. In many cases at the higher stress level Stage I was not detected. The area immediately below the bracketed region "A" is sometimes designated as Stage I also and is attributed to be formed principally by severe deformation by slip (19,20). However, this region is probably more accurately described as a transition between Stage I and Stage II fatigue.

Longitudinal and short-transverse fracture surfaces at the edge of the specimens (fracture origin) are shown for $\varepsilon = 0$ in Figure 21 and for $\varepsilon = 0.7$ in Figure 22. The coarse texture in all four fractographs indicates that the fracture progressed along many paths. The regions in these fractographs are marked as follows:

A - Narrow plateaus
B - Smooth textured zones
C - Tear ridges

It can be seen from these figures that all four specimens exhibited similar fracture characteristics but the relative size and texture of the characteristics varied. In particular, in Figure 22a the $\varepsilon = 0.7$ - longitudinal specimen exhibited much smaller and finer characteristics than the others. The coarseness of the various structures can be rationalized by referring to Figures 4 and 5 where the "y" plane represents the structure for the fracture plane of longitudinal specimens and the "z" plane for short-transverse specimens. It should be noted that no fatigue striations were observed in the above fracture areas. There were many particles visible in all the fracture surfaces. This adds evidence that the fatigue cracks follow the heavy precipitation regions.

Examples of the fracture surfaces at the origin for specimens tested in a salt-fog are shown in Figure 23. There were several important differences in the salt-fog fractures. First, no Stage I regions was observed for any of the specimens. Second, at each of the multiple initiation points there was evidence of corrosion products that appear to be concentrated at the grain boundaries. Third, most of the fracture was composed of large smooth plateau regions.

The structures of Stage II air fatigue fractures (from the center of the fatigue fracture regions) are shown in Figure 24. In all cases the fatigue crack propagates on multiple plateaus that are at different elevations. The plateaus are joined either by walls that have striations too, or tear ridges. It should be noted that there are a large number of spherical type particles in the fractures shown in Figure 24. Nondispersive x-ray analysis of these particles indicated they were enriched in cobalt. Therefore, they are probably the Co_2Al_9 dispersoids. Some of these regions are indicated as "Co" in Figures 24. The results of analyzing the smaller particles marked "P", indicated essentially the same composition as the matrix and therefore they are probably either oxides or Co_2Al_9 that are too small to give any enhanced Co counts. It was found that Co_2Al_9 particles were at the edges of tear ridges or at the walls with striations, while the smaller particles are distributed over the striated plateaus. These smaller particles do not seem to have any significant affect on the striation spacing.

Fracture surfaces of corrosion-fatigue (salt-fog) specimens are shown in Figure 25. Multiple plateaus are observed again but some plateaus have ductile striations (as when tested in air) while others have brittle striations (caused by the salt-fog environment). The two types of striations are the result of differences in orientation from grain to grain.

Thin Film TEM

Thin film transmission electron microscopy studies were made for $\varepsilon = 0.7$ in the heat treated and solution treated conditions and for $\varepsilon = 0$ in the heat treated condition. The structures are shown in Figures 26 through 28.

In Figure 26 the structure of the $\varepsilon = 0.7$, solution treated and quenched structure is shown. The black (and white) circular precipitates are the Co_2Al_9 phase. Most of these particles are within the grains at subgrain boundaries and near subgrain triple points. The subgrain boundaries are free of precipitate other than the Co_2Al_9. From these pictures it is evident that the subgrains exist in the solution treated condition although they cannot be revealed in conventional light microscopy. Some of the main grain boundaries are relatively free of second phases while others are decorated with almost continuous undissolved second phases. These second phases are probably mostly oxides plus possibly some small amounts of $S(Al_2CuMg)$ that did not dissolve on solutionizing. In general, the structure appears to be a relatively homogeneous solid solution of the elements that contribute to the age hardening process.

The heat treated structure of $\varepsilon = 0.7$ is shown in Figure 27. The structure consists of a homogeneous distribution of GP zones within the subgrains with a zone of η' about 0.3μm wide at some of the subgrain boundary regions. Platelets of η(0.05μm long) are at other subgrain boundaries. Larger precipitates of η(∿0.2μm) are at the main grain boundaries in a closely spaced but discontinuous elipsoid form. Of course, the undissolved Co_2Al_9 spherical particles (up to 0.3μm) were carried over from the solution treated condition as were the oxides in some of the grain boundaries. A similar structure was observed for the $\varepsilon = 0$ heat treated material as shown in Figure 28. However, the Co_2Al_9 particles were somewhat larger (∿0.7μm) in the $\varepsilon = 0$ material.

DISCUSSION

Local deformation or flow in forming a hot compacted P/M high strength aluminum alloy has both beneficial and detrimental effects. Increasing the amount of flow increases the longitudinal axial fatigue properties for materials tested in air. However, when fatigue tests are carried out in a salt-fog environment increasing the amount of flow causes a slight decrease in the short transverse properties. Therefore, increasing the amount of deformation during forming increases the amount of anisotropy in the fatigue behavior. In air, the beneficial effect of enhancement of longitudinal properties is responsible for the increase in anisotropy. This is very much in contrast to what occurs in I/M wrought products. Anisotropy of I/M wrought products increases with increasing flow or deformation too, but this is caused by a degradation of the short transverse property for material tested in air.

Certain basic structural changes occur when a metal is subjected to a cyclic stress and these have been classified into the following categories (21):

1. Crack initiation - early fatigue damage that can occur in the first 10% of the life. This damage can be removed by a thermal anneal.

2. Slip-band crack growth - Stage I crack growth - deepening of initial crack on planes of high shear stress.

3. Crack growth on planes of high tensile stress - Stage II crack growth - growth of a well defined crack in a direction normal to maximum tensile stress.

4. Ultimate ductile fracture - fatigue crack long enough so that remaining cross section will not support the applied maximum load.

The relative proportion of each of the above categories involved in the total cycles to failures depends on the material in question and the operating or test conditions.

Some preliminary results of cobalt additions and qualitative deformation effects on the fatigue properties of 7XXX P/M forged aluminum have been presented by Rafalin, et al. (22). It was found in this study that a single upset (called A processing) had little effect on enhancing the fatigue properties for cobalt additions of 0, 0.4 and 0.8%. However, with a triple upset in three mutually perpendicular directions (ABC processing) there was a significant enhancement in longitudinal fatigue properties. The amount of improvement in properties increased as the cobalt content increased. At the time of this presentation detailed fractographic studies were in progress and no correlation was made between structure and properties, nor among performance and the various stages of fatigue listed above. Fatigue specimen geometry, preparation and test conditions used by Rafalin were identical to those used in this work, and his results definitely indicate some beneficial effect from sufficient deformation during forming. His fatigue properties for the ABC process are less than those for $\varepsilon = 0.7$, indicating that the ABC process is not optimum.

The fatigue properties reported in this study have Stage I or crack initiation as the controlling factor,owing to the size and geometry of the specimens used. Therefore, any observed effect must be related principally to Stage I or crack initiation. Thus the enhancement in longitudinal fatigue properties with increasing deformation must be related to an increased resistance to crack initiation for material tested in air. Furthermore, it has been found that crack growth rate in air (Stage II) increases with increasing deformation (23). Therefore, any increase in fatigue life must be associated with an increased resistance to crack initiation.

Microstructural studies revealed that increased deformation produced a finer subgrain structure and a more uniform precipitation during heat treatment (Figures 13-18). Both smaller grains (or subgrains) and more uniform precipitation would promote finer more homogeneous slip. This condition would make it more difficult for Stage I to occur since crack initiation is dependent on non-uniform localized slip on a few planes of high shear stress. The beneficial effects of fine grain size and homogeneous deformation on improving fatigue properties have been verified by several investigators (24-26).

The sensitivity of the fatigue properties to crack initiation is very much in evidence for fatigue tests carried out in the salt-fog environment. Crack initiation occurs at structural weak points. The aggressive corrosive action of the salt-fog creates weak points rapidly by attacking the grain boundaries. Therefore, crack initiation occurs in a relatively few number of cycles in salt-fog. Even though crack propagation in salt-fog is retarded at high stress levels by increased deformation (23), this beneficial effect cannot be realized owing to the initiation controlling factor of the specimen geometry. Therefore, the fatigue properties are significantly less in salt-fog than in air.

The effect of deformation and environment on the fatigue properties of the P/M forged material would take on added significance if a comparison could be made with some standard or base type material. Buzolits in his study of fatigue properties of I/M 7075 and 7475 used identical specimen preparation and testing techniques as those used in this study (27). Therefore, a direct comparison can be made with the results of this study and that of Buzolits. In Figure 29 the fatigue properties in air of the I/M material are shown in the upper part of the plot. It can be seen that the higher purity (lower Fe and Si, see Table I) 7475 alloy has marginally higher longitudinal properties and significantly higher short transverse properties than the 7075 alloy. The longitudinal and short transverse properties of the P/M material that received a deformation of $\varepsilon = 0.7$ are significantly greater than the properties for the I/M 7475 alloy. In addition, the P/M material in the hot compacted condition ($\varepsilon = 0$) without any deformation has as good or better properties than the I/M material. The short transverse properties of the P/M material are as good as, or superior to, the longitudinal properties of the I/M alloys.

The fatigue properties of I/M and P/M materials in the salt-fog environment are shown in Figure 30. The serious degradation of fatigue properties caused by salt-fog on the P/M material was also realized with the I/M alloys. Once again the P/M material had superior properties to the I/M alloys with the short-transverse P/M properties being as good as the longitudinal properties of the I/M alloys. A direct comparison of the higher fatigue property I/M alloy (7475) and the better P/M material ($\varepsilon = 0.7$) is shown in Figure 31 for air and salt-fog environments. In air at σ_{max} = 60 ksi (414 MPa) the P/M material has a longitudinal median fatigue life more than seven times that of the I/M 7475.

CONCLUSIONS

The following conclusions are made as a result of this study:

1. Increasing the local deformation increases the longitudinal fatigue properties but has little effect on the short-transverse fatigue properties in air.

2. Testing in a 3 1/2% NaCl salt-fog causes a tremendous reduction in fatigue properties in both the longitudinal and short-transverse orientation. This degradation in properties is the result of rapid corrosive attack by the salt-fog on the grain boundaries, creating many local weak spots that greatly facilitates crack initiation. No Stage I region was detected on specimens tested in salt-fog but all specimens had many initiation sites.

3. In salt-fog, longitudinal fatigue properties are not sensitive to local deformation but short transverse fatigue properties decrease with increasing local deformation.

4. Anisotropy in fatigue properties increases with increasing local deformation. In air, this is caused by an enhancement of longitudinal properties with increasing deformation; while in salt-fog, there is a degradation of short transverse properties with increasing deformation.

5. The enhancement of longitudinal fatigue properties in air with increasing local deformation is the result of greater resistance to crack initiation (Stage I); caused by smaller subgrain size and uniform precipitation which promotes more homogeneous slip.

6. The mode of fracture in Stage II is a mixture of intergranular and transgranular fracture for both air and salt-fog tests. Increasing the amount of local deformation reduces the amount of intergranular fatigue fracture for the longitudinal orientation. There is a significant amount of secondary cracking when tests are made in salt-fog.

7. Fatigue cracks tend to follow regions of dense precipitation.

8. The fatigue properties of P/M processed material are significantly greater in air and salt-fog than commercial I/M alloys of similar composition and strength level.

TABLE I

Chemical Compositions of Aluminum Alloys

ALLOY	Si	Fe	Cu	Mn	Mg	Cr	Zn	Ti	Co	Al
7075(IM)	.40	.50	1.2/2.0	.30	2.1/2.9	.18/.35	5.1/6.1	.20	-	Bal
7475(IM)	.10	.12	1.2/1.9	.06	1.9/2.6	.18/.25	5.2/6.2	.06	-	Bal
Powder	.06	.05	1.4	-	2.00	-	5.6	-	.35	Bal

TABLE II

Typical U.S. Standard Screen Analysis of Aluminum

Alloy Powder (Weight Percent)

APD	+30	-30 +50	-50 +100	-100 +200	-200 +325	-325
13	0	0	0.5	4	9	86.5

TABLE III

Typical Powder Particle Size by Coulter Counter

Diameter (μm)	75	60	50	40	30	20	15	10	8	5
Percent < dia.	93	88	81	73	63	44	29	16	11	4

TABLE IV

Preform Sizes and Upset Ratios

Preform Size			Upset Size			Upset Ratio	True Height Strain $\varepsilon = -\ln(h_f/h_o)$
$h_o(z)$	$w_o(y)$	$d_o(x)$	h_f	w_f	d_f		
3"	5"	2"	3"	5"	2"	0	0
4.5"	3.25"	2"	3"	5"	2"	1.5:1	0.4
6"	2.5"	2"	3"	5"	2"	2:1	0.7

TABLE V

Location of Center of Gage Length for Tensile and Fatigue Specimens (Reference Figure 5)

Upset Ratio	ε	Spec. Orientation	
		Long	S. Trans
0	0	var	var
1.5:1	.3	B-2	B-2
1.5:1	.6	C-4	C-4
2:1	.6	C-3	C-3
2:1	.7	B-4	B-4

LIST OF REFERENCES

1. H. W. Antes: "Deformation of Porous Materials", Progress in Powder Metallurgy, vol. 28 MPIF Princeton, N.J. pp 5-40, 1972.

2. P. L. Stockl and H. W. Antes: "The Effect of Deformation on Tensile and Impact Properties of Hot P/M-Formed Nickel-Molybdenum Steels", Powder Metallurgy, 17 (33), pp 178-192, 1974.

3. J. P. Cook: "The Effect of Sintering Temperature and Flow on the Properties of Ni-Mo Steel Hot P/M Formed Material", SAE Report 7409, October 1974.

4. H. W. Antes: "P/M Hot Formed Gears", Metals Engineering Quarterly, pp 8-15, Nov. 1974.

5. G. Bockstiegel and U. Bjork: "The Influence of Preform Shape on Material Flow, Residual Porosity and Occurrence of Flaws in Hot Forged Powder Compacts", Powder Metallurgy, 17 (33) pp 126-139, 1974.

6. K. H. Moyer: "A Comparison of Deformed Iron-Carbon Alloy Powder Preforms with Commercial Iron-Carbon Alloys", Modern Developments in Powder Metallurgy, V7: P/M Forging and Copper P/M, Eds.: H. H. Hausner and W. E. Smith, Metal Powder Industries Federation, Princeton, N. J. pp 235-254, 1974.

7. B. L. Ferguson, S. K. Suh and A. Lawley: Int. J. of Powder Metallurgy and Powder Technology, vol. 11, p 263, 1975.

8. B. L. Ferguson, H. A. Kuhn and A. Lawley: Modern Developments in Powder Metallurgy, Eds: H. H. Hausner and P. W. Taubenblat, MPIF, Princeton, N.J., vol. 10, p 15, 1977.

9. M. J. Koczak, C. L. Downey and H. A. Kuhn: Powder Metallurgy Int., vol. 6, p 13, 1974.

10. W. S. Cebulak, E. W. Johnson and H. Markus: Metals Engr. Quarterly, vol. 16, #4, 37, 1976.

11. W. L. Otto, Jr., "Metallurgical Factors Controlling Structure in High Strength P/M Products", AFML Report #TR-76-60, Aluminum Co. of America, Alcoa Center, Pa.; Contract #F 33615-74-C-5077, May 1976.

12. W. L. Otto, Jr. and R. W. Sawtell: "Fatigue Behavior of High Strength Powder Metallurgy Aluminum Alloys", AIME Annual Meeting, Atlanta, Ga., March 1977.

13. D. M. Corbly, Jr.: "Fatigue Crack Growth and Retardation in a Powder Aluminum Alloy", AIME Annual Meeting, Atlanta, Ga., March 1977.

14. K. E. Buchovecky and M. R. Rearick: Forging of Powder Metallurgy Preforms, Eds: H. H. Hausner, K. H. Roll and P. K. Johnson, MPIF, New York, N.Y., p 319, 1973.

15. J. P. Lyle, Jr. and W. S. Cebulak: "Properties of High Strength Aluminum P/M Products". Metals Engr. Quarterly, vol. 14 no. 1, pp 52-63, Feb. 1974.

16. C. Lipson and N. J. Sheth: Statistical Design and Analysis of Engineering Experiments, McGraw-Hill pp 36-48 and pp 84-87, 1973.

17. T. H. Sanders, Jr., R. R. Sawtell, J. T. Staley, R. J. Bucci and A. B. Thakker: "Effect of Microstructure on Fatigue Crack Growth of 7XXX Aluminum Alloys Under Constant Amplitude and Spectrum Loading", Alcoa Laboratories for Naval Air Systems Command; Contract #N00019-76-C-0482 Final Report, April 1978.

18. J. Waldman, H. Sulinski and H. Marcus: "The Effect of Ingot Processing Treatments on the Grain Size and Properties of Al Alloy 7075", Metallurgical Transactions, vol. 5, pp 573-584, March 1974.

19. R. Henderson, G. F. Pittinato, V. Kerlins, A. Phillips and M. A. Russo: SEM/TEM Fractography Handbook, Materials and Ceramics Information Center, Battelle Columbus, Dec. 1975.

20. ASM Metals Handbook: Fractography and Atlas of Fractographs Vol. 9, American Society for Metals, Metal Park, Ohio 1974.

21. G. E. Dieter: Mechanical Metallurgy, second edition, McGraw-Hill, pp 374-376, 1976.

22. M. Rafalin, M. J. Koczak and A. Lawley: "Fatigue of High-Strength Aluminum Alloy P/M Forgings", Fall-Meeting, Met. Soc. (AIME), St. Louis, 1978.

23. H. W. Antes: Role of Deformation Processing on the Fatigue Behavior of a High Strength Powder Metallurgy Aluminum Alloy, Ph.D. Thesis, Drexel University, June 1979.

24. R. E. Sanders, Jr. and E. A. Starke, Jr.: "Effect of Intermediate Thermo-mechanical Treatments on the Fatigue Properties of a 7050 Aluminum Alloy", Metallurgical Transactions A, vol. 9A, pp 1087-1100, Aug. 1978.

25. T. H. Sanders, Jr. and E. A. Starke, Jr.: "The Relationship of Microstructure to Monotonic and Cyclic Straining of Two Age Hardening Aluminum Alloys, Metallurgical Trans. A, vol. 7A, pp 1407-1418, Sept. 1976.

26. R. E. Sanders, Jr. and E. A. Starke, Jr.: "Effect of Grain Refinement on the Low Cycle Fatigue Behavior of an Aluminum-Zinc-Magnesium-(Zirconium) Alloy", Materials Science and Engineering, vol. 28, pp 56-68, 1977.

27. S. Buzolits: unpublished research in progress on fatigue properties of I/M 7075 and 7475 in air and 3 1/2% NaCl salt-fog environments, Drexel University 1979.

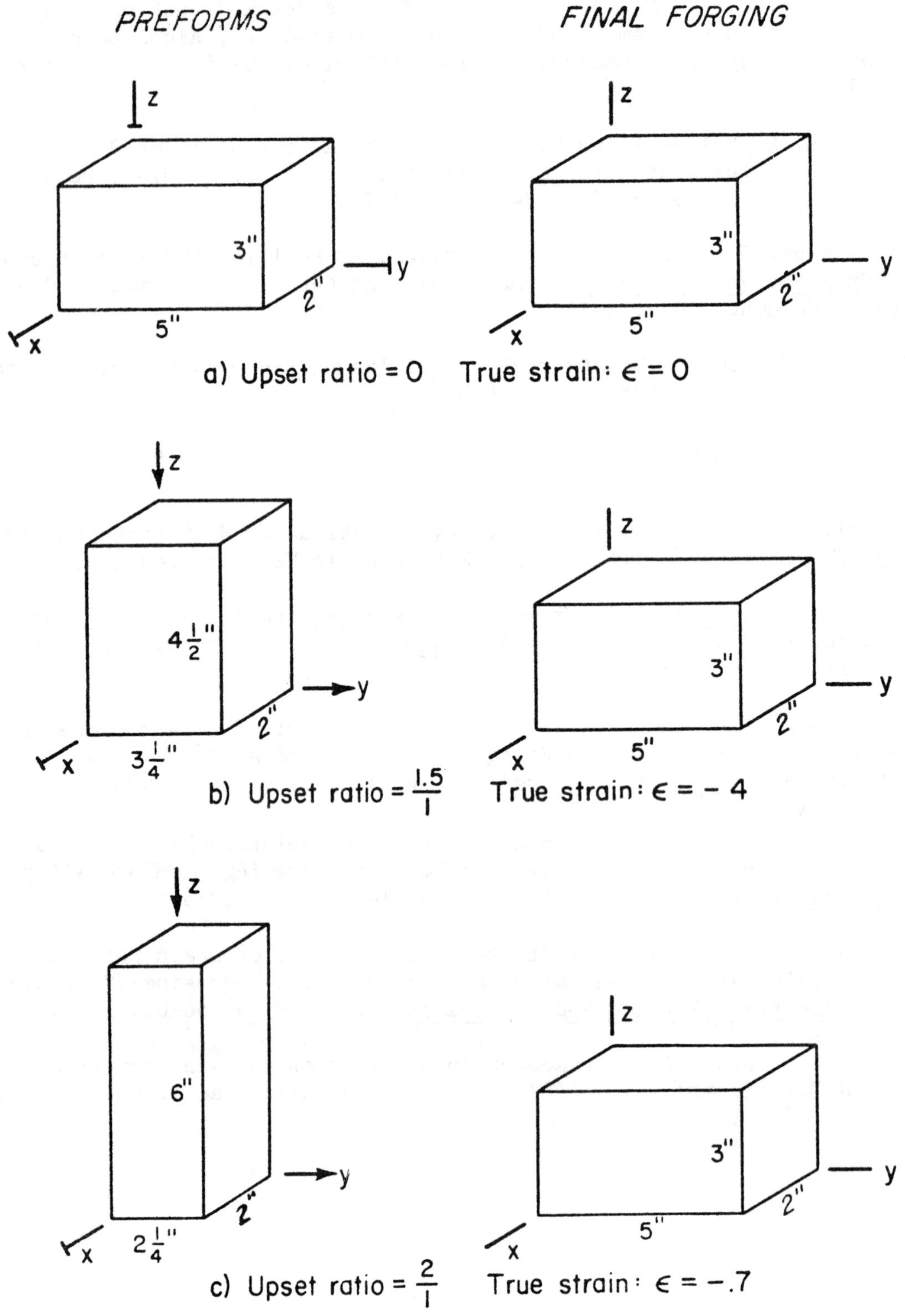

Figure 1. Forging preform sizes and orientations for final forged size.

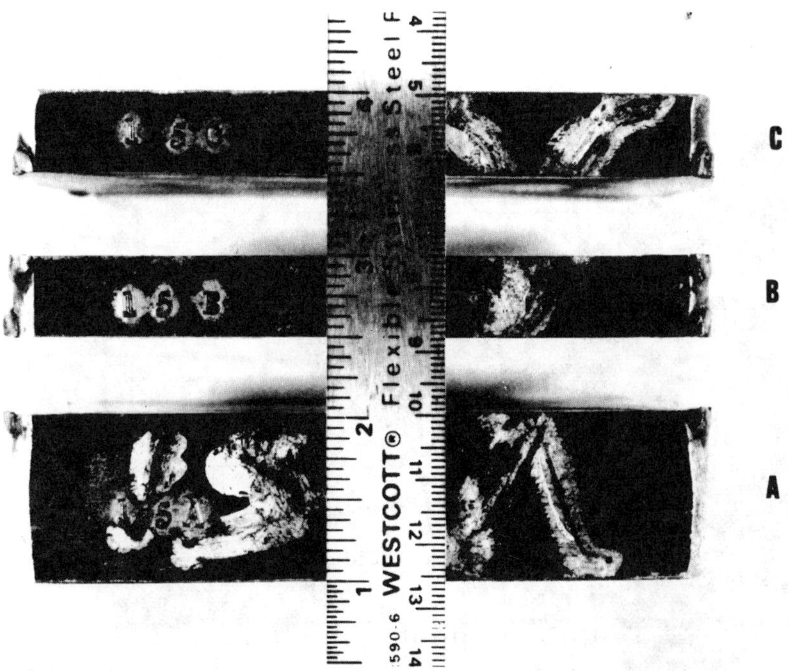

Figure 2. Top view of preform slabs (A=1 in. thick, B=C=1/2 in. thick) after forging.

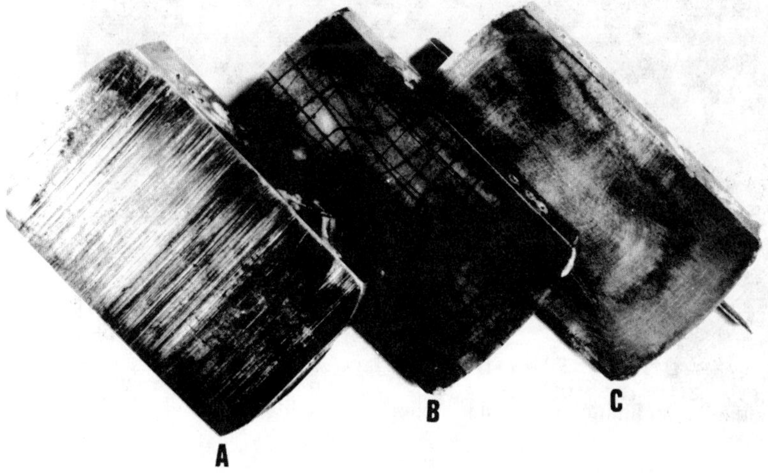

Figure 3. Segments of laminated preform after forging.

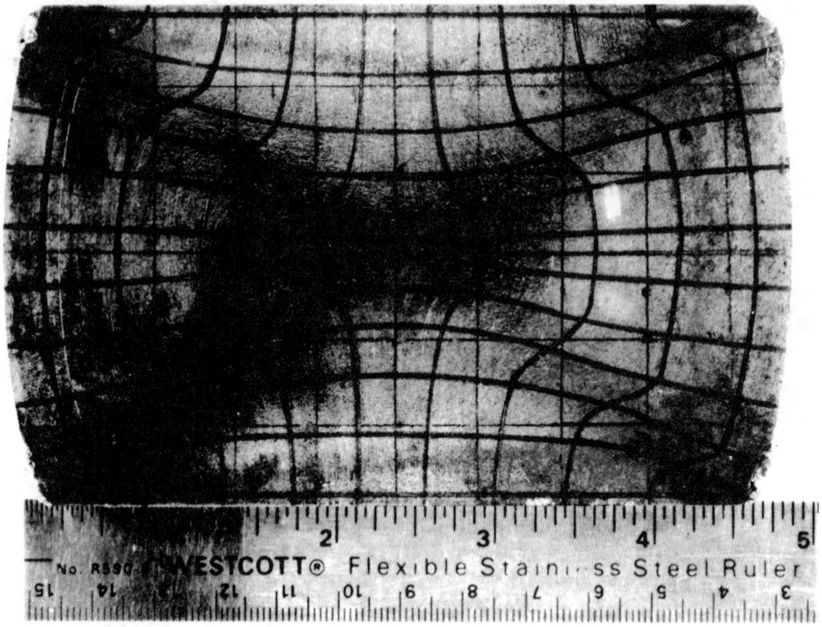

a

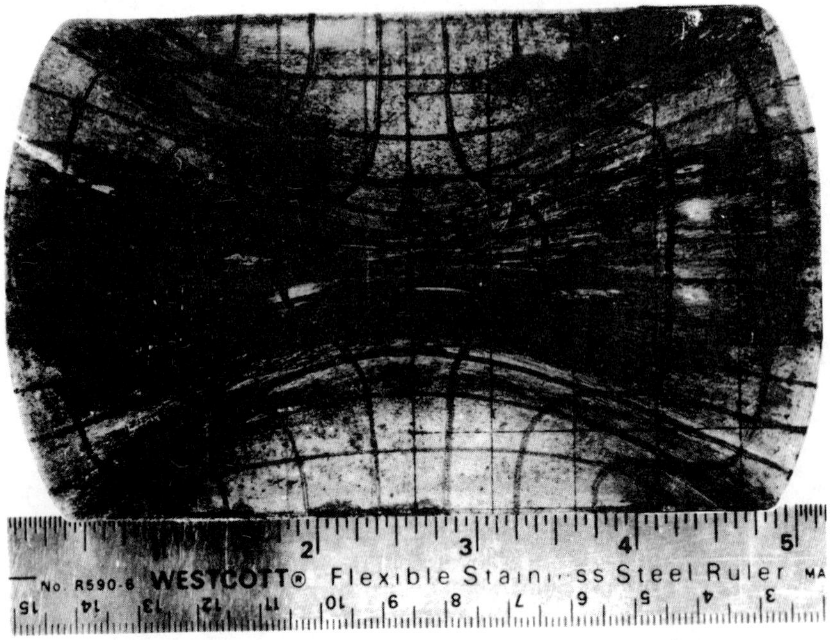

b

Figure 4. Deformed grid pattern in: (a) 1.5:1 upset, (b) 2:1 upset. Note: grids were originally 10 mm square.

1	2	3	4	5	5	4	3	2	1	
					.10	.28	.35	.69		A
					.27	.43	.51	.31		B
					.84	.60	.36	.21		C
										C
										B
										A

a

1	2	3	4	5	5	4	3	2	1	
					.11	.25	.63	.63		A
					.39	.69	.60	.36	.30	B
					1.27	1.14	.60	.36	.30	C
										C
										B
										A

b

Figure 5. Local upset true strains for the center of the area indicated: (a) 1.5:1 upset, (b) 2:1 upset.

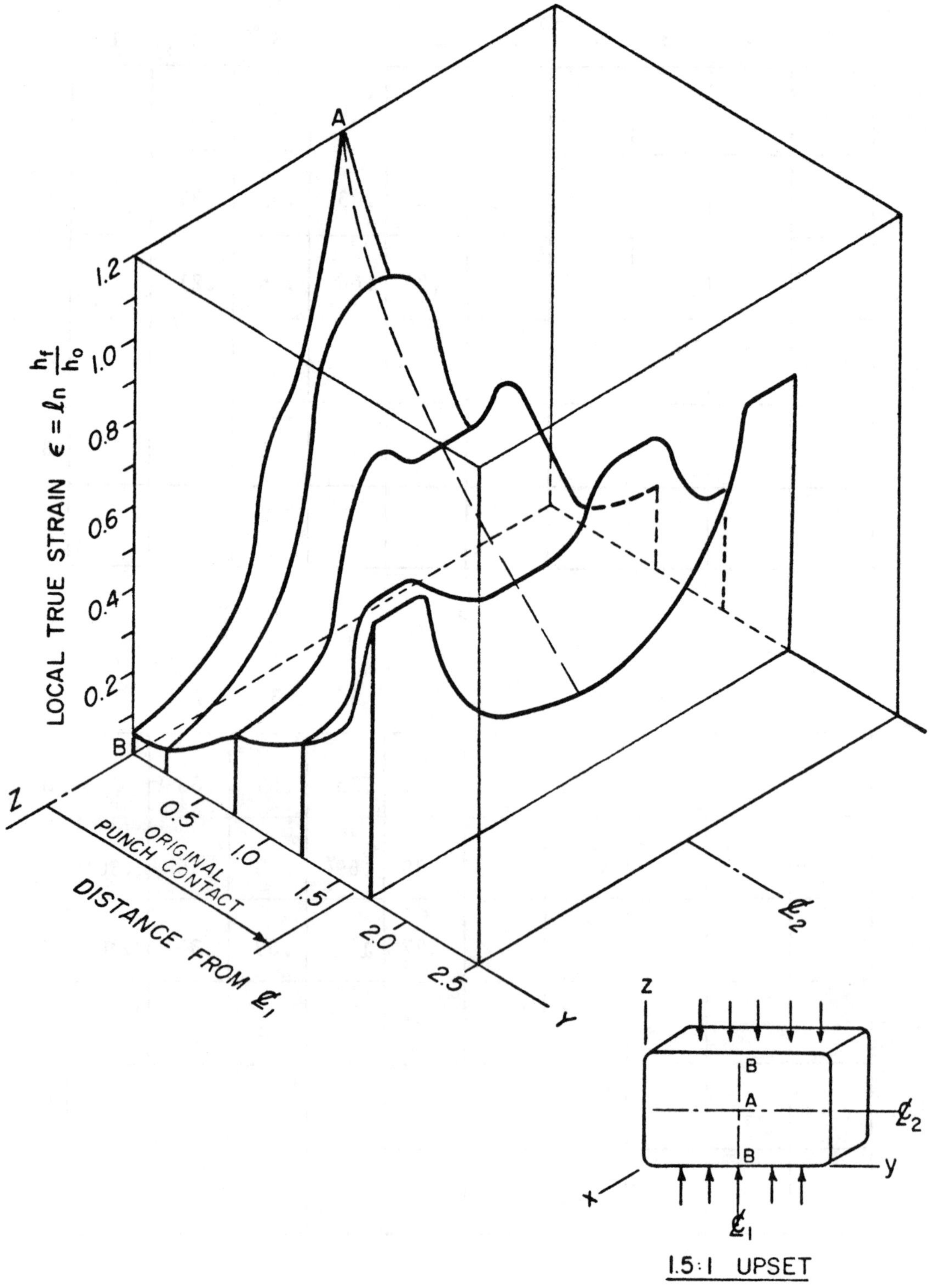

Figure 6. Local true strain as a function of position in the forging for a 1.5:1 upset.

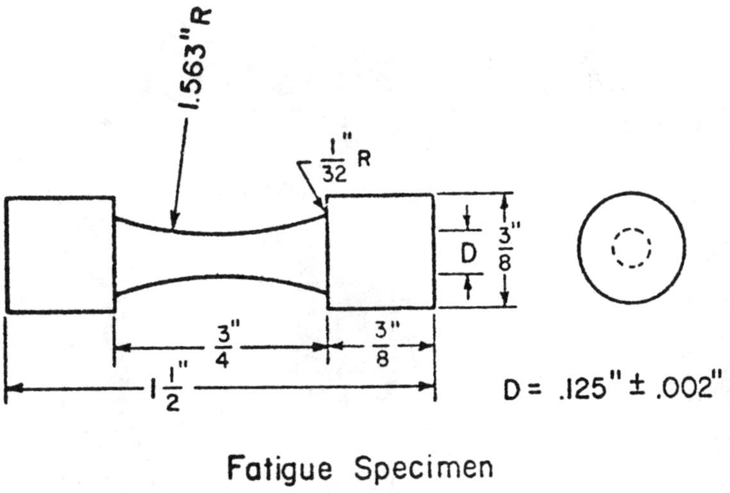

Figure 7. Dimensions of axial fatigue test specimens.

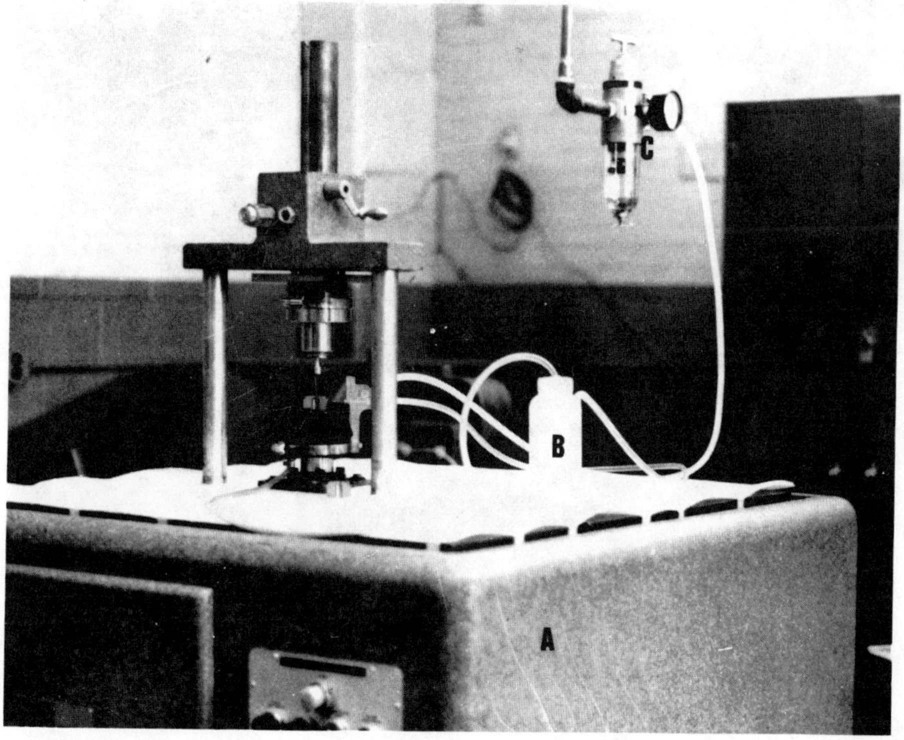

Figure 8. Axial fatigue testing apparatus: A - testing machine, B - salt solution reservoir, C - filtered compressed air.

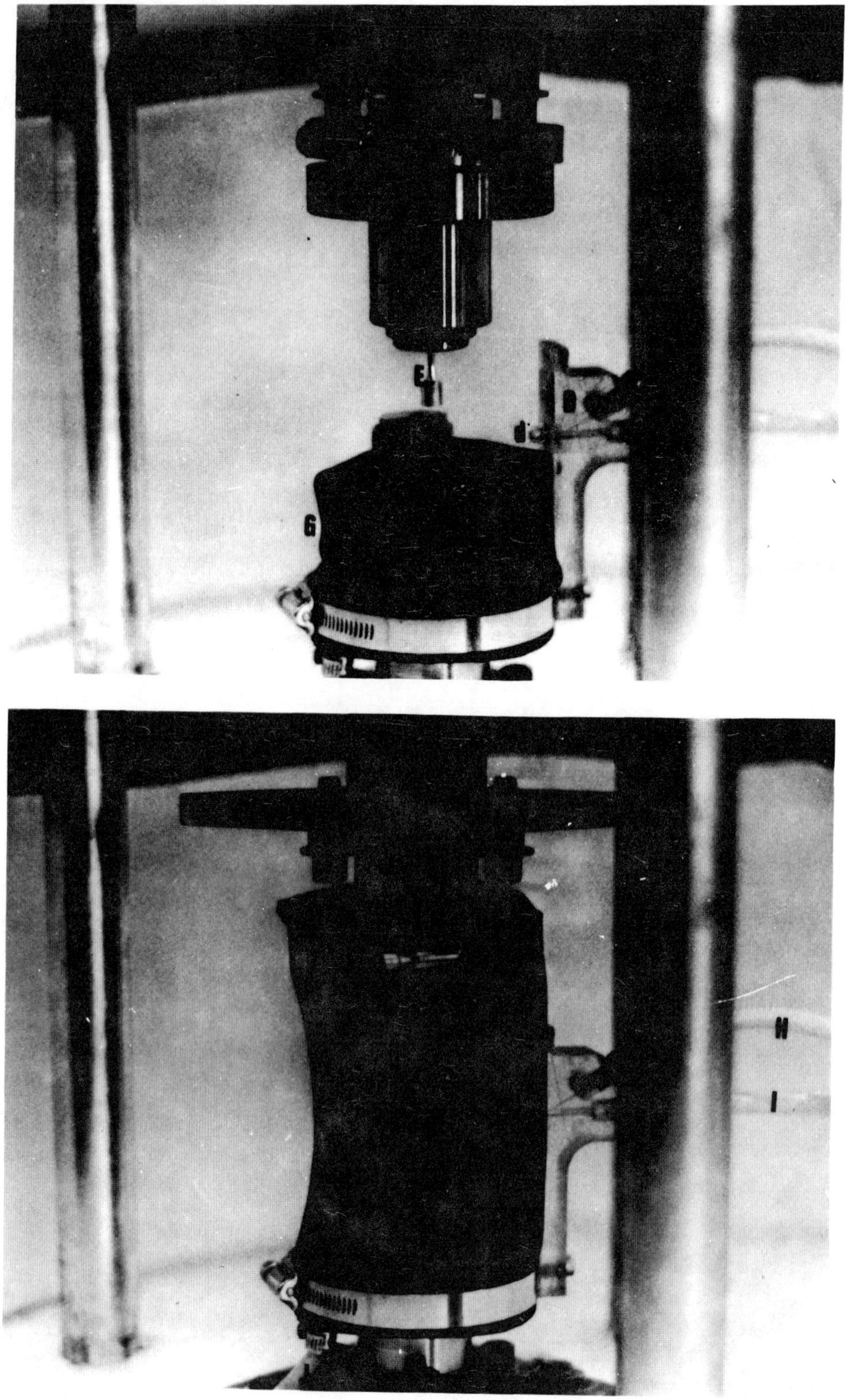

a

b

Figure 9. Axial fatigue salt-fog apparatus close-up (a) D, fog nozzle; E, specimen; F, lower grip; G, rubber chamber. (b) H, air line; I, salt-solution line; J, drain.

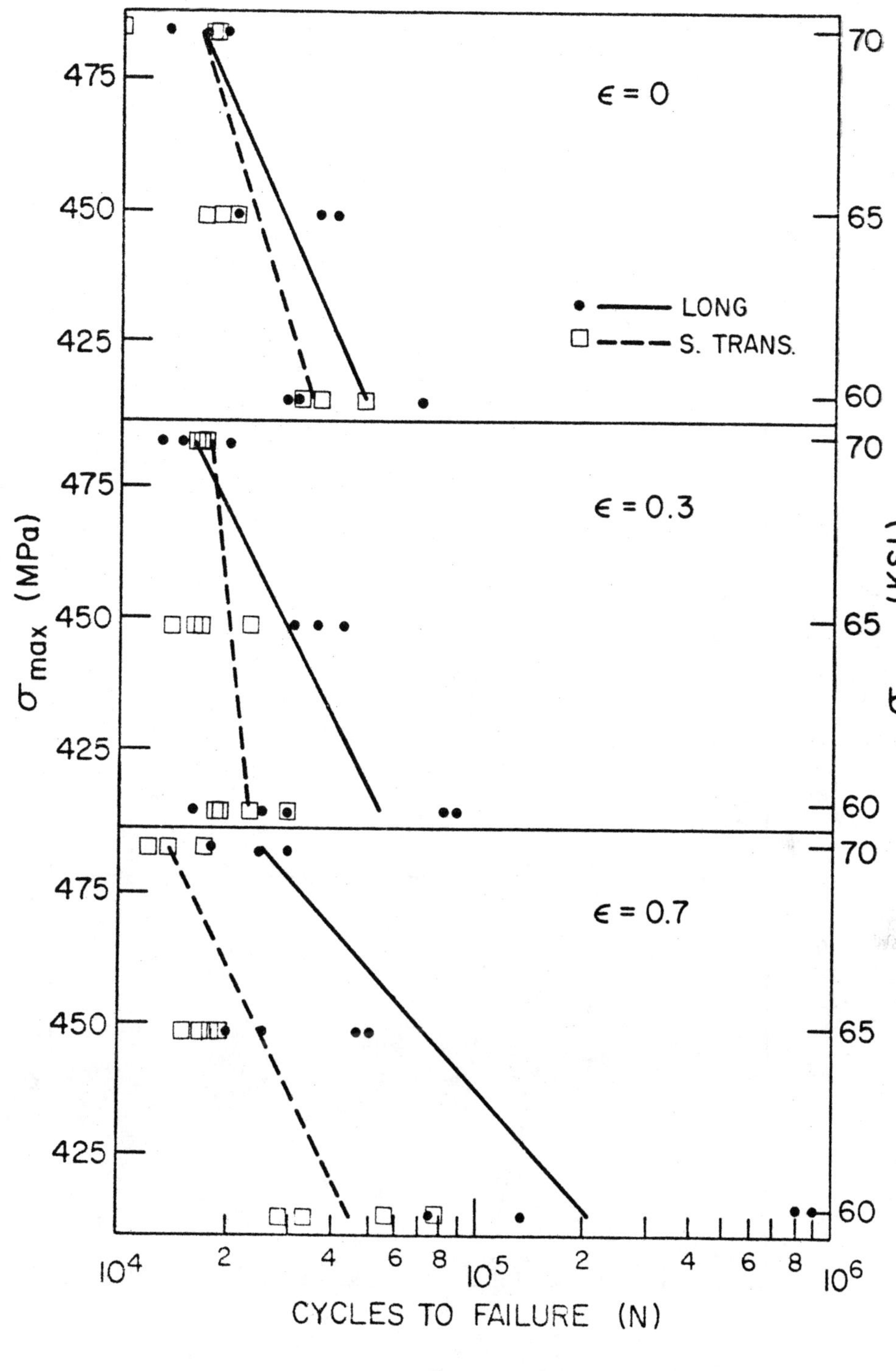

Figure 10. Median fatigue life vs. σ_{max} for P/M forged material tested in air.

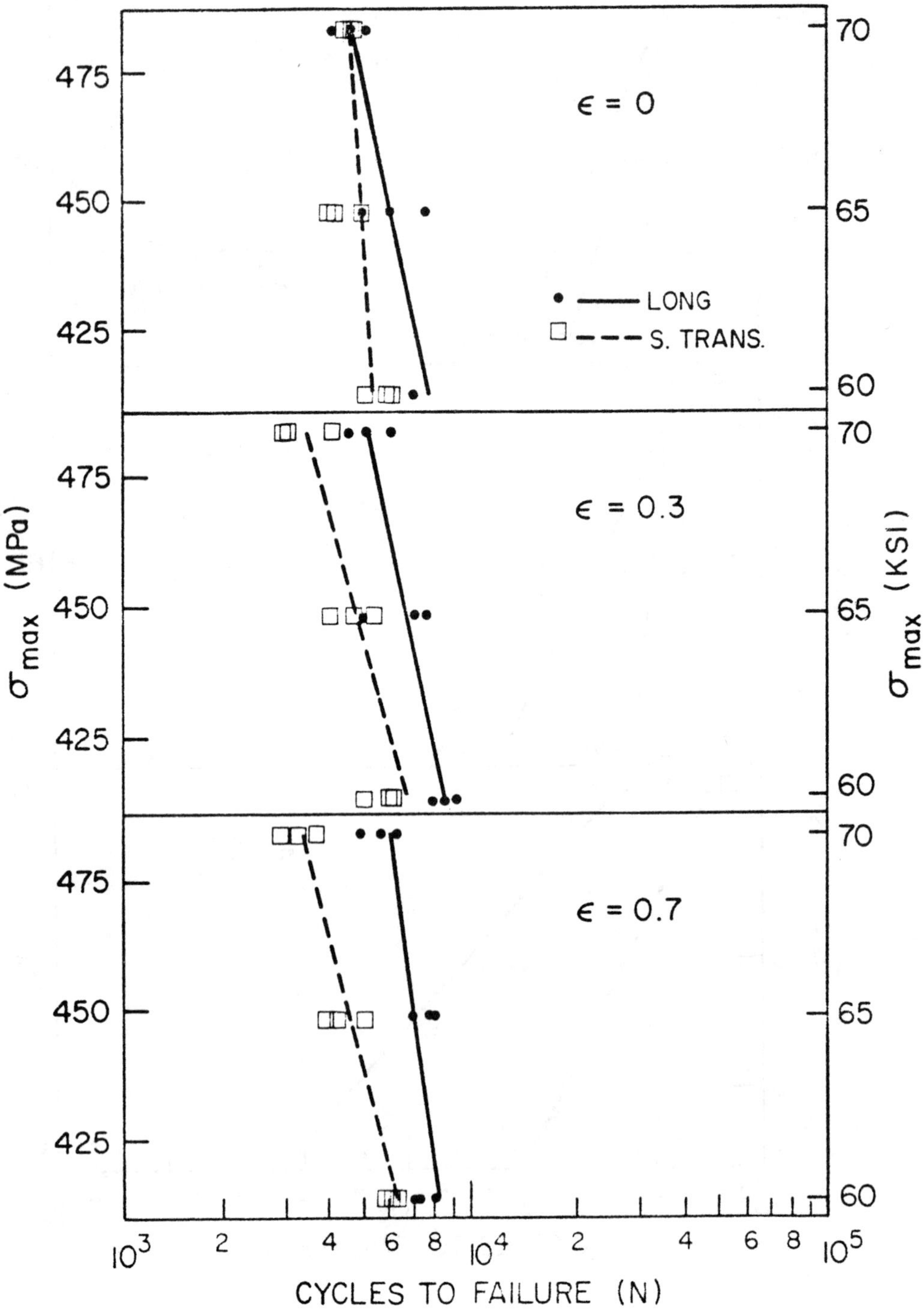

Figure 11. Median fatigue life vs σ_{max} for P/M forged material tested in 3 1/2% NaCl solution salt-fog.

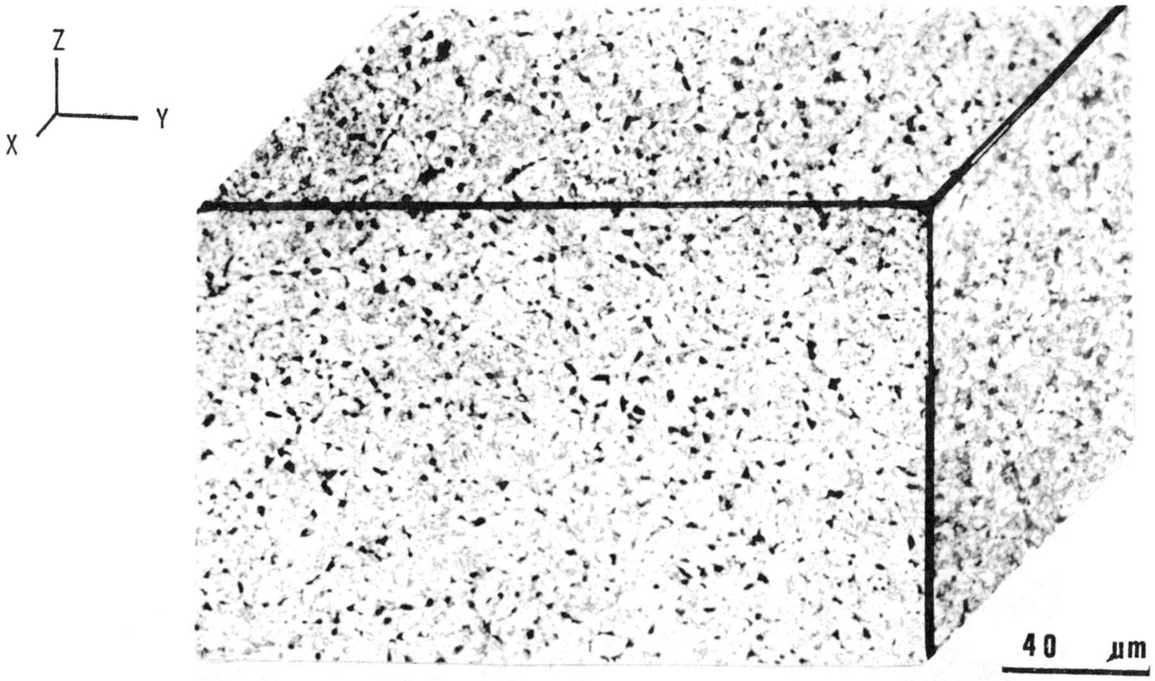

Figure 12. Three dimensional microstructures for hot compacted P/M Alloy, HNO_3 etch, 500X.

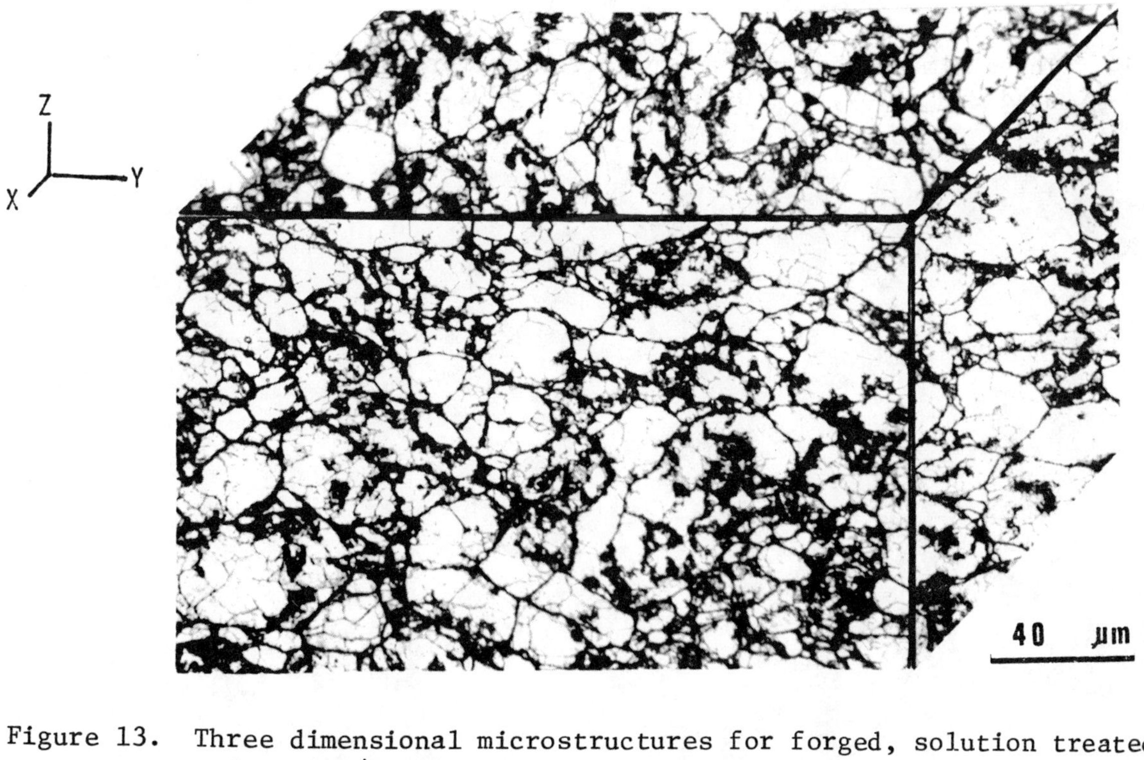

Figure 13. Three dimensional microstructures for forged, solution treated and and aged P/M alloy, ε = 0, HNO_3 etch, 500X.

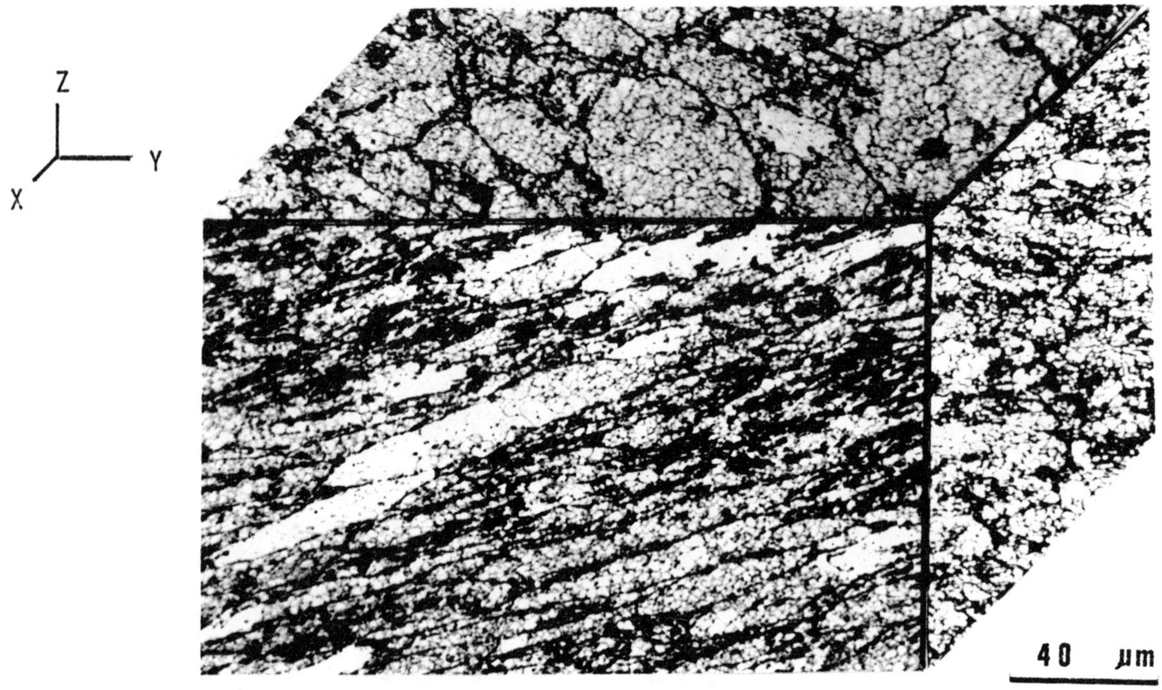

Figure 14. Three dimensional microstructures for forged, solution treated and aged P/M alloy, $\varepsilon = 0.6$, HNO_3 etch 500X.

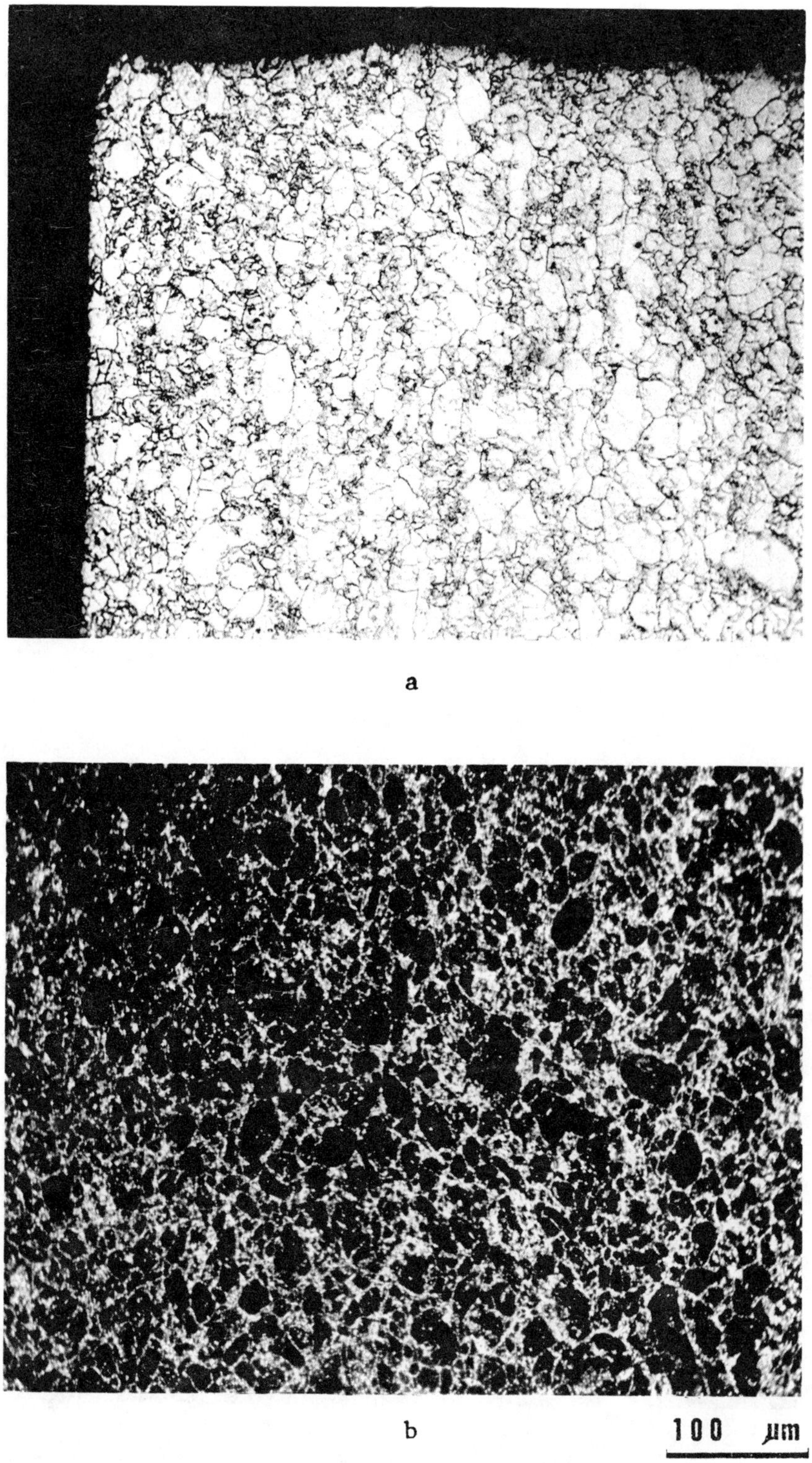

Figure 15. Microstructure, longitudinal fatigue fractured specimen, $\varepsilon = 0$, tested in air, crack runs from left to right: (a) bright field, (b) polarized light, Keller's etch, 200X.

a

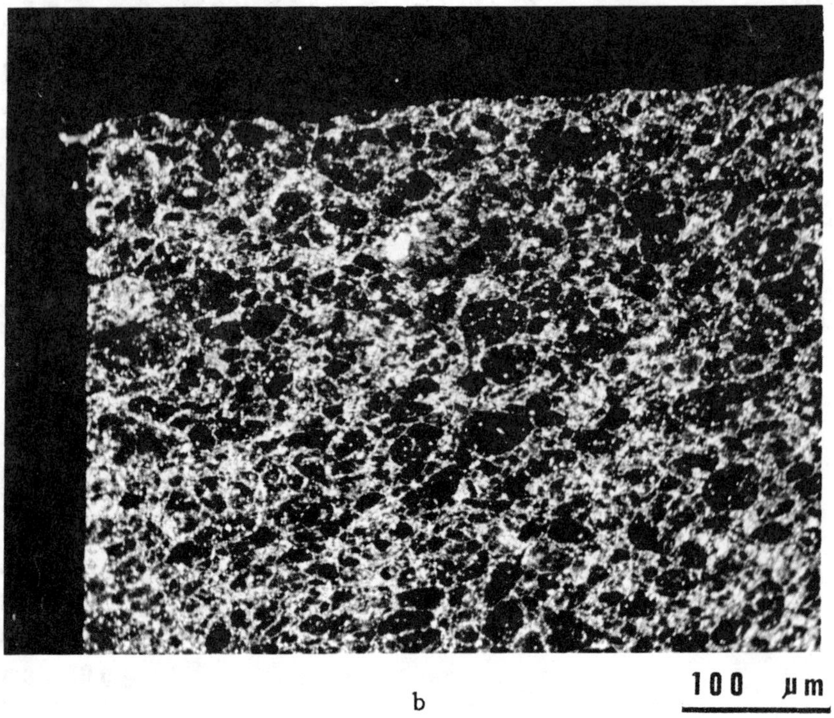

b

Figure 16. Microstructure, short transverse fatigue fractured specimen, $\varepsilon = 0$, tested in air, crack runs from left to right: (a) bright field, (b) polarized light, Keller's etch, 200X.

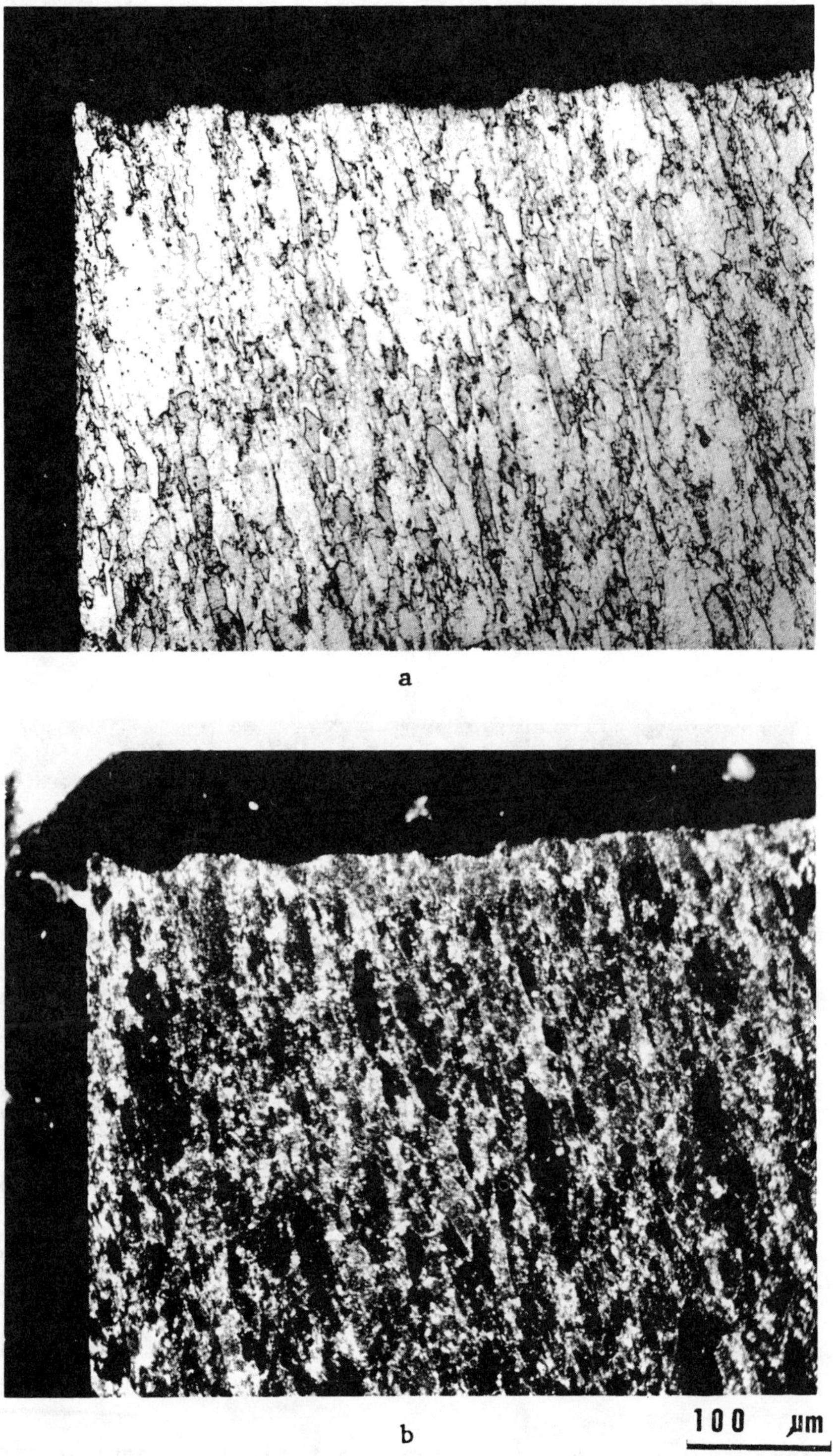

Figure 17. Microstructure, longitudinal fatigue fractured specimen, $\varepsilon = 0.7$, tested in air, crack runs from left to right: (a) bright field, (b) polarized light, Keller's etch, 200X.

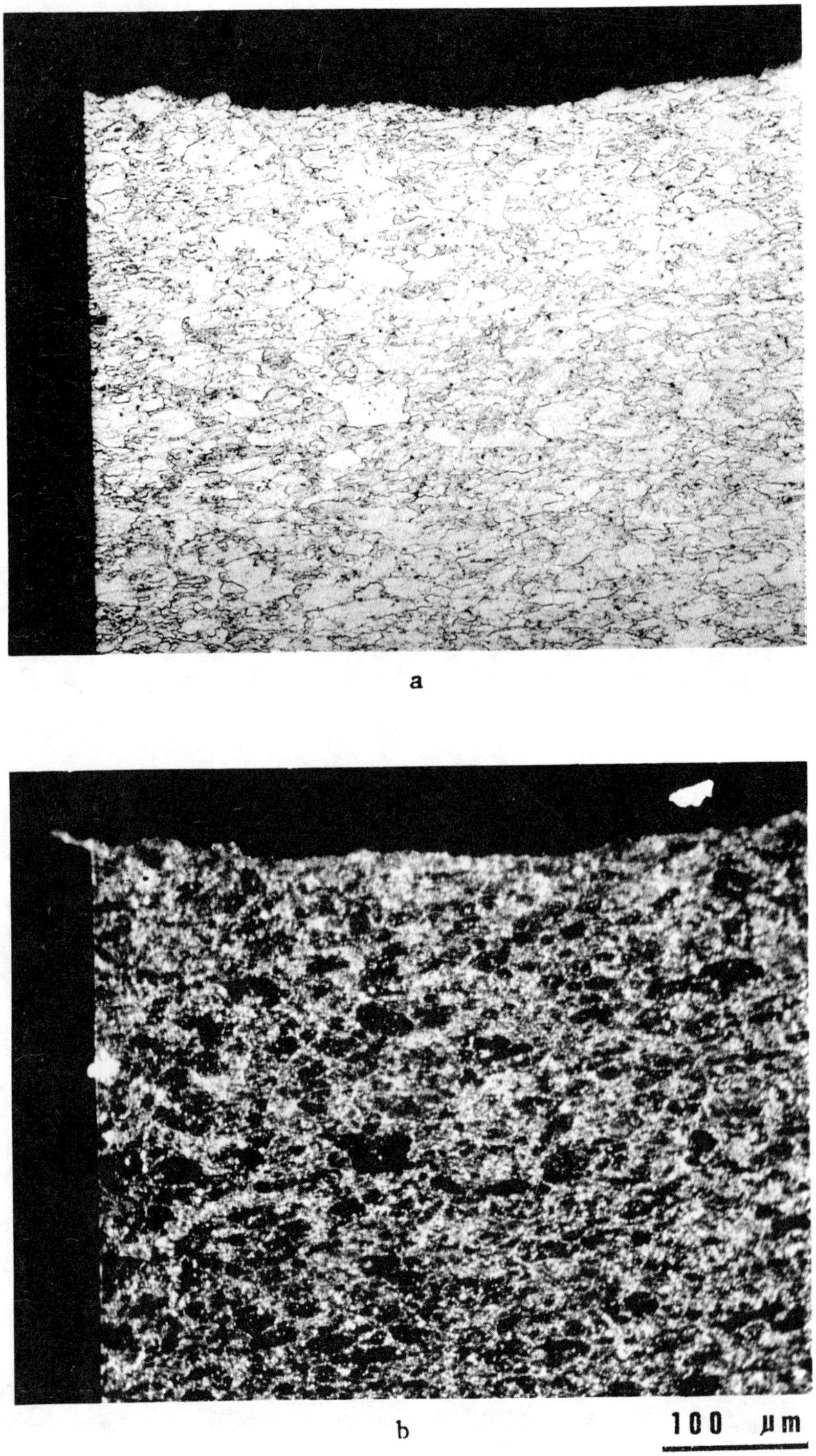

Figure 18. Microstructure, short transverse fatigue fractured specimen, ε = 0.7, tested in air, crack runs from left to right: (a) bright field, (b) polarized light, Keller's etch, 200X.

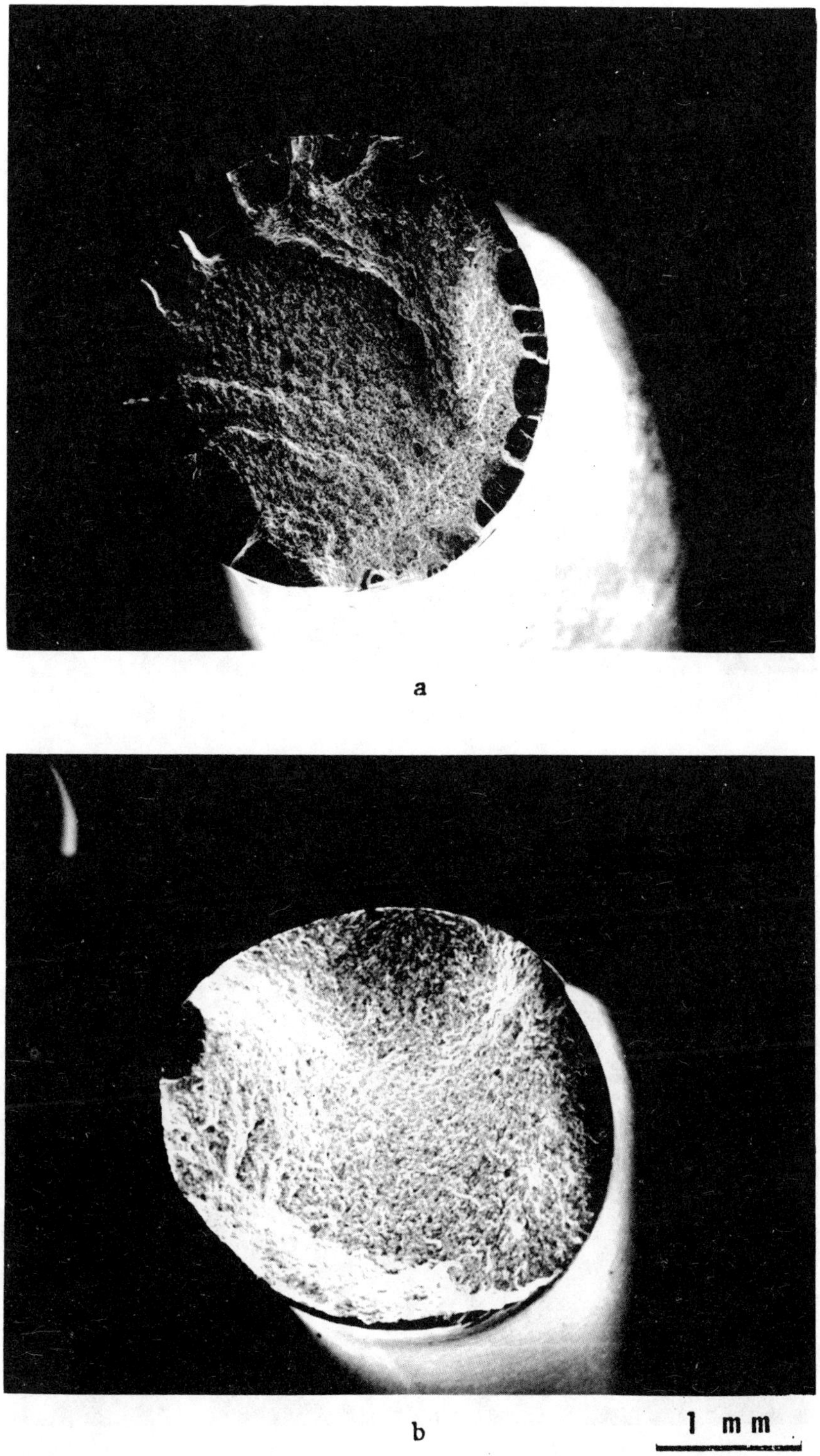

Figure 19. SEM, axial fatigue fracture surfaces, $\varepsilon = 0$, longitudinal, σ_{max} = 60 ksi (414 MPa): (a) salt-fog, multiple initiation sites almost completely around edge of specimen, N = 7,000 cycles; (b) air, single initiation site, N = 30,000 cycles. 20X.

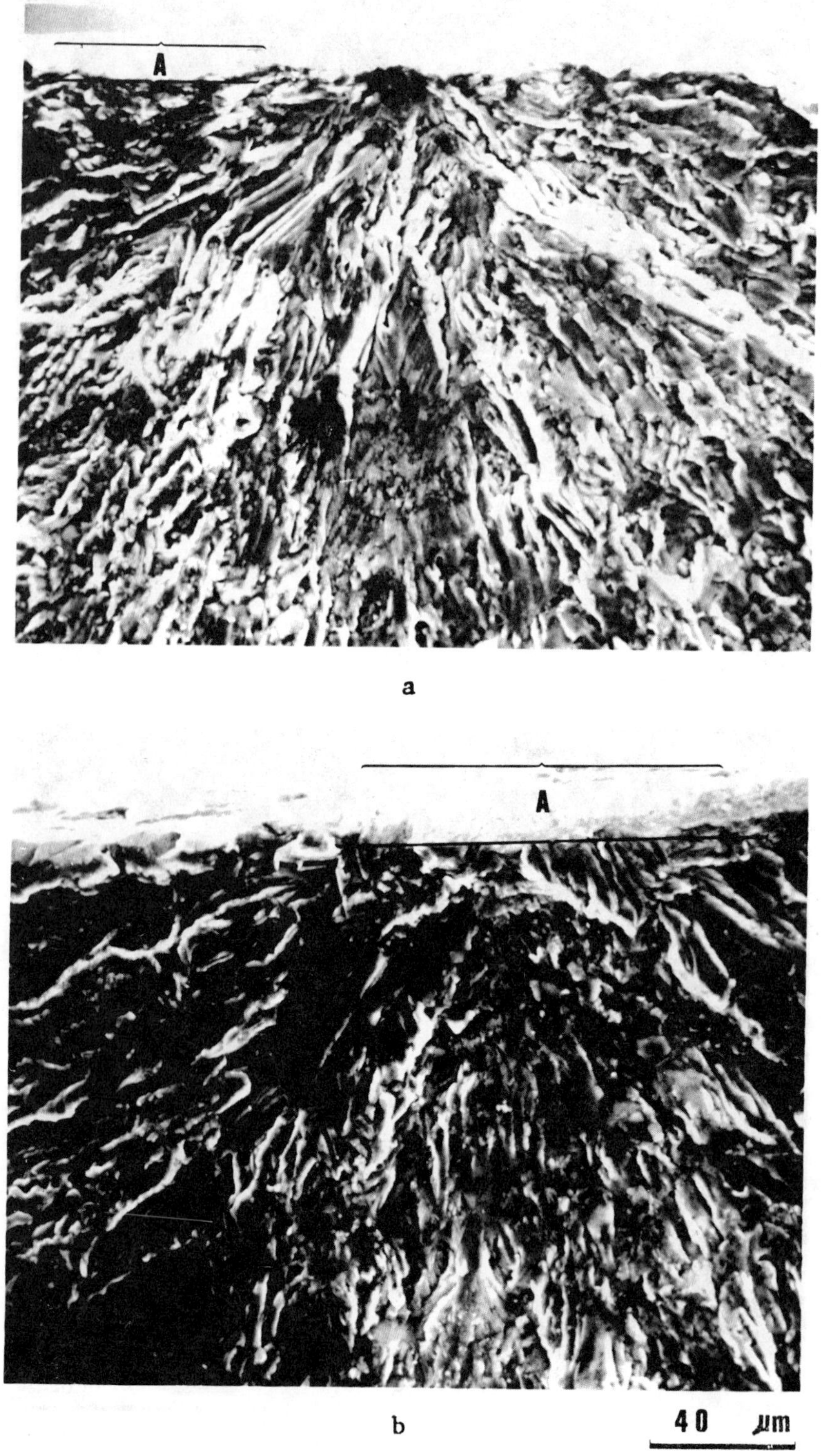

Figure 20. SEM, origin of fatigue fracture surfaces, ε = 0.3, short transverse, air, "A" is Stage I fatigue region: (a) σ_{max} = 70 ksi (483 MPa), (b) σ_{max} = 60 ksi (414 MPa). Direction of crack growth from top to bottom. 500X.

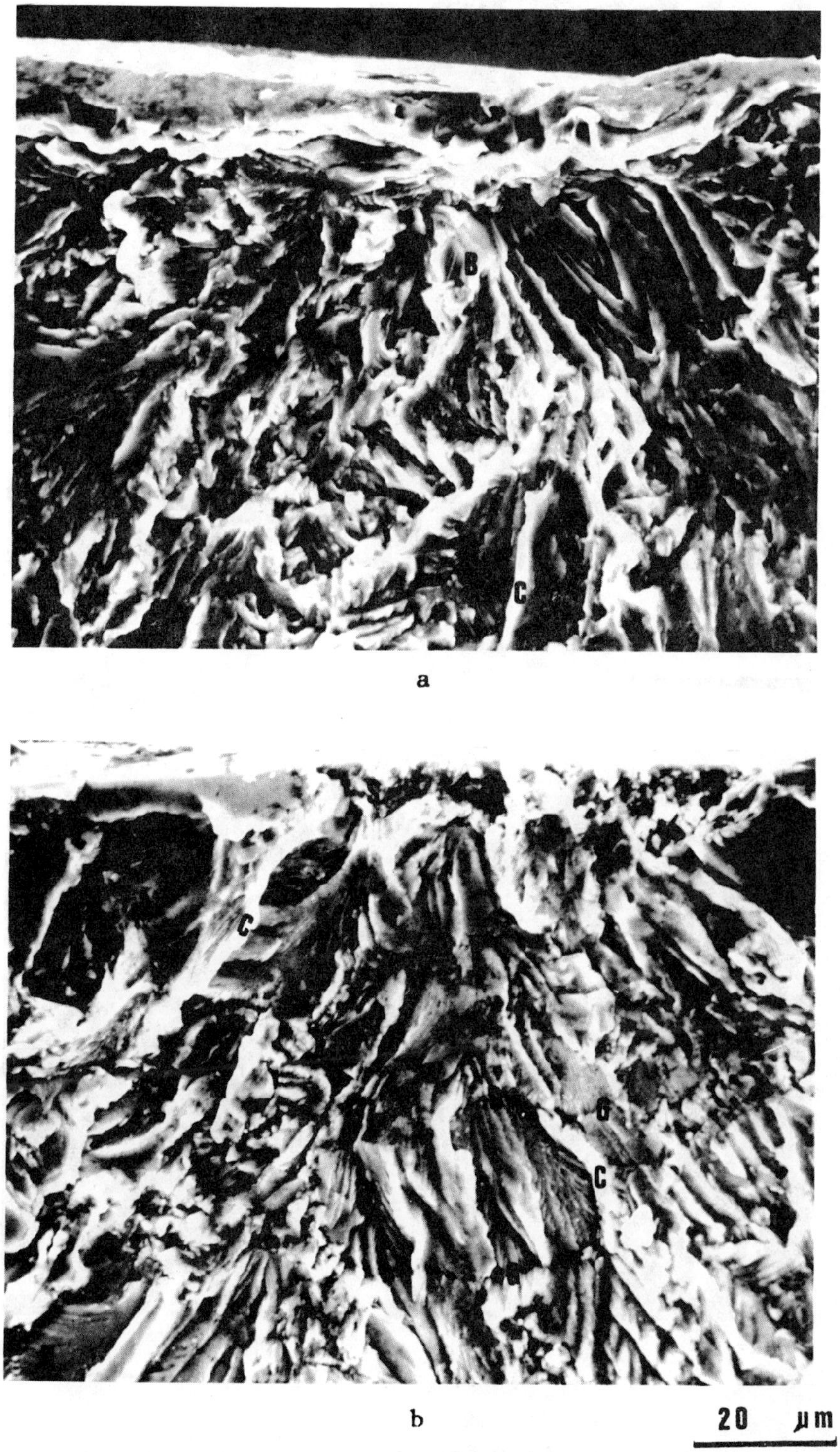

Figure 21. SEM, origin of fatigue fracture surfaces, $\varepsilon = 0$, $\sigma_{max} = 60$ ksi (414 MPa); A, narrow plateaus; B, smooth textured regions; C, tear ridges; G, grain boundaries; S, initiation site: (a) longitudinal, (b) short transverse. Direction of crack growth from top to bottom. 1000X.

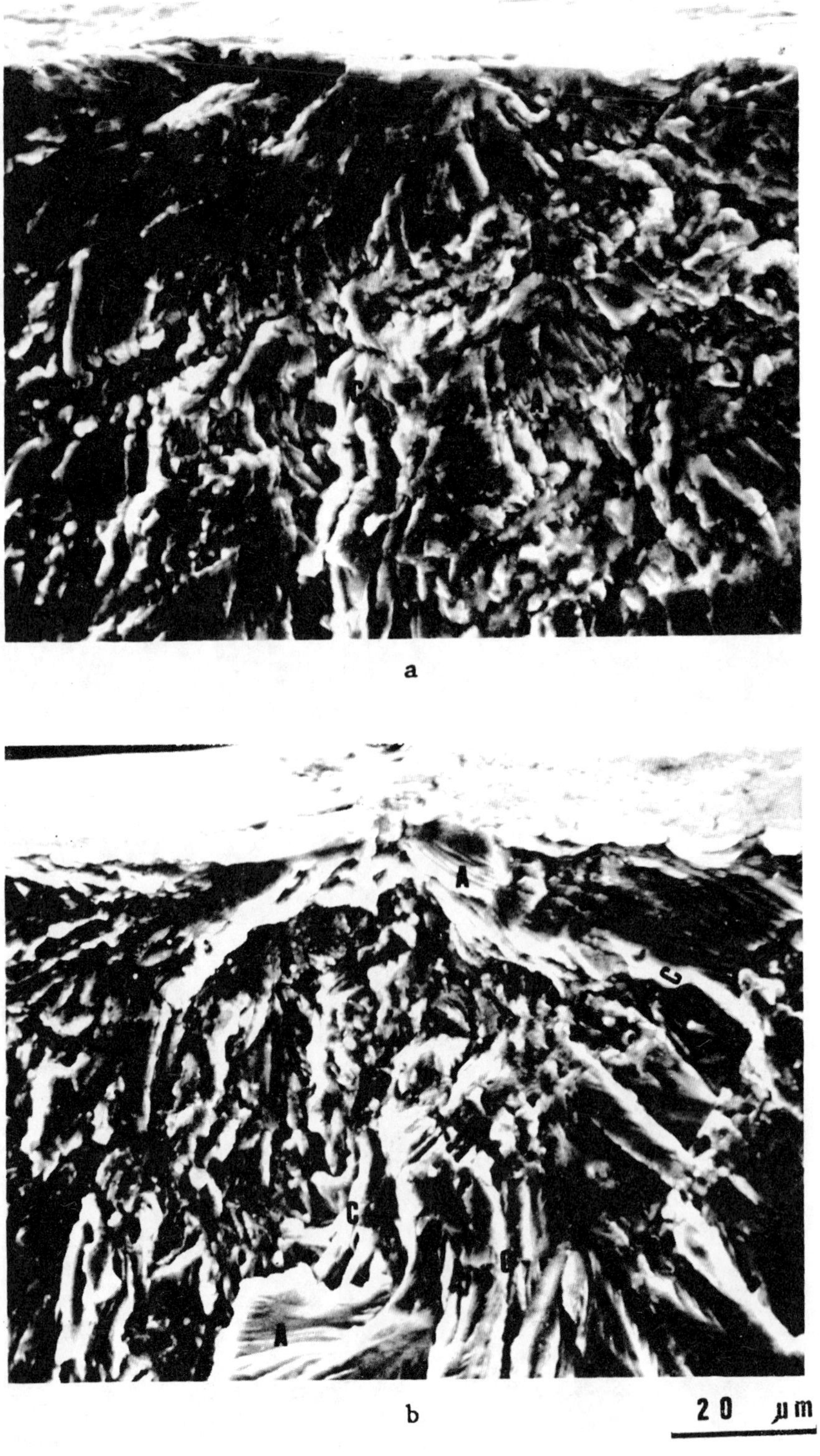

Figure 22. SEM, origin of fatigue fracture surfaces, ε = 0.7, air, σ_{max} = 60 ksi (414 MPa); A, narrow plateaus; B, smooth textured regions; C, tear ridges; G, grain boundaries: (a) longitudinal, (b) short transverse. Direction of crack growth from top to bottom. 1000X.

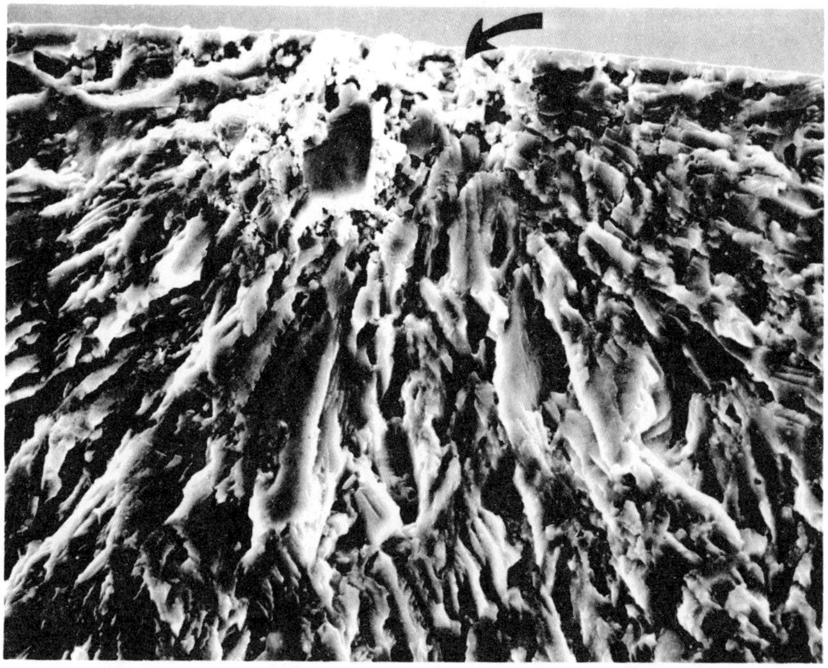

a

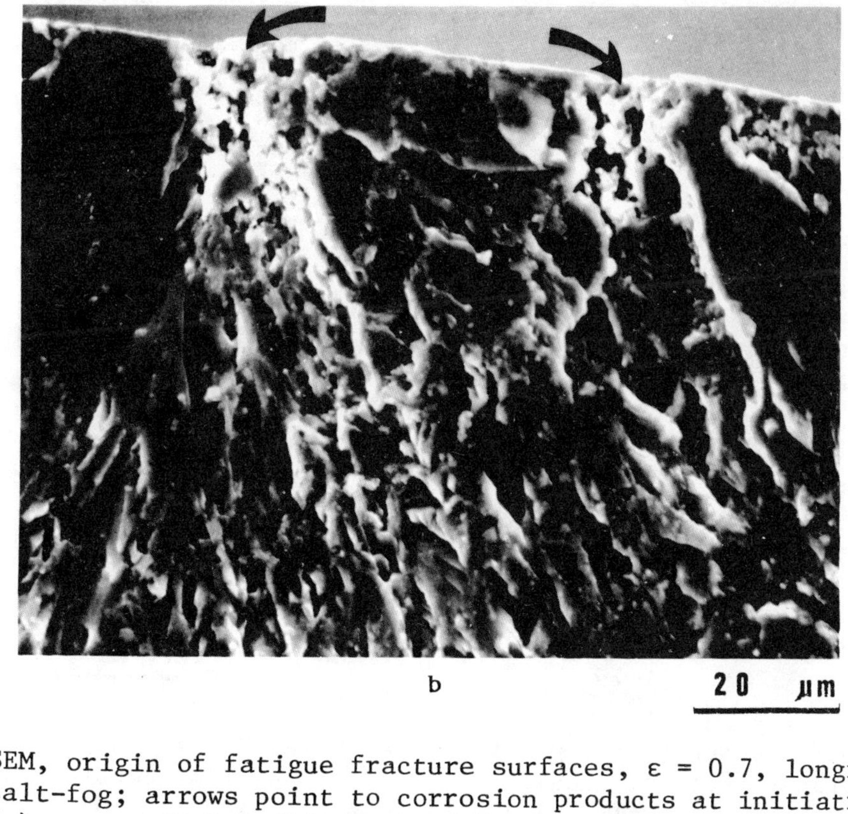

b

Figure 23. SEM, origin of fatigue fracture surfaces, $\varepsilon = 0.7$, longitudinal, salt-fog; arrows point to corrosion products at initiation sites: (a) $\sigma_{max} = 70$ ksi (483 MPa), (b) $\sigma_{max} = 60$ ksi (414 MPa). Direction of crack growth from top to bottom. 1000X.

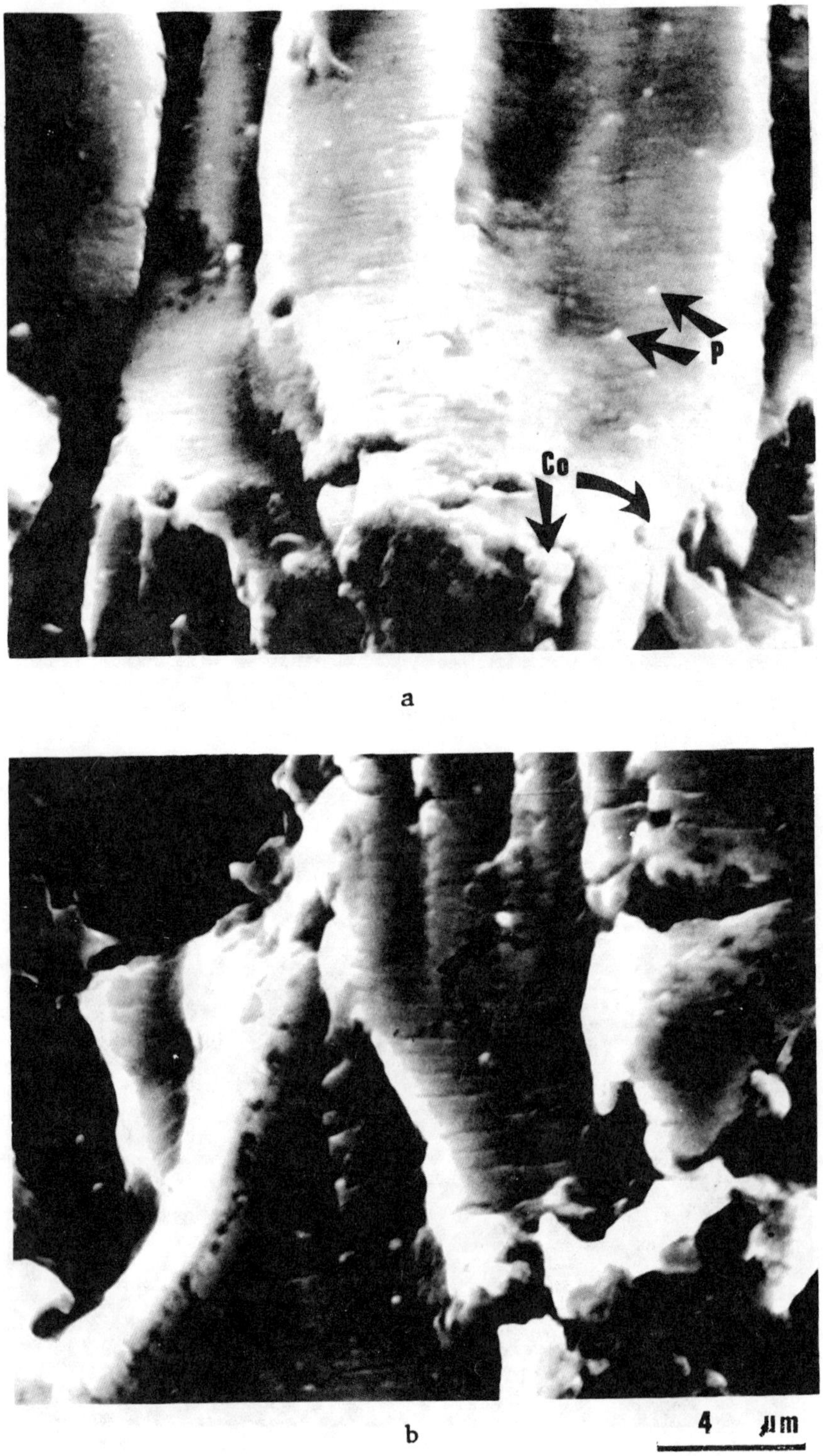

Figure 24. SEM, center of fatigue fracture, Stage II, $\varepsilon = 0.3$, air, short transverse; Co, cobalt rich regions; P, fine particles: (a) σ_{max} = 70 ksi (483 MPa), (b) σ_{max} = 60 ksi (414 MPa). Direction of crack growth from top to bottom, 5000X.

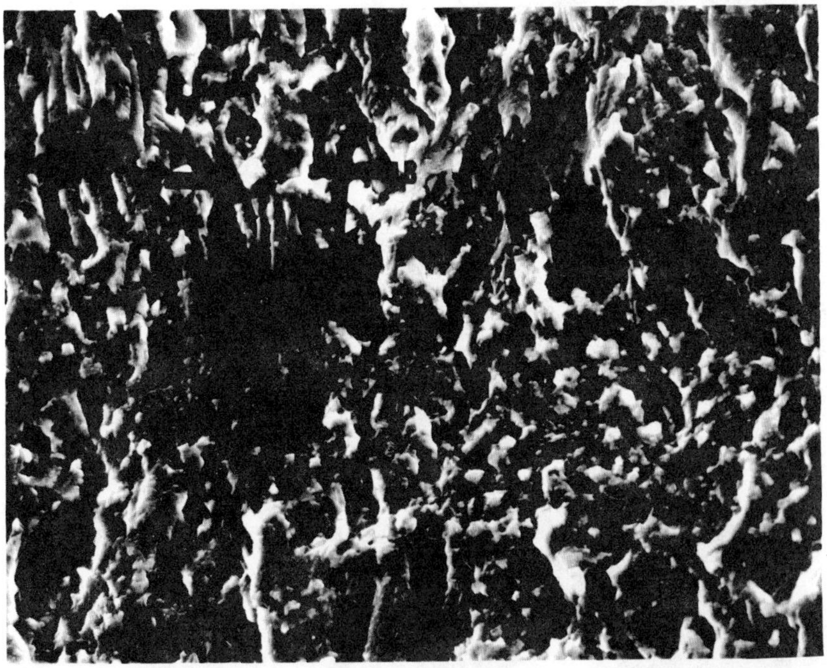

a

b

Figure 25. SEM, center of fatigue fracture, Stage II, $\varepsilon = 0$, salt-fog, longitudinal, σ_{max} = 70 ksi (483 MPa); A, brittle striations; B, ductile striations: (a) 1000X, (b) 5000X. Direction of crack growth from top to bottom.

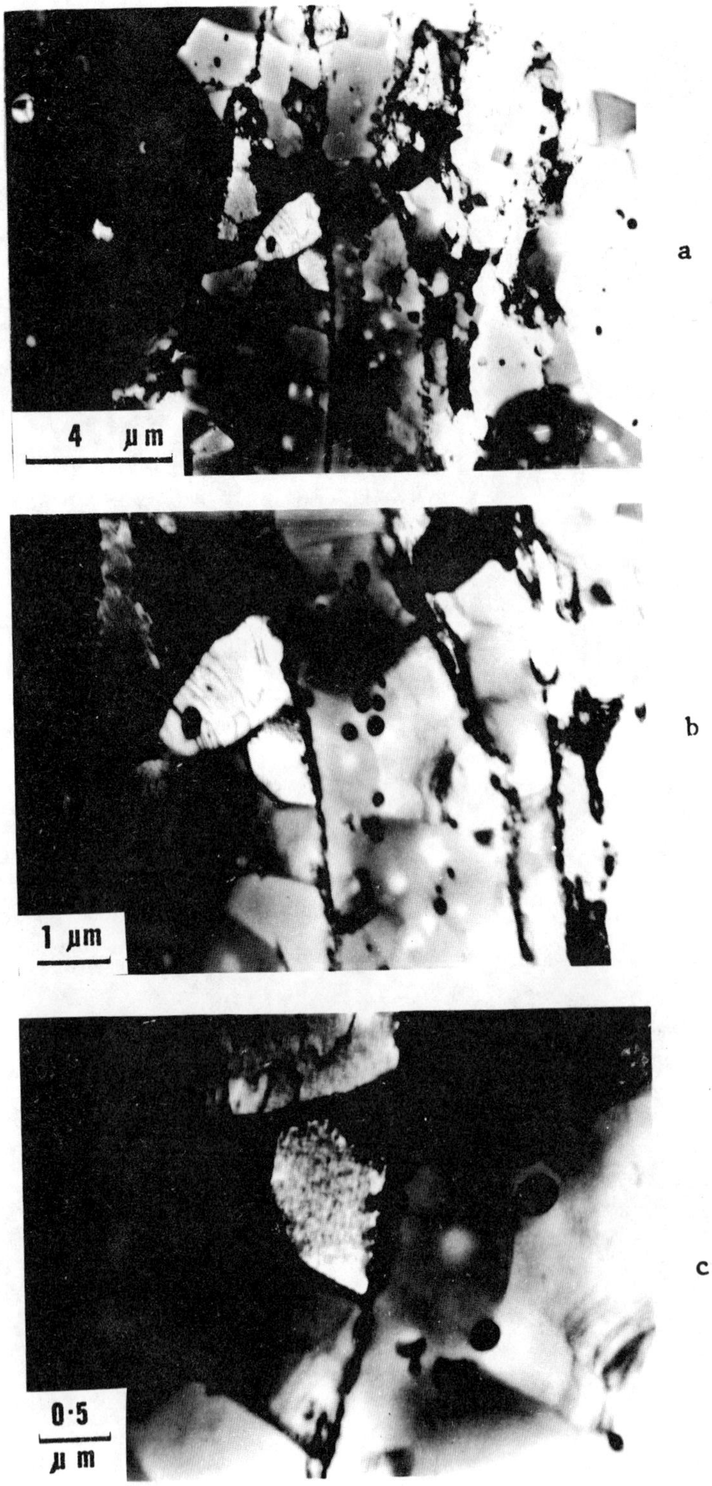

Figure 26. TEM, forged P/M alloy solution treated and quenched, ε = 0.7: (a) 5000X, (b) 10,000X, (c) 20,000X.

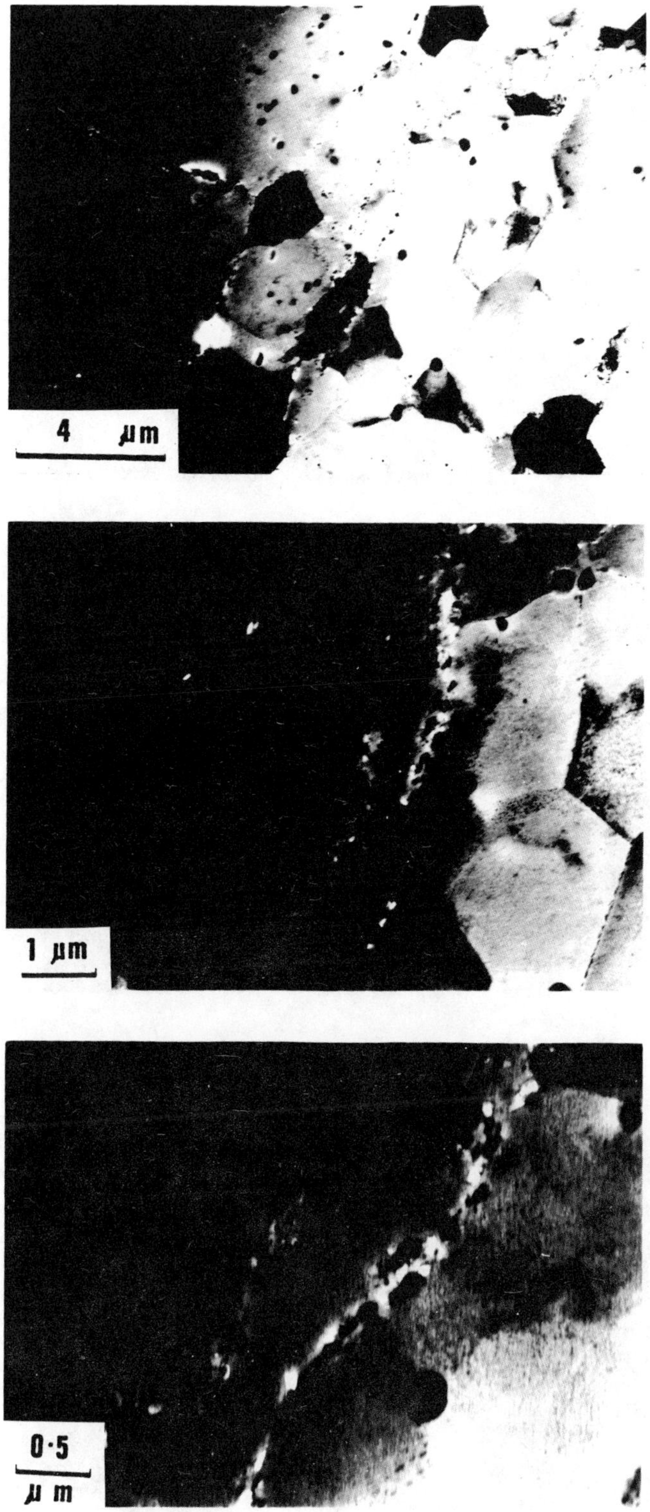

a

b

c

Figure 27. TEM, forged P/M alloy solution treated and aged, ε = 0.7: (a) 5000X, (b) 10,000X, (c) 20,000X.

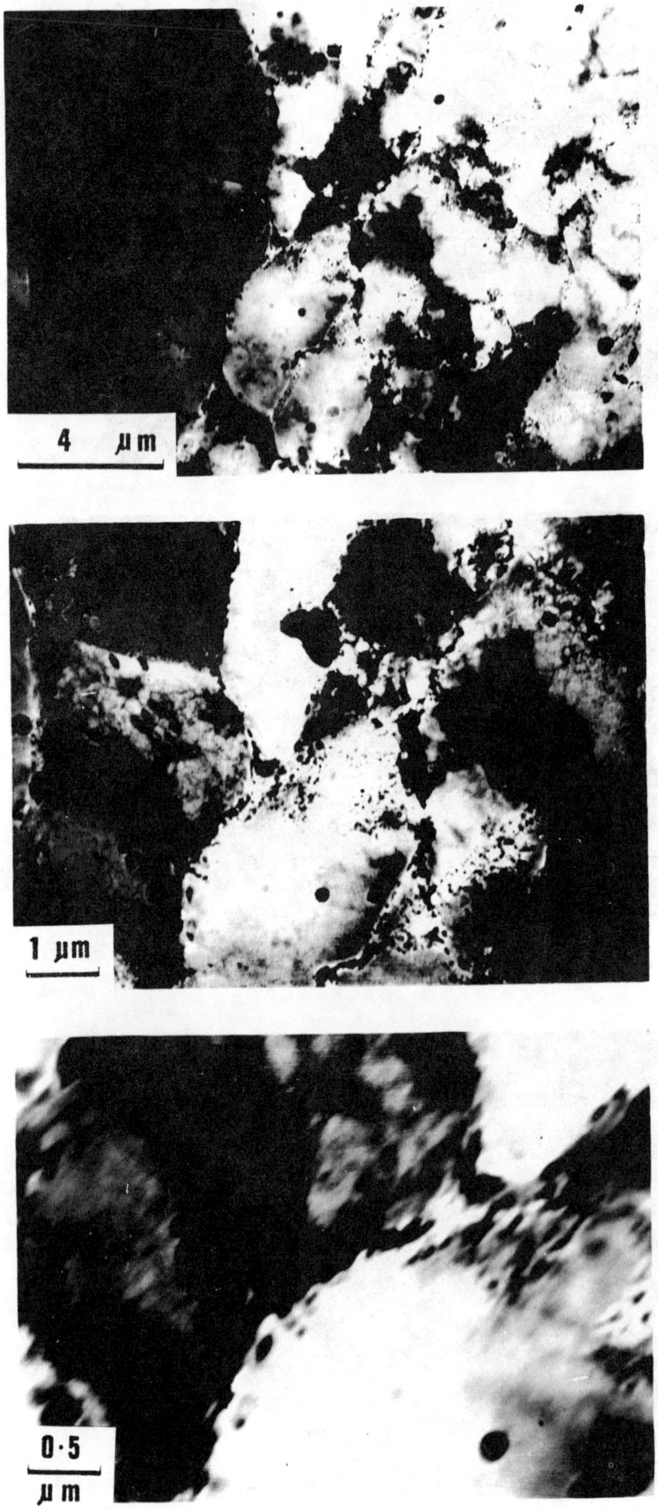

a

b

c

Figure 28. TEM, forged P/M alloy solution treated and aged, ε= 0: (a) 5000X, (b) 10,000X, (c) 20,000X.

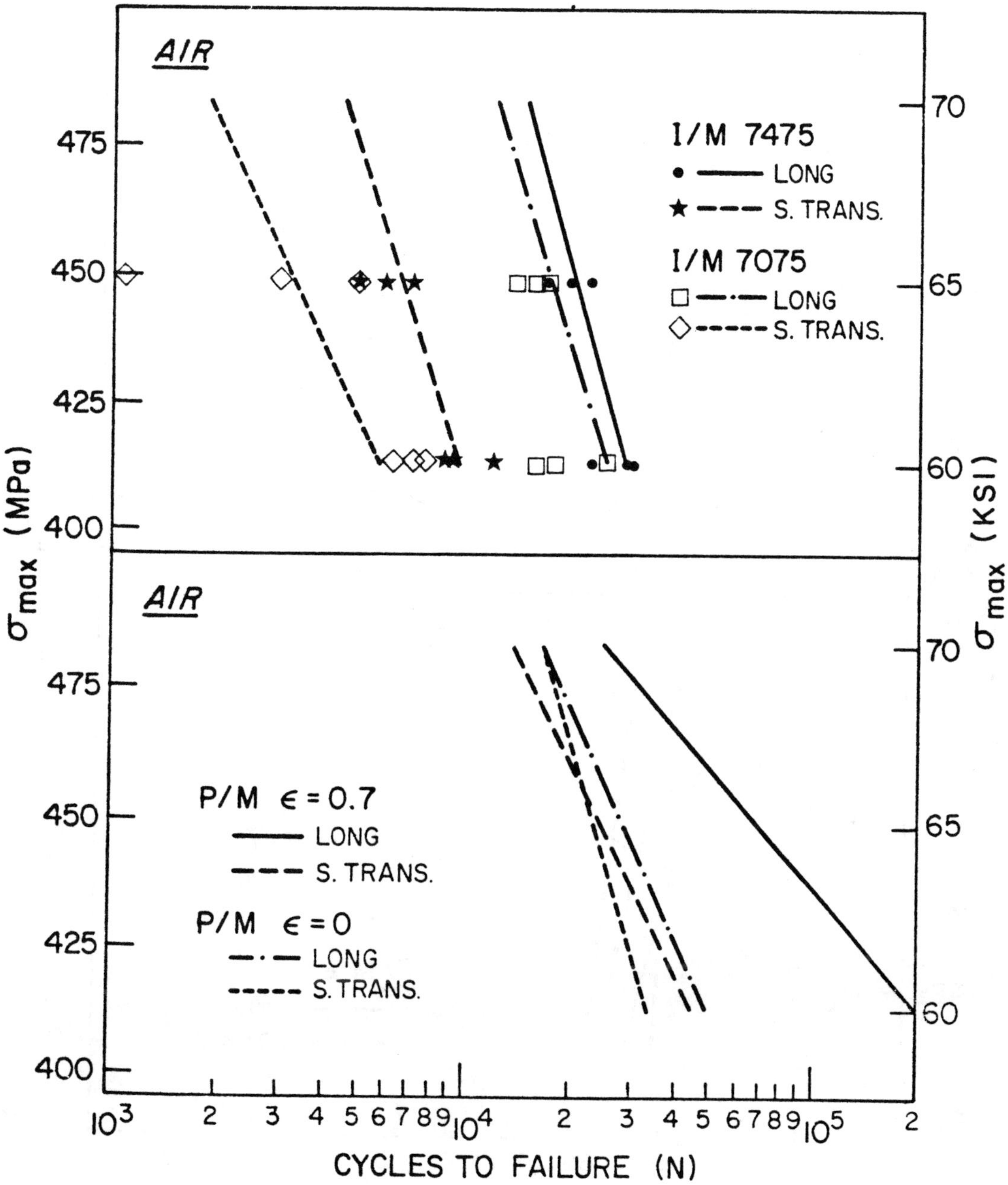

Figure 29. Median fatigue life vs σ_{max} for I/M 7075 and 7475, solution treated and aged material tested in air(27).

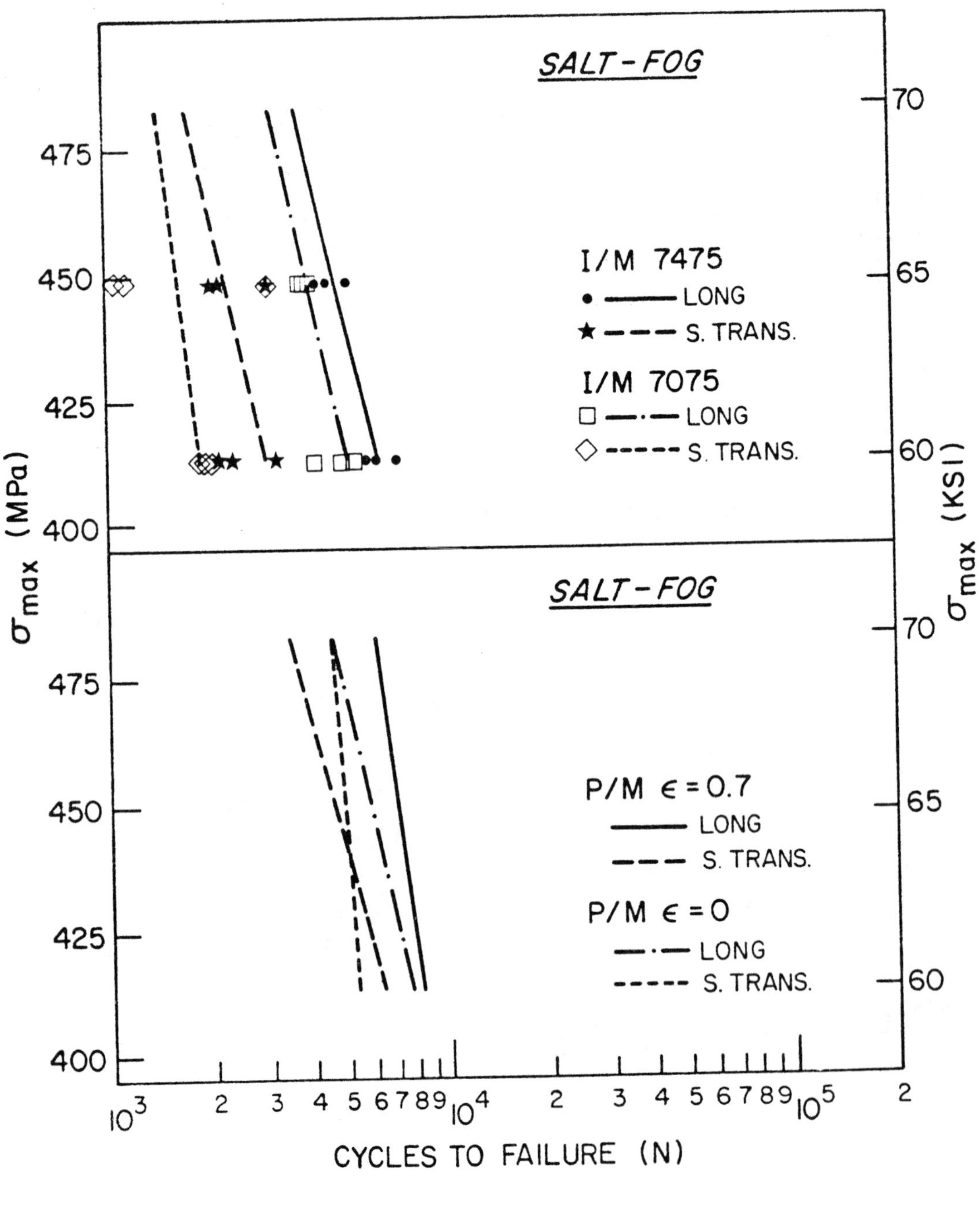

Figure 30. Median fatigue life vs σ_{max} for I/M 7075 and 7475, solution treated and aged material tested in 3 1/2% NaCl solution salt-fog (27).

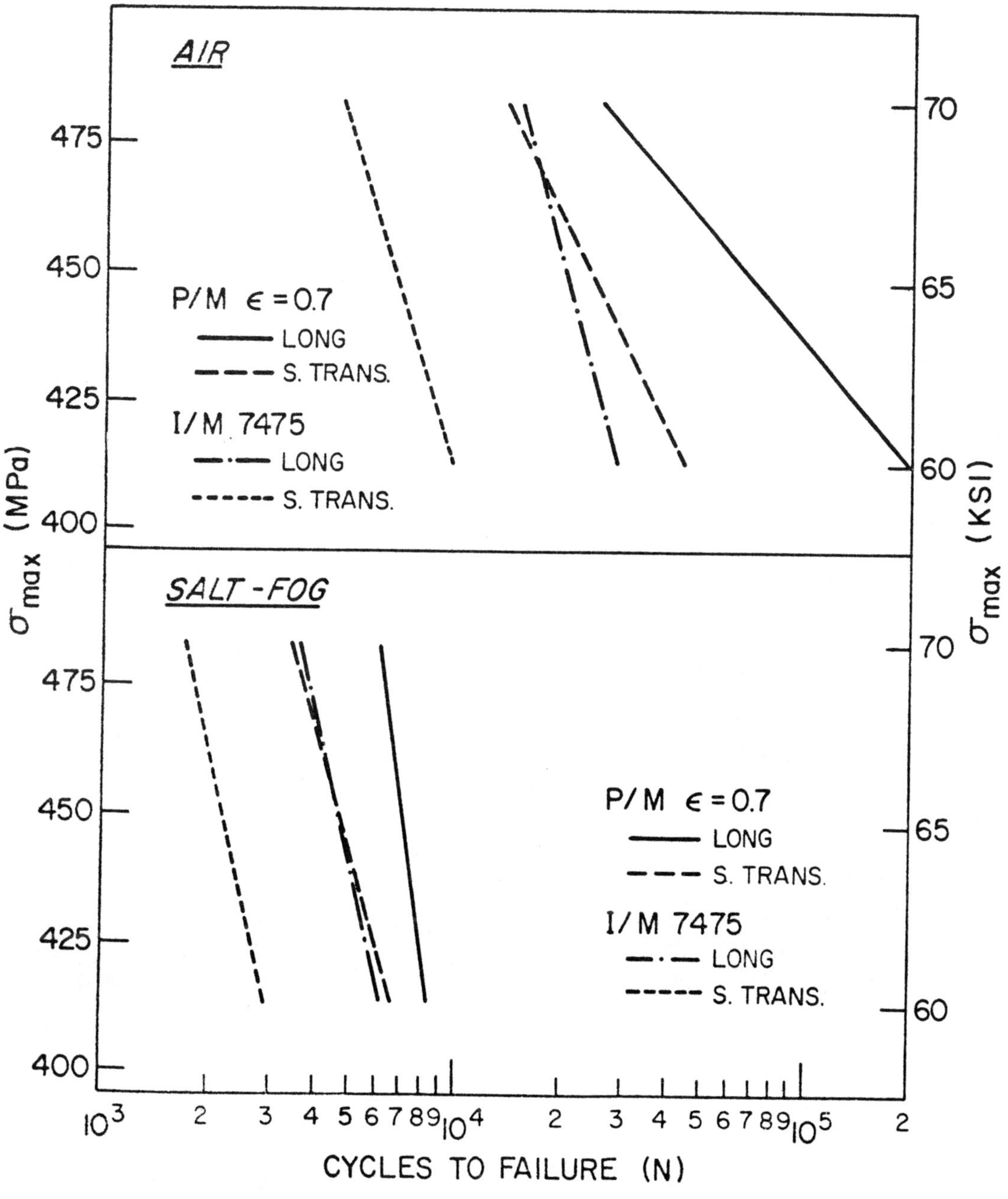

Figure 31. Median fatigue life vs σ_{max} for P/M forged (ε = 0.7) and I/M 7475, solution treated and aged material tested in air and 3 1/2% NaCl solution salt-fog (27).

JOINING OF NICKEL-BASE SUPERALLOYS BY HOT ISOSTATIC PRESSING

C.C. Law and M.J. Blackburn

Materials Engineering and Research Laboratory
Pratt & Whitney Aircraft - Commercial Products Division
United Technologies Corporation
Middletown, CT 06457 and East Hartford, CT 06108

Abstract

Selection of alloys for hybrid (dual-alloy) components must take into consideration the interfacial compositional reactions, heat treatment compatibility and mechanical integrity of the interface between the two alloys. Joining of nickel-base alloys by hot isostatic pressing (HIP) was studied to assess the joint properties of selected alloy combinations. Three different types of joints which involve alloys with chemically similar and dissimilar compositions: modified IN 100 powder/AF 115 powder, modified IN 100 powder/pre-HIP'ed AF 115 solid, and IN 100 powder/$\gamma/\gamma'-\delta$ eutectic casting were studied. It was found that the integrity of the powder/powder joint was excellent. Mechanical testing showed that neither stress-rupture nor tensile failures occurred by debonding. Control of joint location, shape and mixing of powders at the joint are major problems in the fabrication of a powder/powder assembly. Much lower integrity for the powder/solid joint was found. Most test specimens tested failed by debonding at the interface resulting in degradation of both stress-rupture and tensile properties. No new phases or unusual microstructures were identified in either of the above examples. In contrast, in the IN 100 powder/eutectic bond, diffusion of alloying elements across the interface resulted in formation of new phases and rather complex structures. In this case, the properties of the joint were very poor with failures occurring in the diffusion zone at very low strains.

Introduction

In components for advanced systems it is becoming increasingly difficult to satisfy all the mechanical property requirements with one material. For example, the microstructure and processing needed for high temperature applications often result in reduction in lower temperature properties. One method of solving such problems is by selecting suitable materials for each part location and joining them together to form the component. The use of powder alloys may facilitate fabrication of such a hybrid object with the desired characteristics in each location, and the construction of hybrid parts by hot pressing and by pressing and sintering have been described recently (1,2). This article describes the joining by hot isostatic pressing (HIP), of two powder alloys, a powder alloy to a pre-consolidated solid and to a eutectic casting. The microstructures developed at the joints and the integrity of the joints were the major features studied. Several recent government contracts have been awarded to examine the feasibility of producing turbine disk components with graded properties (3,4), using materials similar to some of those discussed here.

The alloys IN 100, modified IN 100 (5) and AF 115 (6) used for the present study belong to the family of gamma prime precipitation hardened nickel-base superalloys. The high strength and creep resistance of these alloys, coupled with severe macrosegregation in cast ingots, make them very difficult to forge. These alloys are, therefore, fabricated by powder metallurgy techniques. Consolidation of these powder alloys is usually accomplished by extrusion or by hot isostatic pressing. To achieve good interparticulate bonding, the lowest effective consolidation temperatures are slightly below the gamma prime solvus which is over ninty-five percent of the melting temperature.

The cast material used in the joining experiments was a recently developed bivariant nickel-base lamellar eutectic, a gamma/gamma prime-delta alloy. When directionally solidified under suitable growth conditions, this alloy freezes with an aligned composite structure containing alternating lamellae of gamma/gamma prime and the intermetallic compound Ni_3Nb (delta phase) (7,8,).

Experimental Details

Powder alloys used for this investigation, AF 115, IN 100 and modified IN 100, were vacuum induction melted and argon atomized. The powder particle size fraction selected was the -80 mesh. The eutectic alloy was produced in the form of 0.6 cm x 8 cm x 20 cm slabs directionally solidified parallel to the long dimension at a rate of 0.6 cm/h. Compositions of the consolidated powder alloys and the eutectic alloy are given in Table I.

Alternating layers of modified IN 100 powder and AF 115 powder were produced by sequentially pouring the powders into a stainless steel powder container (13 cm diameter x 18 cm high) through a fill-tube. To spread out the powder, the container was simultaneously rotated and vibrated after a predetermined quantity of each powder was poured into the container. The assembly was then outgassed at 260°C. The evacuated powder container was sealed and then HIP-consolidated at 1160°C/100 MPa/3 Hr. The consolidation was given a heat treatment of 1175°C/2 Hr./AC + 760°C/8 Hr./AC.

Material for the modified IN 100 powder/AF 115 solid joint evaluation was prepared in the following manner. Flat surfaces on a section of HIP-consolidated AF 115 were prepared by machining. The machined surfaces were cleaned by macroetching, followed by rinsing in distilled water, absolute ethyl alcohol, and drying. The cleaned piece was then embedded in modified IN 100 powder in a stainless steel can. Consolidation and heat treatment of the assembly were the same as those described above.

TABLE I

ALLOY COMPOSITIONS
WEIGHT PERCENT

	AF 115	Modified IN 100	IN 100	$\gamma/\gamma'-\delta$ Eutectic
Ni	Bal.	Bal.	Bal.	Bal.
Cr	10.5	12.3	12.5	5.9
Co	15.0	17.9	18.6	-
V	-	-	0.8	-
Mo	2.8	3.3	3.2	-
W	6.0	-	0.02	-
Al	3.9	5.0	5.0	2.5
Ti	3.9	4.3	4.3	-
Nb	1.8	1.70	0.01	20.1
Hf	2.10	0.74	-	-
B	0.020	0.018	0.020	-
Zr	0.05	0.06	0.06	-
C	0.150	0.025	0.055	0.003
O	88 ppm	94 ppm	61 ppm	10 ppm
N	34 ppm	30 ppm	15 ppm	10 ppm

The IN 100 powder/eutectic solid joint was similarly prepared, but consolidation temperatures selected were 1230°C, 1180°C and 1135°C. The powder/eutectic solid joint was evaluated in the as-consolidated condition.

The modified IN 100/AF 115 joints were evaluated by tensile, stress-rupture and notch-rupture tests for which the specimens are shown in Figure 1. The notch-rupture specimens had an elastic stress concentration factor (K_t) of 3.8 with the joint located at the notch root. The mechanical tests were conducted at several temperatures up to 730°C. The IN 100 powder/eutectic solid joints were evaluated by three-point bend tests at 20°C and 760°C. The dimensions of the specimens were 3.3 cm x 0.6 cm x 0.25 cm and the joint was located at the specimen midpoint where the bending condition is the most severe.

Results

Modified IN 100 Powder/AF 115 Powder Joint

The appearance of the powder/powder joint in a compact after macroetching is shown in Figure 2, from which three potential fabrication problems can be observed. The first is the control of the joint location and shape at a macroscopic level. In the present case curved interfaces instead of the aim planar surfaces was obtained. The second is the shape change after compaction, which is related to tap densities of the powders. AF 115 had a higher tap density than the modified IN 100 and therefore, resulted in less dimensional changes than the modified IN 100. The third is the inevitable mixing of powders at the joint during the can filling and subsequent warm outgassing, which as Figure 2b shows can result in substantial interpenetration of the powders.

The microstructure at the modified IN 100 powder/AF 115 powder joint is shown in Figure 3. The location of the joint is barely discernible at high magnification although macroscopically the interface is easily distinguished. Formation of new phases or unusual microstructure at the joint were not found by either optical or electron microprobe techniques.

In the tensile, smooth stress-rupture and notch stress-rupture tests no failure occurred by debonding at the joint and the properties measured were either those of the modified IN 100 or the AF 115. Typical appearance of a smooth stress-rupture or tensile tested specimen is shown in Figure 4. It was concluded therefore, that the integrity of the powder/powder joint was excellent.

Modified IN 100 Powder/AF 115 Solid Joint

The appearance of the HIP'ed compact and that of the powder/solid joint are illustrated in Figures 5a and b, respectively. As can be seen from the latter figure, this alternative joining technique eliminates the joint location and powder mixing problems described above.

The microstructure at the joint is shown in Figure 6. Precipitation of carbide particles at the joint is obvious, but the density of interfacial particles is lower than that at the prior particle boundaries in AF 115. No porosity, new phases or unusual microstructures were detected at the interface.

In spite of the favorable joint microstructure, the integrity of the solid/powder interface was lower than the powder/powder joint as can be seen from the stress-rupture and tensile results presented in Tables II and III. The stress-rupture and tensile specimens tested failed by debonding with reduction in both properties compared with the powder/powder joints. Based on earlier data it may be noted that the life of the smooth stress-rupture specimens was reduced by only a factor of two, the major effect being on notch-rupture life and tensile ductility. Scanning electron micrographs of the matching fracture surfaces are shown in Figures 7a and b, respectively. The AF 115 fracture surface is characterized by crater-like depressions, matched by protuberances on the modified IN 100 face, which suggest that intimacy of the contacts between the powder particles and the solid surface was not a problem.

TABLE II

TENSILE PROPERTIES OF
MODIFIED IN 100 POWDER/AF 115 SOLID JOINT

Temp. (°C)	0.2% YS (MPa)	UTS (MPa)	EL (%)	RA (%)	Fracture Origin
650	1033	1270	6.2	9.3	Debonding
650	1000	1290	5.3	7.4	Debonding

TABLE III

STRESS-RUPTURE PROPERTIES OF
MODIFIED IN 100 POWDER/AF 115 SOLID JOINT

Temp. (°C)	Stress (MPa)	K_t = 1.0				K_t = 3.8	
		Life (Hr.)	El (%)	RA (%)	Fracture Origin	Life (Hr.)	Fracture Origin
650	965	63.7	1.1	0.6	Debonding	0.0	Debonding
675	935	25.1	5.6	0.9	Debonding	0.1	Debonding
705	760	54.05	3.6	1.6	Debonding	1.2	Debonding

To further investigate the cause for the poor joint integrity, a cube of AF 115 with electropolished surfaces was exposed at the consolidation temperature 1160°C for 3 hours in a vacuum to determine if a carbide or oxide film formed on the surfaces. Following this exposure, extraction replicas were taken from the surfaces and studied by electron microscopy. Electron micrographs of the thermally exposed and unexposed surfaces are shown in Figure 8a and b, respectively. The exposed surface was thermally etched but no carbide or oxide was detected. The amount of carbide particles on both surfaces appears to be the same. Therefore, the observed deficiency of interfacial integrity in this case may be attributed to a lack of sufficient contact-surface area and/or inadequate interdiffusion at the bonding condition used.

IN 100 Powder/Eutectic Casting Joint

The microstructure of the joint produced by hot isostatic pressing at 1230°C is shown in Figure 9. The interface consists of globular regions within which gamma, gamma prime and Widmanstatten delta phases can be observed. Further, cracks have formed along the Widmanstatten delta interfaces. The microstructure suggests that melting of the interface had occurred during the consolidation, although the consolidation temperature is about 50°C below the eutectic temperature. For verification, a joint produced at a lower consolidation temperature was heated to 1230°C for 1 hour followed by quenching into water. This procedure should reveal the incipient melted regions, if formed, by producing a characteristic dendritic solidification structure. The interfacial microstructures after quenching is shown in Figure 10, which clearly shows that localized melting had indeed occurred during the 1230°C consolidation.

The microstructure of the joint produced by consolidation at 1180°C as illustrated in Figure 11, is distinctly different from that produced at 1230°C. Three regions may be distinguished near the interface: powder diffusion zone (PDZ), interface (I) and eutectic diffusion zone (EDZ). The PDZ differs from the unaffected powder in that it has larger grain size and the grain boundaries are no longer decorated with blocky gamma prime particles. Electron microprobe analyses show that the interface consists predominantly of niobium carbide particles (white) and gamma prime particles (dark) (Figure 11b). The EDZ differs from the unaffected eutectic in that the delta phase becomes less distinct from the matrix. The latter shows an absence of gamma prime particles. Major characteristics of the complex joint microstructure can be qualitatively understood from the compositional differences of the alloys (Table I). The IN 100 composition has higher aluminum and titanium contents which tend to diffuse toward the eutectic side. Aluminum and titanium losses reduce the amount of gamma prime phase and lower the gamma prime solvus, which contributes to the observed grain growth in the PDZ. The niobium content in the eutectic favors diffusion of this element toward the IN 100 side of the couple. The loss of niobium in the delta phase is compensated by removing aluminum from its gamma/gamma prime matrix in addition to the aluminum and titanium transfered from the IN 100. The interchange of niobium with aluminum and titanium in the delta phase results in the loss of gamma prime particles from the matrix and the loss of phase boundary definition. The interdiffusion of the alloying elements also results in an interface decorated with niobium carbide and gamma prime particles. These explanations are also consistent with the electron microprobe phase analyses.

The microstructure of the powder/eutectic joint produced at a lower consolidation temperature, 1135°C, is similar to that given in Figure 11 except that the diffusion zones are smaller and that, because of the higher niobium content, delta phase has reprecipitated in the EDZ as Widmanstatten platelets.

Bend tests conducted at 20°C and 760°C showed that the weakest location in the powder/eutectic joint is at the EDZ. This is true for both consolidation temperatures (1180°C and 1135°C). Appearance of the joint produced at 1135°C after fracture by bending at 20°C is illustrated in Figure 12. The fracture stresses are in the range of 1075 MPa to 1225 MPa, relatively insensitive to the test temperatures and consolidation temperatures (1180°C and 1135°C). Although these stress levels are quite high the powder/eutectic joints fractured in a brittle manner.

Discussion

Results presented above highlight some of the problems that may be encountered in the fabrication of a joint by hot isostatic pressing technique. The two major problems identified are to produce high quality joints and to accurately define joint location and shape.

In the present study, joint integrity was shown to depend on the forms of the alloys (powder or solid) and chemical compositions of the alloys. The integrity of the powder/powder joint was found to be excellent, and no failure caused by debonding at the interface was found in mechanical tests. However, the powder/solid joint of the same alloys fabricated at the same temperature, failed predominantly by debonding. The higher integrity of the powder/powder joint is considered to be attributable to the deformation in the powder particles during compaction, which enhances bonding by breaking up surface films, and mixing of powders which provides mechanical interlocking between the two alloys at the interface. The latter factor seems to be very important since deformation of the AF 115 solid interface also took place during the powder compaction (Fig. 7a), albeit to a lesser extent than that in the powder particles. Another potential problem which affects the integrity of a powder/solid joint, is the cleanliness of the solid surface. However, surface cleanliness did not appear to be an obvious problem in the modified IN 100 powder/AF 115 solid joint, neither was the formation of oxide or carbide film on the AF 115 solid surface. The poor integrity of the modified IN 100 powder/AF 115 solid joint is attributed, to a lack of mechanical interlocks and/or inadequate bonding conditions used. Thus bond quality may be improved by using a higher temperature and a longer hot isostatic pressing time.

The importance of alloy compositions on the integrity of the joint is amply demonstrated by the IN 100 powder/eutectic solid joint. In contrast to the above case, rapid interdiffusion of alloying elements occurs during hot isostatic pressing of alloys with significantly different chemical compositions, resulting in changes in chemical and mechanical property characteristics in the diffusion zones. In the case of the IN 100 powder/eutectic solid joint, the loss of niobium from the delta phase in the diffusion zone apparently weakened the material such that it became the preferential site for cracking. The interdiffusion also resulted in the lowering of the incipient melting temperature in the eutectic diffusion zone (Figure 10) and grain growth in the IN 100 powder diffusion zone (Figure 9). This condition did not arise in the case of the modified IN 100/AF 115 joints because the chemical compositions of the two alloys are rather similar. The main difference is the 6 percent of tungsten in AF 115, while the modified IN 100 contains no tungsten. However, this difference was not sufficient to produce any detectable microstructural change in either alloy, because tungsten enters into solid solution with nickel, a common feature of both alloys, and tungsten also diffuses very slowly.

Another problem which may affect the integrity of a joint in nickel-base superalloys is the formation of topologically close packed phases such as sigma, mu, chi and Laves. These intermetallics are hard and brittle and form as Widmanstatten platelets after long time exposure at intermediate temperatures (760°C to 980°C). Such precipitation usually causes a deterioration in tensile ductility and impact properties and also in stress-rupture life and ductility. Many advanced nickel-base superalloys are so saturated with solutes that they are on the verge of instability with respect to the formation of the intermetallics. Thus, the mixing of powders at the joint, or the interdiffusion of alloying elements, can change the long time stability of the material near the interface. It is possible to predict the phase stability using electron-vacancy calculations (9, 10, 11, 12) if the interfacial composition can be assumed to be an average of the two alloy compositions. However, for compositions based on IN 100, which are near the borderline of the (gamma + gamma prime) and (gamma + gamma prime + sigma) phase fields, it is known that losses of aluminum and titanium contribute to phase stability. Both AF 115 and the eutectic are stable alloys (6, 7, 8). Therefore, no interfacial sigma formation, as a result of powder mixing and/or interdiffusion, is expected of the alloy couples chosen for this study.

In the case of the powder/powder joint, the definition of the location and shape of the interface was shown to be a potential problem. However, this problem can be circumvented either by designing the joint in a noncritical location of an object or by using a specially designed powder filling equipment and procedures. Since hot isostatic pressing results in a fully dense consolidation, the shape of a hybrid component obviously depends on the tap densities of the powders. Therefore, basic parameters of the powder particles such as size distributions, shape and flow characteristics are important considerations and must be carefully controlled.

References

1. H.M. Skelly and J.T. McGoey: Int. J. of Powder Metallurgy and Powder Technology, vol. 14, p. 99, 1978.
2. H. Schreiner: Powder Metallurgy Int., vol. 10, p. 198, November 1978.
3. J.M. Marder: "Development of Materials and Process Technology for Dual Property Disks" NAS 3-21351, June 1978.
4. D. Wilson: "Development of Process Technology for Dual Property Disks" AFML Contract No. S-33615-77-C-5253.
5. M.J. Blackburn and R.A. Sprague: Metal Technology, vol. 4, p. 388, 1977.
6. J.L. Bartos: "Development of a very High Strength Disk Alloy for 1400°F Service" AF-TR-74-187, June 1974.
7. F.D. Lemkey and E.R. Thompson: NMAB 308-II, vol. 2, p. 105, January 1973.
8. F.D. Lemkey and G.P. McCarthy: "Proceeding of the Conference on In Situ Composites II" p. 77, Xerox Individualized Publishing, May 1976.
9. C.C. Law, W. Wallace, C.P. Ashdown and D.A. Grey: Accepted for publication in Metal Science J., 1979.
10. W. Wallace: Metal Science J., vol. 9, p. 547, 1975.
11. R.G. Barrow and J.B. Newkirk: Met. Trans., vol. 3, p. 2889, 1972.
12. L.R. Woodyatt, C.T. Sims and H.J. Beattie Jr.: Trans. TMS-AIME, vol. 236, p. 519, 1966.

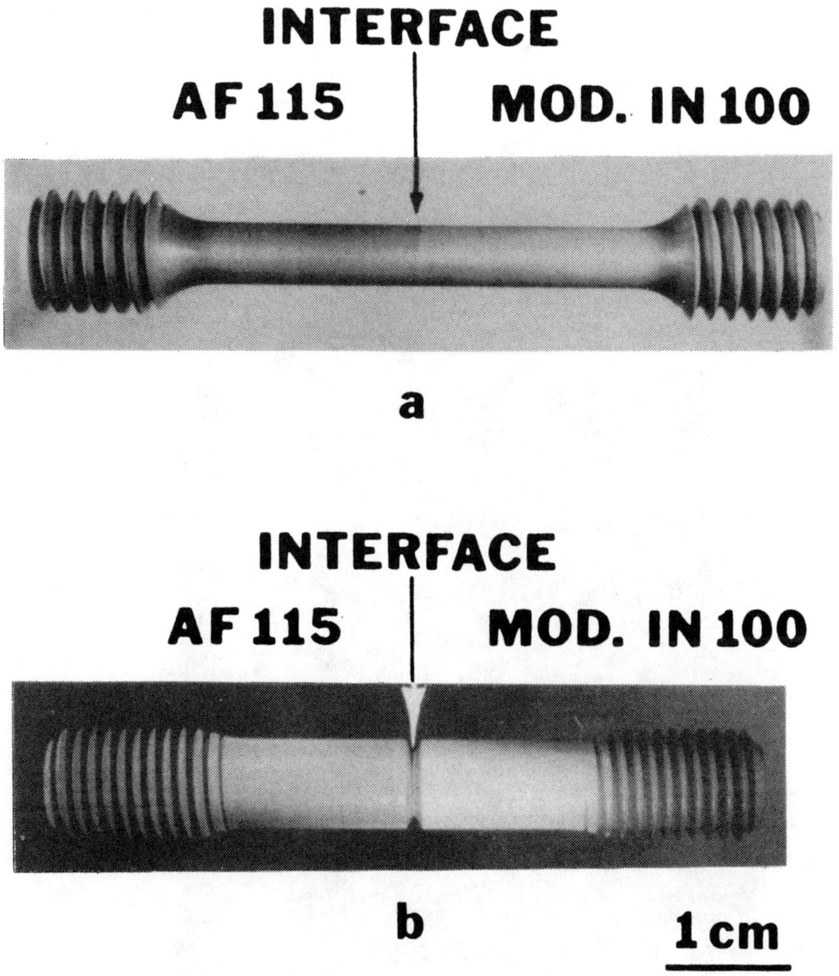

Specimens for the evaluation of joint integrity, (a) smooth stress-rupture/tensile specimen, (b) notched stress-rupture specimen.

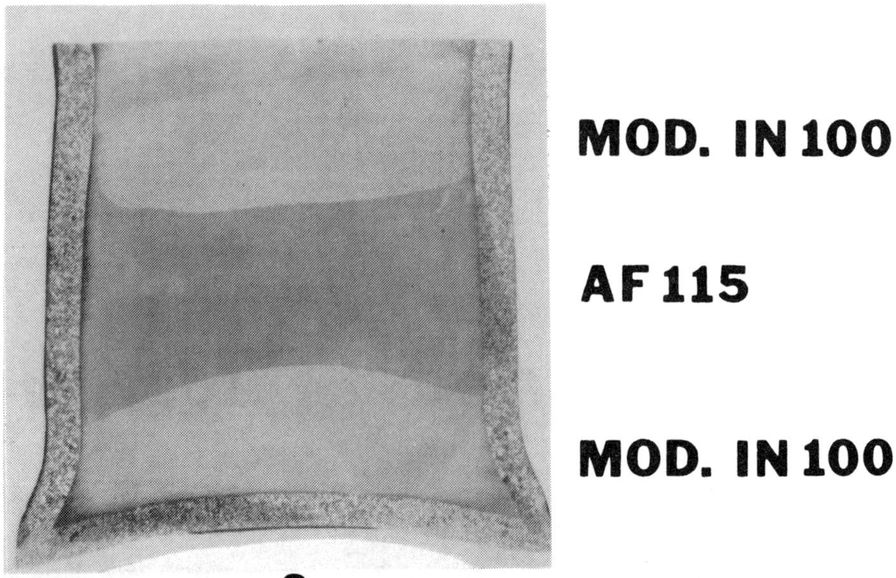

a

b

(a) Appearance of the modified IN 100 powder/AF 115 powder joint in a HIP-compact, (b) Mixing of powders at the joint.

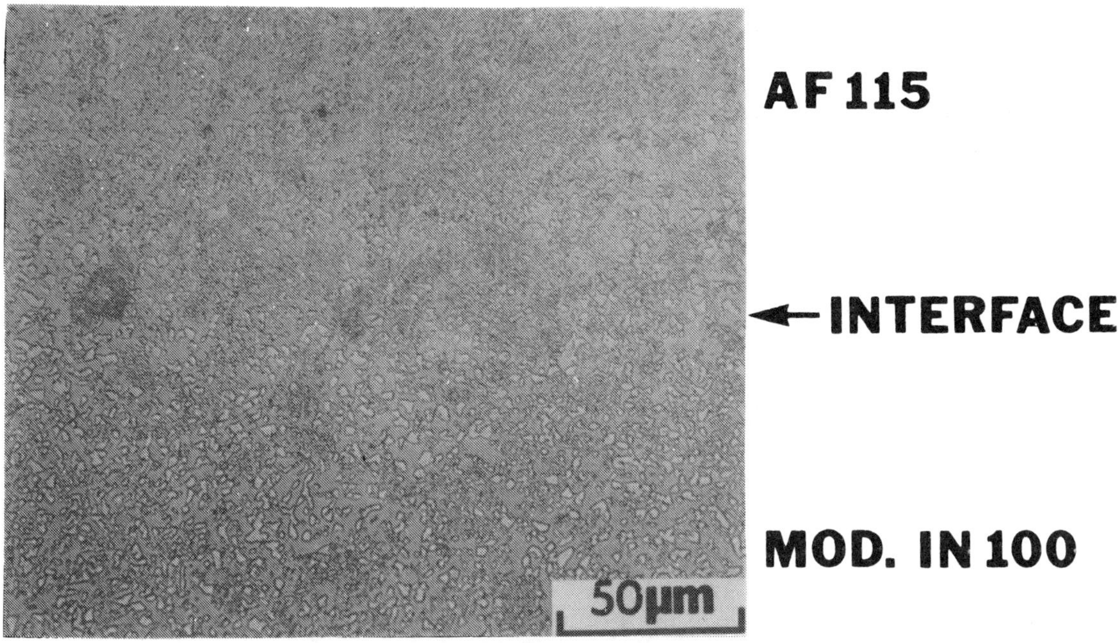

Microstructure of the modified IN 100 powder/AF 115 powder joint.

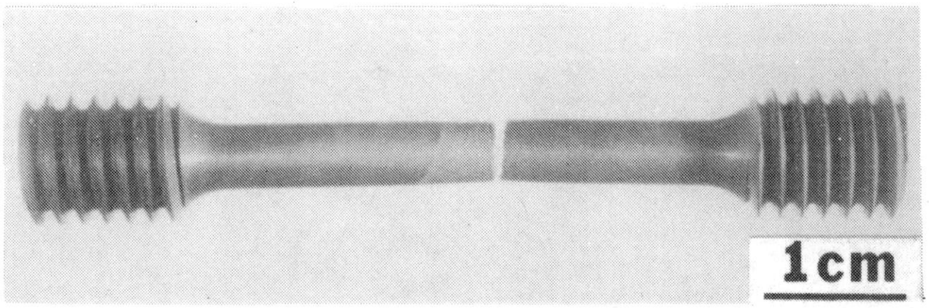

Appearance of a modified IN 100/AF 115 joint specimen after testing.

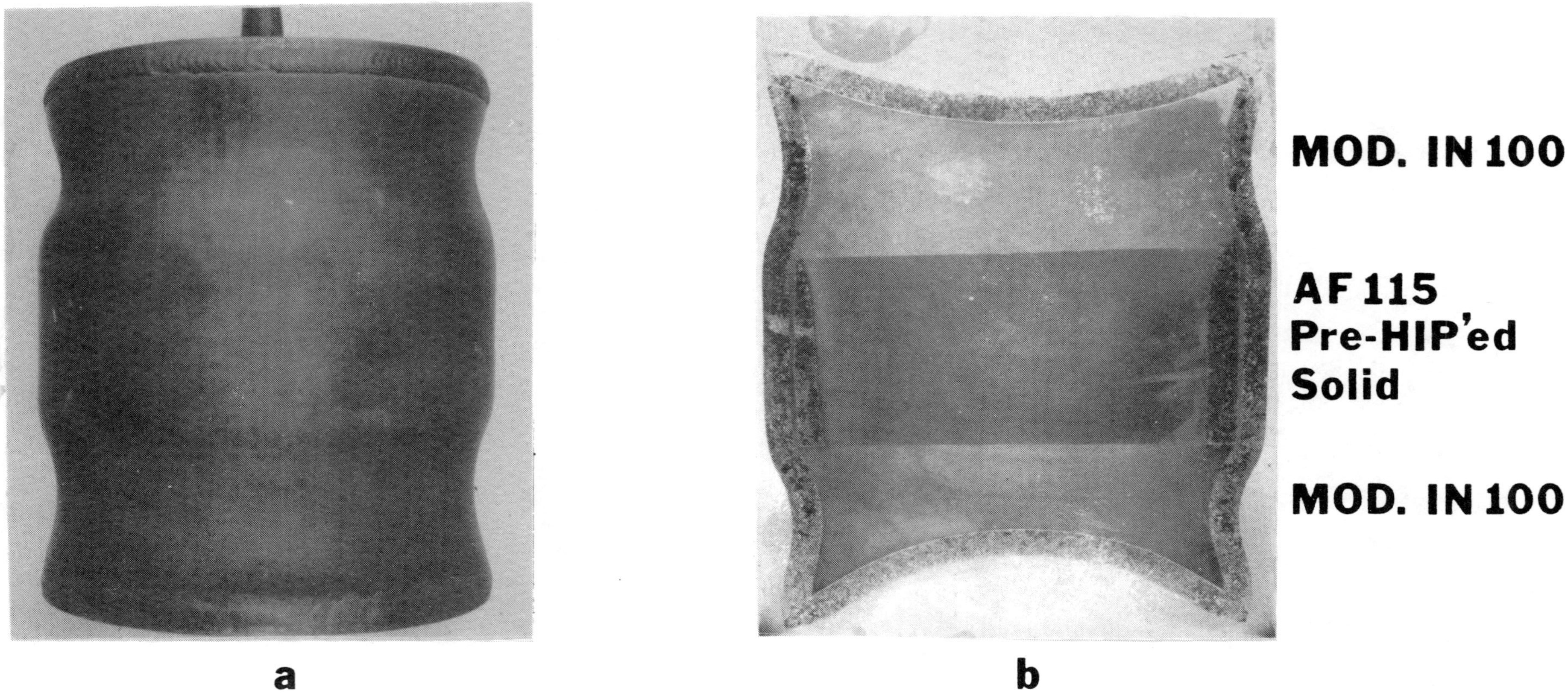

Appearance of the modified IN 100 powder/AF 115 pre-HIP'ed solid HIP-compact (a) before and (b) after sectioning and macroetching to reveal the powder/solid joint.

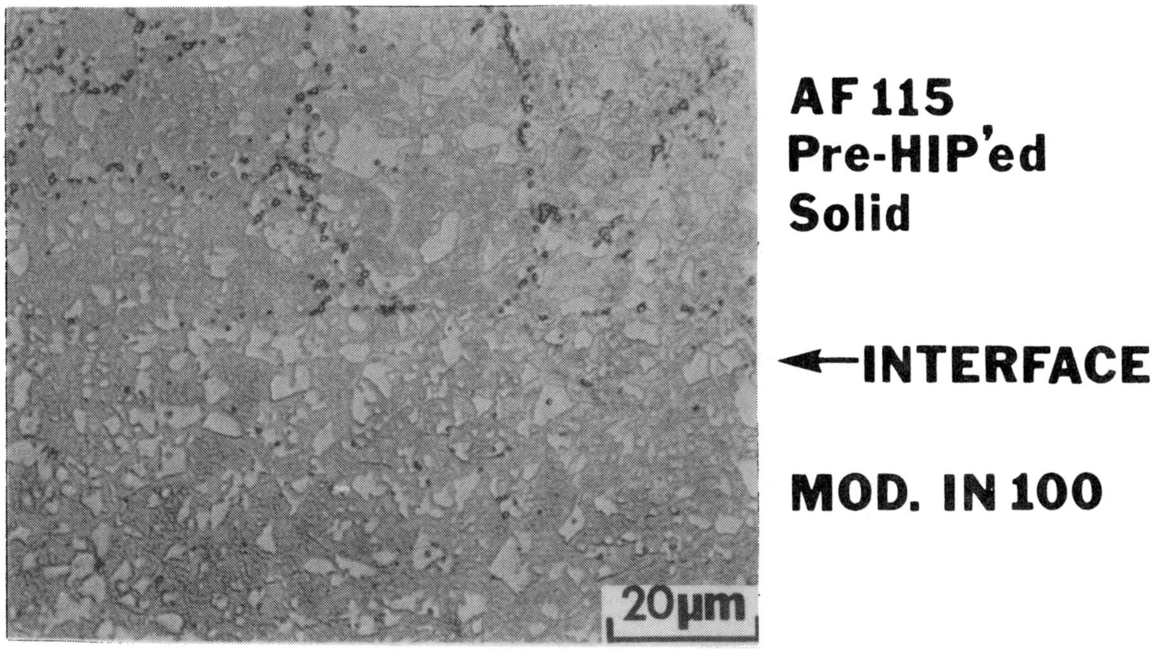

Microstructure of the modified IN 100 powder/AF 115
pre-HIP'ed solid joint.

Appearance of the fracture surfaces between (a) pre-HIP'ed
AF 115 solid and (b) modified IN 100 powder.

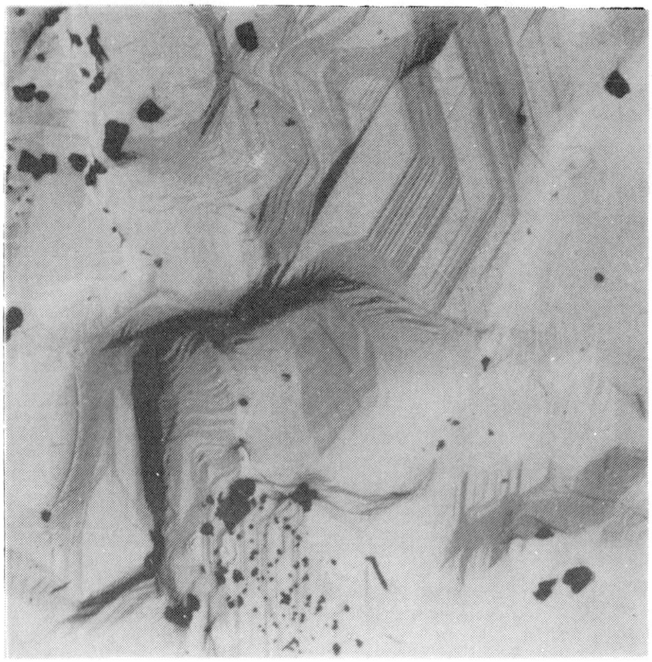

a

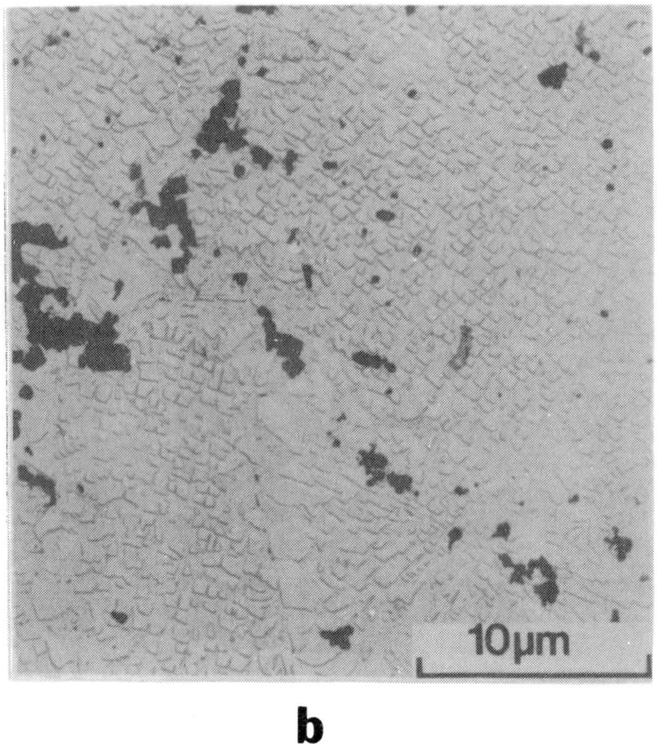

b

Extraction replica of AF 115 surfaces (a) after 1160°C/3 Hr. in vacuum (b) unexposed.

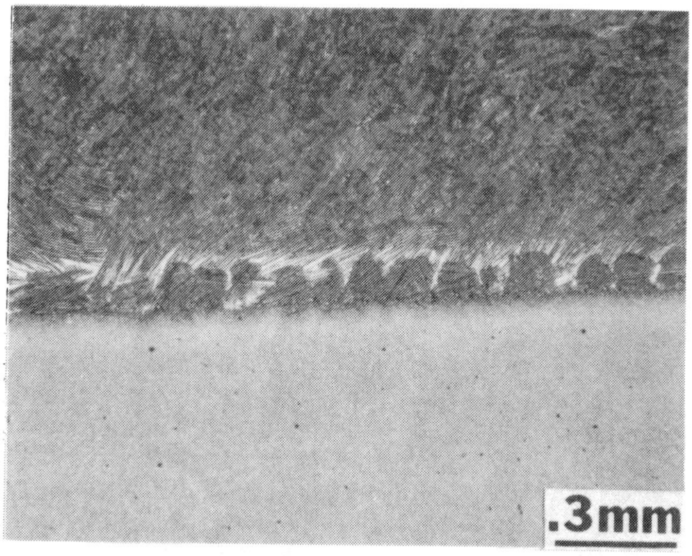

a

b

Microstructure of an IN 100 powder/eutectic solid joint produced by hot isostatic pressing at 1230°C.

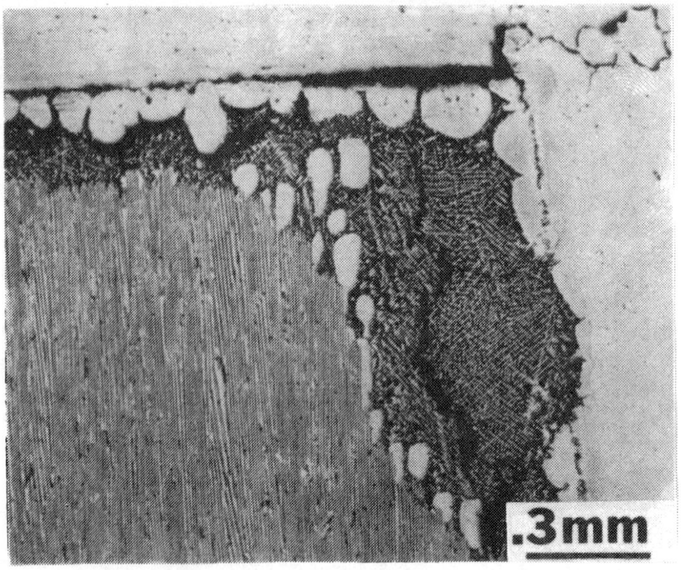

Microstructure of an IN 100 powder/eutectic solid joint after a heat treatment of 1230°C for 1 hour followed by quenching in water.

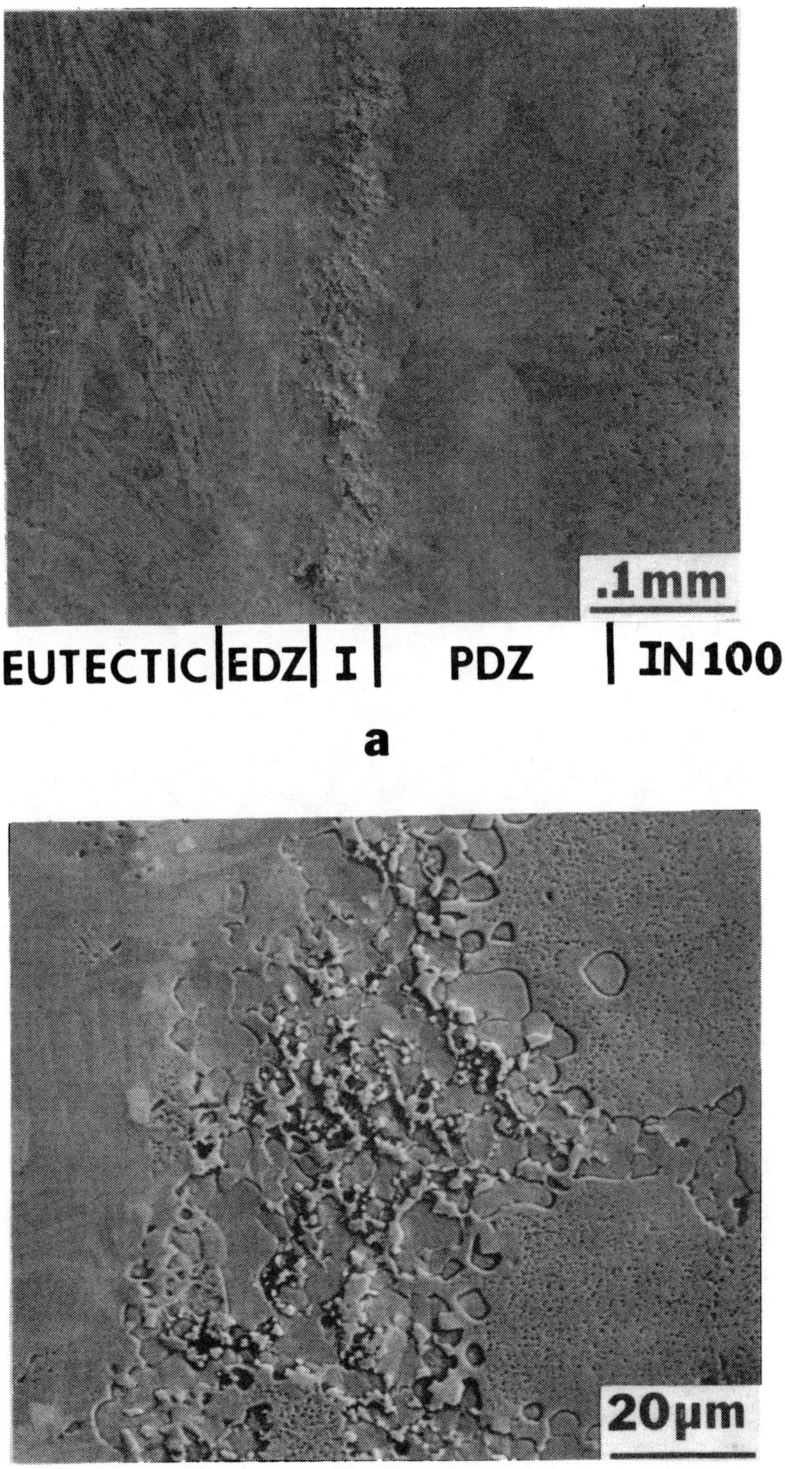

Microstructure of an IN 100 powder/eutectic solid joint produced by hot isostatic pressing at 1180°C. (a) The joint consists of an eutectic diffusion zone (EDZ), an interface (I) and a powder diffusion zone (PDZ). (b) Detailed microstructure of the joint.

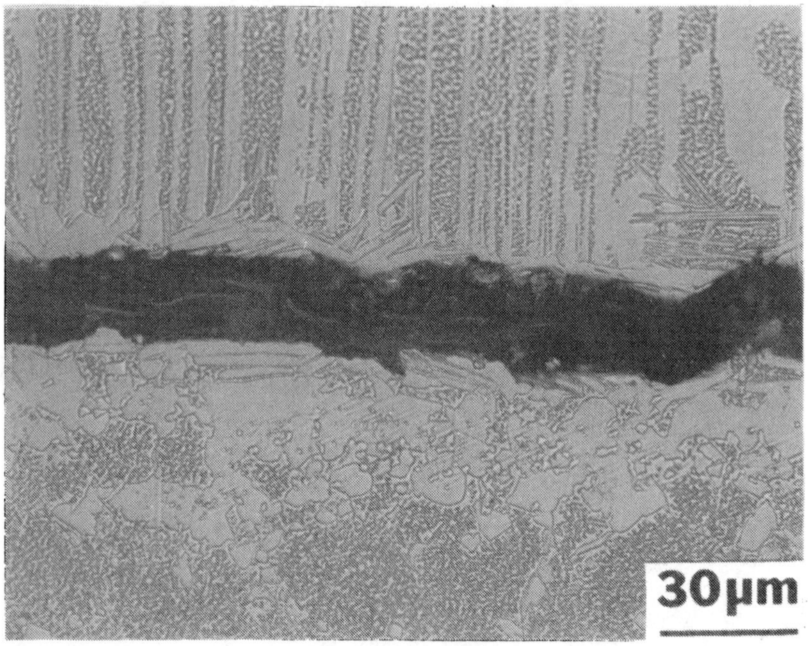

Appearance of the fracture in an IN 100 powder/eutectic solid joint produced by hot isostatic pressing at 1135°C. The specimen was fractured by bending at 20°C.

HIGH TEMPERATURE SINTERING OF FERROUS POWDER METAL IN NITROGEN BASE ATMOSPHERES

S. Mocarski and D. W. Hall

Ford Motor Company
Detroit, Michigan

ABSTRACT

Automotive P/M components and samples of commercial powders at 6.2 to 6.8 g/cc density were sintered at the conventional sintering temperature of 1120°C (2050°F) and also at high temperatures of 1232 and 1288°C (2250 and 2350°F) using different time cycles in an industrial walking beam furnace. For water atomized low alloy steel powder, the unnotched Charpy impact value increased from 14 joules (10 ft. lbs.) at 1120°C (2050°F) to 26 joules (19 ft. lbs.) at 1288°C (2350°F) after 30 min. sintering which is attributed mainly to the rounding of the pores. Morphology of the pores is shown in a series of photomicrographs. Spheroidization of the pores is rapid at high temperatures and the rate of spheroidization increases with the complexity of powder particles. Rounding of the pores at high temperatures is also associated with the elimination of prior particle identity as shown in a series of fractographs. High temperature sintered parts, due to improved impact strength and ductility, have potential for expanding the field of powder metallurgy, because parts so processed can be substituted for all but the most demanding forgings and also for nodular iron castings. High temperature sintering, due to reduction of oxides of manganese and chromium, promises increased application for the field of P/M forgings, as these least costly alloying elements can become fully effective to increase hardenability. Also, alloying by the use of master alloy premixes becomes possible. It was found that parts stacked 5 high sintered satisfactorily and separated easily after light tapping. Thus thermal efficiency of P/M processing at high temperature can be improved by heavier furnace loading. A nitrogen-base atmosphere of 96:4 N_2/H_2 composition with a small addition of CO gas performed satisfactorily during high temperature sintering, and such an atmosphere is considered superior to an endothermic gas for this application. Dimensional changes were studied and problems connected with dimensional control at high temperature are discussed.

INTRODUCTION

Sintering of metal powders has been the subject of numerous investigations and publications with various models being developed for material transport and shrinkage during sintering. The classical paper of Kuczynski (1) marked the beginning of analyses by many investigators. Excellent summaries of the present knowledge of sintering were made by Thümmler and Thomma (2), Moon (3), and Hirschhorn (4). Despite the fact that over 500 papers on the subject can be quoted, the mechanism of sintering is very complex and still not fully understood. The sources of driving forces for sintering are now widely accepted. In contrast several mechanisms of material transport have been advanced but disagreement exists among leading researchers as to which are involved at the various stages of sintering. The most commonly accepted theory is that at least one or sometimes several mechanisms are active simultaneously as the sintering process progresses. It has been known for many years that stronger parts with superior overall properties can be obtained by high temperature sintering (5,6). Interest in the industial use of high temperature sintering of ferrous parts has decreased in the last decade, mainly due to the limitation of belt-type furnaces to maximum temperatures of 1120 - 1150°C (2050 - 2100°F). This type of furnace, despite its disadvantages, is most commonly used in the U.S. for mass production and little data are contained in commercial and technical literature on the production of parts or compacts sintered above 1120°C (2050°F). European sintering practice was at first confined, to a considerable extent, to small batch sintering at temperatures up to 1325°C (2415°F) (7) but with the introduction of mass production of automotive parts and improvements in powder quality a trend to lower sintering temperatures and to belt-type furnaces has been noted, particularly in view of the high maintenance cost of the older type furnaces and difficulties encountered with dimensional control.

With the advent of modern high temperature furnaces, particularly the walking-beam, and the need for improvement of mechanical properties, weight reduction and energy savings it was decided to conduct an industrial experiment in a production type furnace. The purpose of the experiment was the evaluation of advantages and problems involved in high temperature sintering using a nitrogen based atmosphere. High temperature sintering using vacuum furnaces is also gradually finding more applications but vacuum sintering studies are not included in the present investigation.

EXPERIMENTAL PROCEDURE

a) Description of Facility

A production high temperature furnace, see Figure 1, walking beam type, 0.30 m (12") wide and 6.77 m (22.2') long, as described in commercial catalogues (8), was used for this industrial scale experiment. The furnace was heated by electric molybdenum elements and had sufficient power (188 kW) to operate at 1800°C (3300°F). The furnace was equipped with gas tight doors which operated automatically during the furnace cycle, a flame curtain being switched on while the door was opened to admit a tray. Trays, 30 x 23 cm, (12 x 9") were made from mild steel sheet 3/16" thick, see Figure 2. The furnace was equipped with nitrogen gas barriers which protected the molybdenum elements from oxidation by the accidental entrance of air. The furnace was gas tight and only 12.2 cubic meters/hour (430 cfh) of gas atmosphere were used.

The furnace was designed for sintering of products other than ferrous metal parts and, therefore, lubricant burn-off was performed in a separate conventional belt-type furnace at a temperature of 760°C (1400°F) under a protective nitrogen gas atmosphere. The high temperature sintering furnace had a comparatively short cooling zone which resulted in several loads being slightly oxidized after being prematurely pulled from the furnace.

b) Furnace Loading

The trays were loaded into the furnace as shown in Figure 2. Each experimental run consisted first of 2 trays carrying dummy loads followed by several trays with experimental production parts, test samples and a 75 mm (3 inch) diameter, 75 mm high compact drilled to its center to accommodate a travelling thermocouple which monitored each run. Again two trays with dummy loads were at the end of each test run.

c) Sintering Time and Temperature

The experiments were designed to study three temperatures and three sintering times at each temperature. The matrix for these tests is given in Tables I and II. Table I shows the times at temperature for relatively thick parts as monitored by the travelling thermocouple and plotted on a time-temperature recorder. Table II shows the slightly longer times at temperature for thin transverse rupture bar specimens.

d) Atmosphere

Gasified liquid nitrogen was used as a base atmosphere with the addition of dissociated ammonia to provide a reducing gas having a composition of 96% nitrogen and 4% hydrogen; to this mixture various amounts of methane were added to the first 10 runs. While trying to determine the time-temperature profiles, methane gas additions were varied to achieve a neutral atmosphere, i.e. one which would not cause carburization or decarburization of the samples. The dew point of the furnace was -42° to -48°C (-45° to -55°F); the carbon potential in the furnace was determined after each run by metallographic examination. While the final setting of 0.8% methane was found to provide satisfactory results as to the carburizing-decarburizing condition, the methane content had to be controlled within narrow limits. An increase of methane to 1.2% caused excessive sooting in the furnace and on the parts. Due to the critical nature of the methane addition and the undesirable sooting condition which could result in damage to the furnace it was decided to add bottled carbon monoxide gas rather than methane. An addition of 0.8% CO to the base 96% N_2 - 4% H_2 atmosphere resulted in satisfactory carbon control and neither carburization nor decarburization was observed in the test samples despite their variation in carbon content. The microstructure of a test bar sintered for 30 minutes at 1288°C (2350°F) is shown in Figure 3. A uniform carbon content from the surface to the center of the bar can be observed. The dew point of the furnace for CO additions was the same as for the atmosphere with methane additions.

e) Samples and Powders Investigated

Each furnace run consisted of production parts, 75 mm dia slugs of prealloyed iron powder and test bars of the compositions shown in Table III.

T A B L E I

SINTERING TIMES AND TEMPERATURES FOR RELATIVELY THICK PARTS

	Sintering Times (minutes) at temperatures shown		
	$1120^{\circ}C$ * ($2050^{\circ}F$) **	$1232^{\circ}C$ ($2250^{\circ}F$)	$1288^{\circ}C$ ($2350^{\circ}F$)
Time (Minutes)	6	6	6
	12	12	12
	25	30	30

T A B L E II

SINTERING TIMES AND TEMPERATURES FOR THIN SPECIMENS

	Sintering Time (minutes) at temperatures shown		
	$1120^{\circ}C$ * ($2050^{\circ}F$) **	$1232^{\circ}C$ ($2250^{\circ}F$)	$1288^{\circ}C$ ($2350^{\circ}F$)
Time (Minutes)	8	8	8
	16	16	16
	30	35	35

* $\pm 14^{\circ}C$

** $\pm 25^{\circ}F$

T A B L E III

SAMPLES USED FOR HIGH TEMPERATURE SINTERING

Description of Sample	Description of Powder	Density g/cc	Chemical Composition-Weight %								
			C	Si	Mn	S	P	Cu	Ni	Mo	Fe
Test Bars	Water atom. Cast Iron Atomet 28	6.8	.51	.04	.01	.004	.027	-	-	-	Bal
Test Bars	Sponge Iron MH 100	6.4	.72	.09	.03	.010	.002	-	-	-	Bal
Test Bars	H_2 Red. Mill Scale P-100	6.2	.03	.10	.46	.003	.008	2.8	-	-	Bal
Test Bars	Water Atom. Mn-Mo Low Alloy Steel	6.8	.15	.04	.45	.018	.010	-	.05	0.49	Bal
Test Bars 75 mm slugs	Water Atom. Mn-Ni-Mo Low Alloy Steel	6.8	.20	.03	.35	.016	.010	-	.43	0.65	Bal

RESULTS

All sample dimensions were measured after pressing, after lubricant burn-off and after sintering. Representative specimens were sectioned and prepared for metallographic examination. The large, 75 mm diameter cylinders were normalized, cut in sections and machined into Charpy impact bars.

a) Mechanical Test Results

Unnotched Charpy impact test bars were prepared from 75 mm dia slugs made from MDO-101 Ni-Mn-Mo low alloy steel powder, density: 6.8 g/cc in a normalized condition. The test results showed a big increase in impact strength with increasing sintering time and temperature, see Figure 4. Conventional sintering at 1120°C, for 25 minutes resulted in an impact strength of 16 joules (12 ft. lb.) and 30 minutes at 1288°C 26 joules (19 ft. lb.). The impact strength shows little variation with impact test temperature, the average decrease in impact strength from room temperature to -51°C (-60°F) being only 3 joules. This is an extremely important benefit of high temperature sintering, as in many applications the low ductility and impact strength of porous iron alloys prevent designers from specifying parts from P/M despite the economic advantages offered by the process. No other powders were impact tested; it would be of great interest to obtain impact strength data for other powders and for various heat treat conditions.

b) Fractography of Impact Bars

SEM fracture photographs of impact bars of water atomized low alloy Mn-Ni-Mo steel at 6.8 density taken originally at a magnification of 500x are shown in Figs. 5a-d. It can be seen that conventional sintering for 25 minutes at 1120°C does not result in sufficient transport of material to eliminate prior powder particle boundaries and the fracture looks almost like loose powder with some dimples here and there where the necks between the powder particles were formed. Six minutes sintering at 1232°C (Fig. 5b) moves enough material to almost obliterate the identity of powder particles; complete elimination of prior particle boundaries occurs after sintering for 6 minutes at 1288°C (Fig. 5c). Extending the time at the latter temperature coarsens the texture (Fig. 5d) of the fracture. This behaviour can be attributed to grain growth and coarsening of the pores.

c) Metallographic Examination - Pore Morphology

The polishing of metallographic specimens of porous sintered iron is a difficult task. Only with proper techniques of specimen preparation can the study of pore morphology be accomplished. Procedures outlined by Pease [9] for the mounting and polishing of porous specimens were rigorously followed. Specimen porosities as determined by the wet density method were compared with pore areas on polished metallographic samples. A QTM microscope with a TV monitor connected to a computer terminal automatically scanned, selected, measured and averaged 20 to 50 random areas on each specimen. Metallographic polishing was continued in one minute intervals until the percentage of pore area as measured on the QTM corresponded to the known percentage porosity of the sample.

The photomicrographs in Fig. 6 through 12 show the effects of sintering time and temperature on the shape, size and distribution of pores in several iron and low alloy steel powders.

Water Atomized Cast Iron, 6.8 g/cc Density. Fig. 6a-f is a series showing the pore structure of water atomized iron, Atomet 28, containing 0.5% carbon at sintering times of 8 and 30 minutes and at sintering temperatures 1120, 1232 and 1288°C. The micrographs were taken originally at 500x. Note in Fig. 6a some of the original particle boundaries are still visible; in Fig. 6b the particle boundaries are breaking up forming small isolated voids but the original particle shape can still be distinguished. A structure similar to that shown in Fig. 6b is found in Fig. 6c indicating that the pore structure is approximately equivalent whether sintering this powder at 1120°C for 30 minutes or 1232°C for only 8 minutes. As we proceed to longer times and higher temperatures Fig. 6d, e, f it is noted that pore rounding and pore spheroidization becomes evident. Prior particle boundaries are no longer present.

Sponge Iron 6.4 g/cc Density. Figs. 7a-i shows, at 200x original magnification, sponge iron MH100 containing about 0.7% carbon. Here again short-time sintering for 8 minutes at 1120°C is insufficient to eliminate the original particle boundaries and even sintering for 16 and 30 minutes does not completely remove them. Significant rounding of pores and some spheroidization begins to appear in Fig. 7e at 1232°C for 16 minutes. Many rounded pores are apparent in 7f and sintering appears to be quite adequate. Speroidization and rounding of isolated pores are clearly shown in Fig. 7i. Fig. 8a-f shows the pore shape and its variation with sintering time and temperature for sponge iron, MH100 at 500x magnification.

Reduced Mill Scale Iron Powder plus 2.8% Copper, 6.15 g/cc Density. The pore morphology of this powder is shown in Figs. 9a-d and 10a-d taken originally at 200 and 500x respectively. The original particle shape of this powder is considered complex and irregular, the compacts have appreciably more interconnected porosity than other powders tested. Sintering at 1120°C for only 8 minutes shows in Fig. 9a and 10a that very little of original particle boundaries remain. This behaviour is due in part to the presence of copper which becomes for a short period of time a liquid phase at the sintering temperature. Note the large irregular pores which were the original sites of the copper phase before melting and solution occurred. The rapid rate of sintering is especially noticeable at 1288°C Fig. 10d, and the spheroidization of the pores is reminiscent of the spheroidal graphite nodules in inocculated cast iron.

Water Atomized Mn-Ni-Mo Low Alloy Steel Powder, 6.8 g/cc Density. The change of pore shape with temperature and time is shown in Figs. 11 and 12. This powder has the least complex particle shape of all powders tested and also shows the least amount of densification and pore rounding. Sintering at 1120°C within the time periods investigated does not eliminate the identity of powder particles, while 16 minutes and longer at 1232°C and 8 minutes at 1288°C already exhibit rounding of the pores. Some coarsening of the pores is observed after 30 minutes at 1288°C.

Grain Size. From Table IV it can be seen that the grain size throughout sintering at times and temperatures indicated is rather stable. A slight increase in grain size is noticeable except for sponge iron which is the same at all temperatures run; sintering for 30 minutes at 1288°C results in a grain size of 7-6 for water atomized cast iron, 7 for sponge iron, 6 for reduced mill scale and 7-6 for low alloy steel powder. Thus the greatest amount of grain coarsening was observed on the most active powder, i.e. on reduced mill scale, and even here the degree of coarsening was not great when compared with conventional steels.

TABLE IV
ASTM GRAIN SIZE

Powder:	Grain size after sintering at:								
	1120°C			1232°C			1288°C		
Minutes	8	16	30	8	16	30	8	16	30
Water Atomized Cast Iron Atomet 28	7-8	7-8	7-8	7	7	7	7	7	7-6
Sponge Iron MH 100	7	7	7	7	7	7	7	7	7
Reduced Mill Scale P-100 + 2.8% Cu	Not Discernible			7-6	7-6	7-6	7-6	7-6	6
Water Atomized Low Alloy Mn-Ni-Mo Steel	8	8	8	7	7	7	7-6	7-6	7-6

d) Dimensional Changes During Sintering

Fig. 13 shows dimensional changes from die size for selected powders after sintering at 1120, 1232 and 1288°C (2050, 2250 and 2350°F). The data are only approximate, as the cooling rates from the sintering temperatures varied. They represent only the order of magnitude of the size change from the die during sintering and are appropriate only for the purpose of further discussion. More statistical data would be required to obtain usable data for production purposes.

The data for sponge iron are incomplete as this powder more than others has shown a greater trend to carbide formation and this possibly in conjunction with varying cooling rates resulted in a considerable scatter of results. The tests could not be repeated under similar conditions at this time. Comparing the dimensional changes shown in Fig. 13 it can be seen that the particle shape and other powder characteristics have a considerable influence on the shrinkage or growth of iron P/M parts. Usually powder manufacturers try to blend their powders in such a way as to obtain a curve representing shrinkage vs. sintering time as flat as possible, and to limit the dimensional change from the die size to less than 0.5%; otherwise size control during sintering becomes difficult (10). It can be shown from Fig. 13 that sintering at 1232°C will give reasonable size control for the majority of powders or premixes studied. Sintering at 1288°C, however, results in rapid shrinkage for most powders, particularly for reduced mill scale powder with 3% copper added. This powder sintered for 35 min. at 1288°C exhibited a shrinkage in excess of 1%. This can be explained by its complex particle shape, large specific surface and the magnitude of interconnected internal porosity. Thus, for conventional P/M applications with close tolerances, sintering at 1288°C could create dimensional control problems and very likely a sizing or cold forging operation would be necessary.

DISCUSSION OF RESULTS

a) Pore Shape and Mechanical Properties

High temperature sintering brings about some shrinkage and densification of the compacts, also a change in pore shape. Both these factors, but particularly pore shape result in a sizeable increase in impact strength.

A recent paper by Merhar (16) shows many advantages of high temperature sintering, among them the increase in impact strength and ductility of stainless steels.

He quotes an unnotched impact strength after 1 hour of sintering at 1260°C (2300°F) of 8 joules (6 ft. lbs.) and 78 joules (58 ft. lbs.) after sintering at 1371°C (2500°F) at a density of 6.5 to 6.8 g/cc; further, 34 joules and 167 joules (25 and 123 ft. lbs.) respectively at 7.0 to 7.2 g/cc density are quoted; the benefits of high temperature sintering being greater at higher densities.

Many investigators point to the importance of pore shape (11-13) with regard to tensile strength and elongation; not much data can be found in the literature on impact strength. Some papers (14,15) compare sintered P/M to cast iron. The notch effect of flake graphite is similar to the notch effect of sharp irregular pores, however, some of the models developed for the stress-strain behaviour of cast iron do not apply to powder metallurgy. Lawley (18) and Ferguson (19) explain the behavior of residual pores in hot formed P/M. Voids behave differently than non-metallic inclusions in metals, and graphite in cast iron can be considered an inclusion. The development of nodular iron by modification of the graphite shape has obviated the undesirable feature of the notch sensitivity caused by graphite flakes. Nodular iron with round spheroidal graphite is now a popular material which has replaced many former steel forgings in the automotive vehicle some 10 to 20 years ago, e.g. crankshaft, slip yoke, companion flange and connecting rods. Similar improvement in the properties of sintered P/M components can be achieved by spheroidization of the pores via high temperature sintering. In many industrial applications P/M is not used because design engineers are deterred by the low ductility and fear of brittle failure particularly with the rise of consumerism and product liability. More work would be required to obtain appropriate fatigue properties, tensile strength and ductility data.

b) Pore Rounding and Spheroidization

Comparing the pore morphologies of different powders it can be seen that the degree of pore spheroidization during the same sintering cycle varies from powder to powder. It would appear that the particle shape, specific surface, internal porosity, the amount of cold work and dislocations have an influence on pore rounding. The greatest degree of spheroidization was observed on hydrogen reduced mill scale powder, Fig. 10d. Gold (17) has described this powder and has shown that conventional sintering results in greater strength for this powder than for water atomized powder. The water atomized low-alloy powder had possibly the least amount of pore spheroidization, nevertheless it still showed an increase in impact strength from 14 joules at 1120°C to 26 joules when sintered at 1288°C for 30 minutes, the density being about 6.8 g/cc.

It is obvious from the metallographic review of several series of powders (Figs. 6-12) and from SEM pictures of fractures (Fig. 5) that there is an interchangeability of time-temperature cycles to achieve the same degree of pore shape which is reflected by an increase in strength and ductility. This equivalency could be concluded from the work of Bockstiegel (20) on the effect of sintering temperature on strength and Jordan and Duvez (21) who studied densification at different temperatures. As the degree of densification and strength appear to be interdependent, so the degree of sintering at different times t and absolute temperatures T could be interrelated by the equation

$$\log\left(\frac{t_1}{t_2}\right) = \frac{Q}{R}\left(\frac{1}{T_1} - \frac{1}{T_2}\right)$$

Where R is the gas constant
Q the molar heat of activation

Q the heat of activation, appears to be much dependent upon the particle shape being lowest for complex shaped powders. The increase in surface tension of complex particles results in forces which cause the transport of material. It would also appear that Q would decrease as the cleanliness of the powder increases. Oxygen rich films, particularly of irreducible oxides in given physico-chemical conditions, are known to slow down or even prevent sintering.

A modification of present-day powders by developing powders of complex shape and higher cleanliness i.e. with less oxygen content and smaller less numerous non-metallic inclusions will facilitate pore sheroidization. Complex particle shape appears to be of major importance in accelerating densification and pore spheroidization during sintering. The greater the amount of surface area in the original material the greater the driving force for sintering [22].

The equivalence of sintering time-temperature relationship was also reported by Gummeson and Stosuy [6] who state that for sponge iron 4 hrs (240 mins) at 1120°C is equivalent to 10 min. at 1204°C. Material transport during sintering depends mainly upon diffusion phenomena; the speed of material transport increases exponentially with temperature and small increases at higher temperature levels may cause the speed of reaction to increase one or two orders of magnitude. It is conceivable that proper development of metal powders for high temperature sintering would result in materials having small evenly distributed pores. Bockstiegel et al. [23] investigated additions of fine and coarse slag to powder specimens to determine the effect of size and distribution of non-metallics in hot formed P/M preforms and extended this further to porous bodies. Small evenly distributed non-metallics or pores result in optimum properties. Similarly very coarse globules of graphite in nodular iron are undesirable. Longer isothermal sintering results in fewer pores [24] but of more rounded shape. The last two stages of sintering i.e. pore rounding and pore coarsening result in the best combination of mechanical properties up to a point. Excessively long sintering times, at lower densities, however, may cause very large pores to form with an adverse effect on mechanical properties.

c) Dimensional Control

Dimensional problems encountered during sintering were discussed by Armstrong et al. [25] giving consideration to all variables of the P/M process. Modern high speed presses with new types of feed shoes and electronic force monitoring devices allow much closer density controls than in the past. One of the most important parameters which can cause problems in production is the dimensional change incurred during sintering. This parameter, usually controlled by the powder manufacturers by changing the particle size distribution, presents a constant challenge to keep dimensions within the specified limits. As shown in Fig. 13 sintering at temperatures above 1232°C for most powders investigated would present a problem for parts with close tolerances; since the process of densification is rather rapid, components of uneven thickness could warp due to the variance in time-at-temperature in different portions of the part. The method of loading parts on the tray is critical. Parts should be loaded in such a way as not to cast a "shadow" on each other. The heat transfer at temperatures above 1120°C occurs mainly by radiation and a part or a section which does not "see" the heat remains cooler, sinters slower and shrinks less than one fully exposed to the radiation.

During one trial run, the parts were stacked on a tray allowing a much heavier loading than would be possible with a belt furnace and a narrow dimensional process spread within the stack was obtained. Of course, the matter depends wholly upon the part tolerance required by design. For very close tolerances, subsequent coining or cold forging operations would be required. If parts contain high percentages of carbon the coining operation may not be practical since high temperature sintered parts have poor re-pressing characteristics as outlined by Bockstiegel (26). For components with wider tolerance, high temperature sintering may give sufficient strength and ductility to substitute P/M parts for steel fogings or nodular iron castings.

d) Use of Nitrogen Base Atmospheres for High Temperature Sintering

The endothermic atmosphere is nominally about 40% N_2, 40% H_2 and 20% CO with variable residual amounts of CO_2 and H_2O. The effectiveness of this gas mixture, however, as a protective atmosphere is very sensitive to the concentrations of these residuals. Fig. 14 shows the dependence of carbon potential on dew point at temperatures up to 1400°C. It can be seen that e.g. when sintering parts with a carbon content of 0.8% C at 1300°C (2372°F) a dew point of -28°C (-18°F) should be maintained in the high temperature zone. It is not practical to generate an atmosphere of such a low dew point in an endothermic generator as during the operation excessive sooting would necessitate very frequent interruptions to burn-off the soot plugging the catalyst and the piping. Thus methane gas additions would have to be made. This atmosphere during cooling from 1300° to say 1000°C would cause severe carburization of the processed parts, see Fig. 14, as the carburization reaction is very rapid at temperatures between 1000° and 1300°C. Say at 1100°C a dew point of -20°C would be required in order to maintain 0.8% carbon and -28°C would cause carburization. This problem has been discussed by Kaufman (27) in a recent paper and the endothermic atmosphere is possibly the least suitable for high temperature sintering. In addition, the endothermic atmosphere requires a lot of energy.

Cryogenic nitrogen gas atmospheres using various minor gas additions have recently been introduced (28); Kaufman(29) has successfully used a mixture of cryogenic nitrogen with endothermic gas. Of the two atmospheres investigated here, i.e. one 96% N_2 plus 4% H_2 with methane additions, and the second using the same base with CO additions, the second appears to be preferred since extremely clean bright parts were obtained and a wide range of carbon compositions was unaffected by this atmosphere. An excessive methane addition resulted in considerable sooting while lean additions caused decarburization. The range of control appears to be very narrow. On the basis of these trial runs it can be stated that nitrogen base atmospheres can be successfully used for high temperature sintering and one gas composition may perform satisfactorily for a wide range of temperatures. More tests, however, would be required before passing judgement on which is the best nitrogen-base atmosphere.

GENERAL COMMENTS

a) Energy Efficiency of High Temperature Sintering

Energy saving can be achieved during high temperature sintering, as the walking beam or vacumm furnace loading can be much higher than in the conventional belt type furnace. Maximum belt loading is approximately 98 kg/sq. meter (20 lbs/sq. ft.) while tray loading of 294 kg/sq. meter (60 lbs/sq. ft.) or more is possible. Part carriers or trays can be made from mild steel and the "parasitic" weight of the trays is estimated to be equal or less than the weight of the belt per unit weight of processed powder metal compacts, taking heavier loading into consideration. Also with higher temperature, the time cycle at temperature can be reduced, thus increasing the furnace throughput. The energy to heat 1 kg of steel to 1120°C is 194 KCal (350 BTU/lb) while a temperature of 1288°C, still considered practical for ferrous P/M processing, requires 216.3 KCal (390 BTU/lb)[30] which is only 11% more than conventional sintering. The principal sources of wasted energy in a high temperature sintering furnace are heat losses to the furnace atmosphere and heat loses through the furnace walls. Wall losses would obviously be higher if sintering temperatures are raised but these could be reduced by increasing the furnace wall thickness. The additional costs incurred would be more than offset by doubling the product throughput via the walking beam principle.

b) High Temperature Sintering of Preforms for Hot Forging

High temperature sintering above the temperature of reduction of oxides of manganese and chromium allows the use of these alloying elements in powders thus permitting a full hardenability contribution from these alloys [31-33]. This is important particularly for parts requiring high hardenability in view of the shortage of molybdenum and the somewhat prohibitive cost of modified 4600 steel powder now commonly used for P/M hot forming. Here short-time high-temperature sintering is sufficient to reduce the oxides [34]. High temperature sintering also allows the use of master alloys for one's own alloy steel making when sintering and also the use of prealloy-premix combinations which permits one to tailor alloys according to hardenability requirements [35-39].

Preforms for hot forging could be heavily stacked for high temperature sintering, as forging is the last operation and this controls the final tolerance. During one test run at 1232°C for 30 minutes, component parts were stacked two, three, four and five high to determine if interface welding between parts would occur at a high sintering temperature. No parting compounds were used and only slight interfacial sintering was noted for parts three or more high. A gentle tap was sufficient to separate them.

CONCLUSIONS

Preliminary industrial scale experiments in high temperature sintering have shown considerable benefits for both porous conventional powder metallurgy parts and for preforms used for dense hot forgings.

Conventional P/M parts, when high temperature sintered, will exhibit rounded, or even spheroidized pores, depending upon sintering conditions. Superior mechanical properties, particularly impact strength and ductility result from this method of processing. It is expected that high temperature sintering will expand the application of powder metallurgy to parts where steel forgings or nodular iron castings are now used. More work, however, is required to determine the mechanical properties of different powders after high temperature sintering.

Preforms for hot forging, when high temperature sintered, besides superior properties attributed to better oxide reduction, can allow the use of powders with low cost alloying elements, i.e. manganese and chromium. High temperature sintering also opens a new way of making one's own steel from iron or steel powder by admixing master alloys.

References

1. G. C. Kuczynski. Self-diffusion in Sintering of Metallic Particles. Trans. AIME 185, 169 1949

2. F. Thümmler and W. Thomma. Metallurgical Review 12. (115) 69 1967

3. J. R. Moon. Sintering of Metal Powders. Powder Metallurgy International Vol. 3 No. 3 p. 147 and No. 4 p. 194, 1971

4. J. S. Hirschhorn. Introduction to Powder Metallurgy. American Powder Metallurgy Institute 1969

5. G. Bockstiegel. Basic Aspects of Sintering. Iron Powder Handbook Hoeganaes Corp. 1957

6. P. U. Gummeson and A. Stosuy. Iron-Carbon Behaviour during Sintering. Hoeganaes Corporation Publication # 1647. 1967

7. S. I. Hulthen and G. Gustavsson. Some Thoughts on Sintering. International Journal of Powder Metallurgy Vol. 4 No. 3 July 1968

8. BTU Engineering Corporation. Walking Beam Furnaces for Sintering of P/M Compacts. Walking Beam Bulletin 601 1978

9. L. F. Pease III. Metallography and Properties of Sintered Steels. Progress in Powder Metallurgy 1977, Vol. 33 Editors S. Mocarski and T. W. Pietrocini

10. P. F. Lindskog and G. F. Bocchini. Development of High Strength P/M Precision Components in Europe. Hoeganaes Corp. P/M Technical Conference Philadelphia, 1978

11. D. Pohl and F. Redlinger. Metallographic Measurement of Pore Shape and Its Significance for the Mechanical Properties of Sintered Iron. Powder Metallurgy International Vol. 9 No. 4 1977 p. 164

12. R. Wanyanabe, T. Miyamoto and Y. Masuda. Quantitative Analysis of the Morphological Changes of Residual Voids in Sintered Materials. Powder Metallurgy International Vol. 6 No. 3 1974

13. A. Salak, V. Miskovic, E. Dudrova and E. Rudnayova. The Dependence of Mechanical Properties of Sintered Iron Compacts upon Porosity. Powder Metallurgy International Vol. 6 No. 3 1974

14. G. Findeisen. Z. Metallkunde 52 (1961) 3

15. G. Push and W. Schatt. Investigations of Elastic-Plastic Deformation of Sintered Iron. Powder Metallurgy International Vol. 3 No. 1 1971

16. J. R. Merhar. The Application of High Temperature Sintering in the Production of P/M Components. Hoeganaes P/M Technical Conference, Philadelphia, Pa. 1978

17. R. B. Gold. Should Density be the Specification of Iron P/M Parts. Precision Metal. June 1977

18. A. Lawley, Analysis of Mech. Properties - Structure Relations in Powder Forgings. Powder Metallurgy Processing Editors H. A. Kuhn and A. Lawley. Academic Press 1978

19. B. L. Ferguson. Ph.D. thesis, Drexel University. 1976

20. G. Bockstiegel. On the Rate of Sintering. Transactions of AIME Journal of Metals May 1956 p. 580

21. C. B. Jordan and P. Duvez. The Densification of Copper Powder Compacts in Hydrogen and Vacuum. AIME Metals Transactions. February 1949 p. 96

22. J. S. Hirschhorn. Introduction to Powder Metallurgy. American Powder Metallurgy Institute 1969 p. 170

23. G. Bockstiegel and C. A. Blande. The Influence of Slag Inclusions and Pores on Impact Strength and Fatigue Strength of Powder Forged Iron and Steel. Powder Metallurgy International. Vol. 8 No. 4 1976 p. 155

24. F. H. Rhines, C. E. Birchenall and L. A. Hughes, Behaviour of Pores during the Sintering of Copper Compacts. Transactions AIME Vol. 188 (1950) p. 378

25. D. A. Armstrong and L. R. Hildebrand. Tolerance Factors in P/M Parts Production. International Journal of Powder Metallurgy 4 (2) 1968 p. 35

26. G. Bockstiegel. Hoeganaes Handbook. Sintering Sect. E. Chapter 20 p. 3 Hoeganaes Corporation.

27. S. M. Kaufman. Three Zone Sintering Funace. Progress in Powder Metallurgy 1977, Vol. 33, p. 185. Editors S. Mocarski and T. W. Pietrocini.

28. H. S. Nayar. Sintering and Infiltration of Steel P/M Parts in Nitrogen-Base Atmospheres. Progress in Powder Metallurgy 1977 Vol. 33 p. 197 Editors S. Mocarski and T. W. Pietrocini.

29. S. M. Kaufman. Nitrogen-Base Atmosphere Sintering. Paper presented at 1979 SAE Congress and Exhibition, Cobo Hall, Detroit.

30. W. Trinks and M. H. Mawhinney. Industrial Furnaces, 2 Volumes, John Wiley & Sons, 1967 edition.

31. P. C. Eloff and S. M. Kaufman. Hardenability Considerations in the Sintering of Low Alloy Powder Preforms. Powder Metallurgy International, Vol. 3 1971 p. 71

32. S. Mocarski. Influence of Process Variables on Properties of Mod. 8600 and Manganese-Nickel-Molybdenum Low Alloy Hot Formed P/M Steels. Modern Developments in Powder Metallurgy Vol. 7 p. 303. Editors H. H. Hausner and W. E. Smith, 1973

33. P. Lindskog and S. E. Grek. Reduction of Oxide Inclusions in Powder Preforms Prior to Hot Forming. Modern Developments in Powder Metallurgy Vol. 7 p. 285, Ed. H. H. Hausner and W. E. Smith, 1973

34. P. J. Guichelaar and R. D. Pehlke. Gas Metal Reactions during Induction Sintering. 1971 Fall Powder Metallurgy Conference Proceedings p. 109. MPIF Editor S. Mocarski

35. H. F. Fischmeister and L. E. Larsson. Fast Diffusion Alloying for Powder Forging Using a Liquid Phase, Powder Metallurgy Vol. 17, No. 33, 1974 p. 227

36. L. Albano-Mueller, F. Thuemmler and G. Zapf. High Strength Sintered Iron-Base Alloys by Using Transition Metal Carbides. Powder Metallurgy 1973, Vol. 16 No. 32 p. 236

37. S. Mocarski. U.S. Patent No. 4,071,354 January 1978. Master Alloy for Powders

38. S. Mocarski and D. W. Hall. U.S. Patent No. 4,069,044 January 1978 Method of Producing a Forged Article from Prealloyed - Premixed Water Atomized Ferrous Alloy Powder

39. S. Mocarski and D. W. Hall. Properties of Hot Formed Mo-Mn-Ni P/M Steels with Admixed Copper. Modern Developments in Powder Metallugy Vol. 9 p. 467. Editors H. H. Hausner and P. W. Taubenblat 1976

FIGURE 1. A Production High Temperature Walking Beam Furnace.

FIGURE 2. Trays of Production Parts Ready to Enter the Furnace.

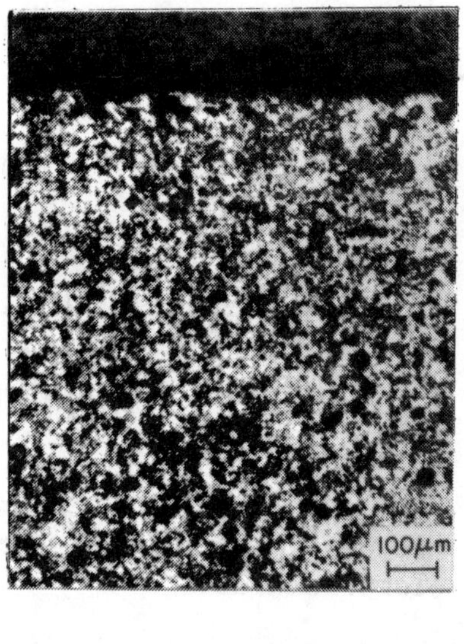

FIGURE 3. Microstructure of a Test Sample at 100x, Nital Etched. Note the Uniform Pearlitic Structure from Edge to Center with no Decarburization.

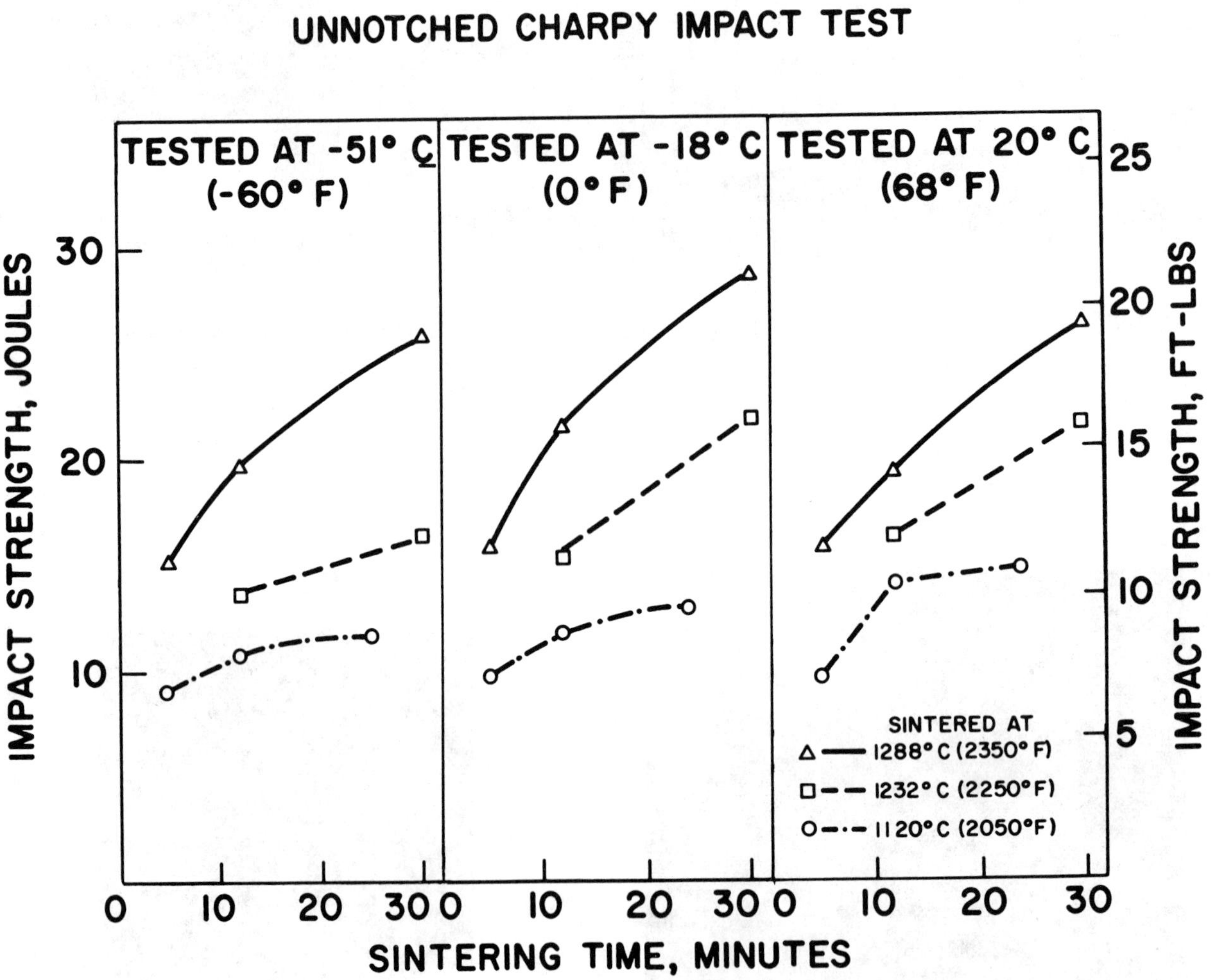

Figure 4. Impact Strength for Normalized Low Alloy Mn-Ni-Mo Steel Powder, 0.2% C, 6.8 g/cc Density, Sintered at Various Times and Temperatures.

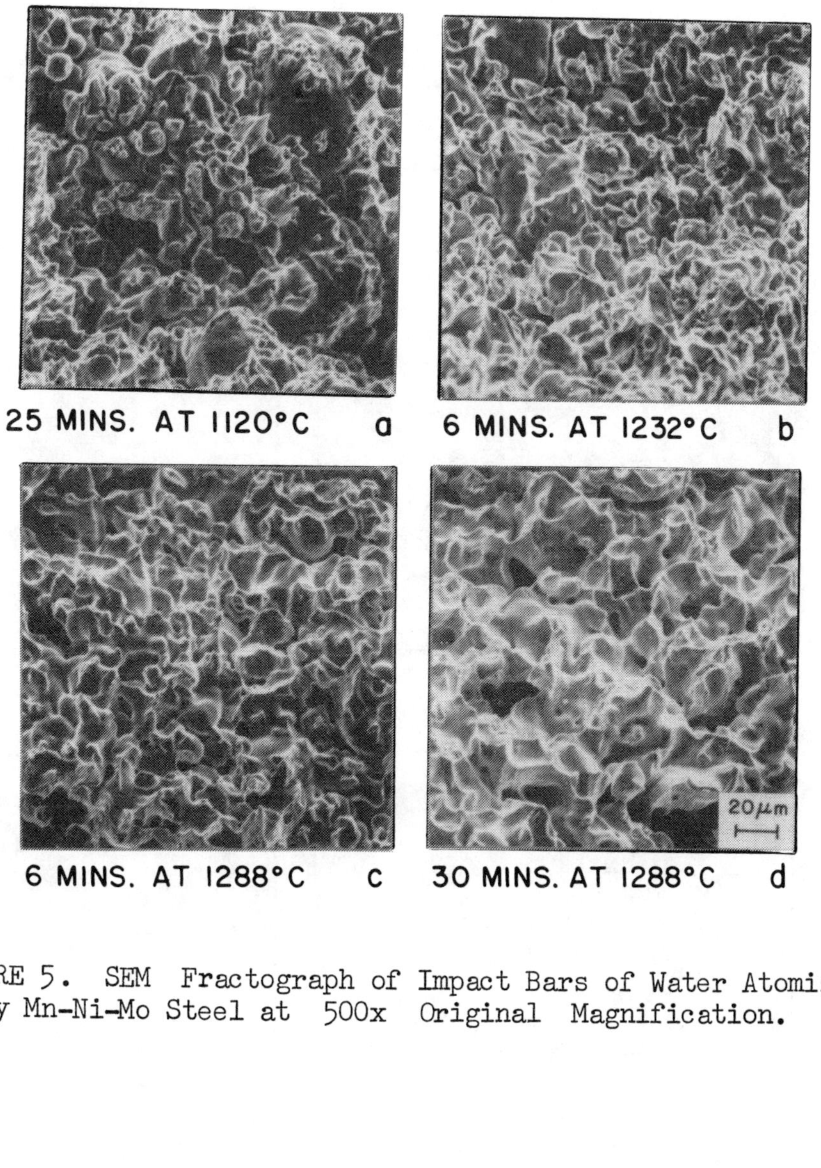

FIGURE 5. SEM Fractograph of Impact Bars of Water Atomized Low Alloy Mn-Ni-Mo Steel at 500x Original Magnification.

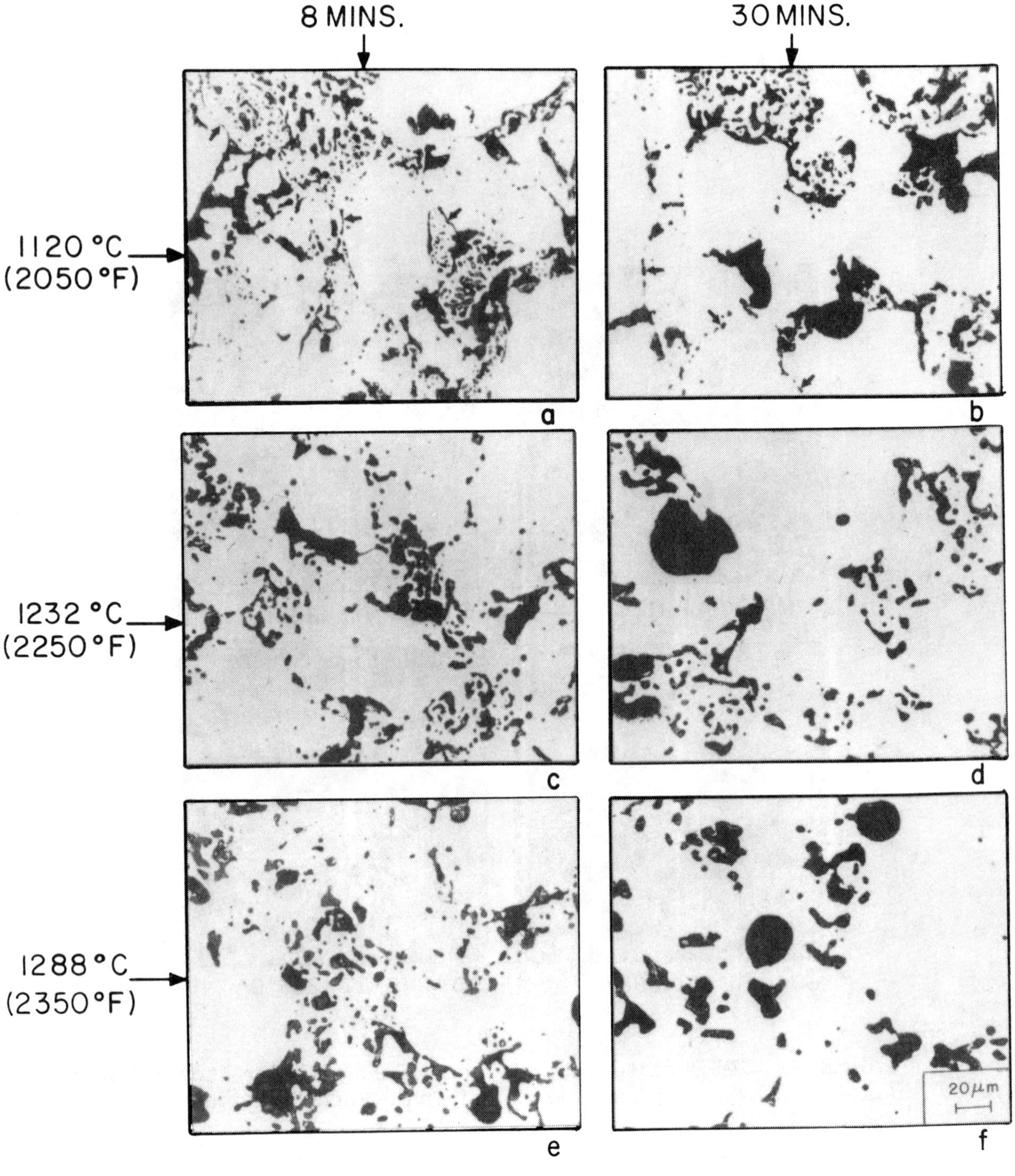

FIGURE 6. Pore Morphology of Water Atomized Cast Iron Powder, Density 6.8 g/cc, Sintered at times and Temperatures Shown. Original Magnification: 500x.

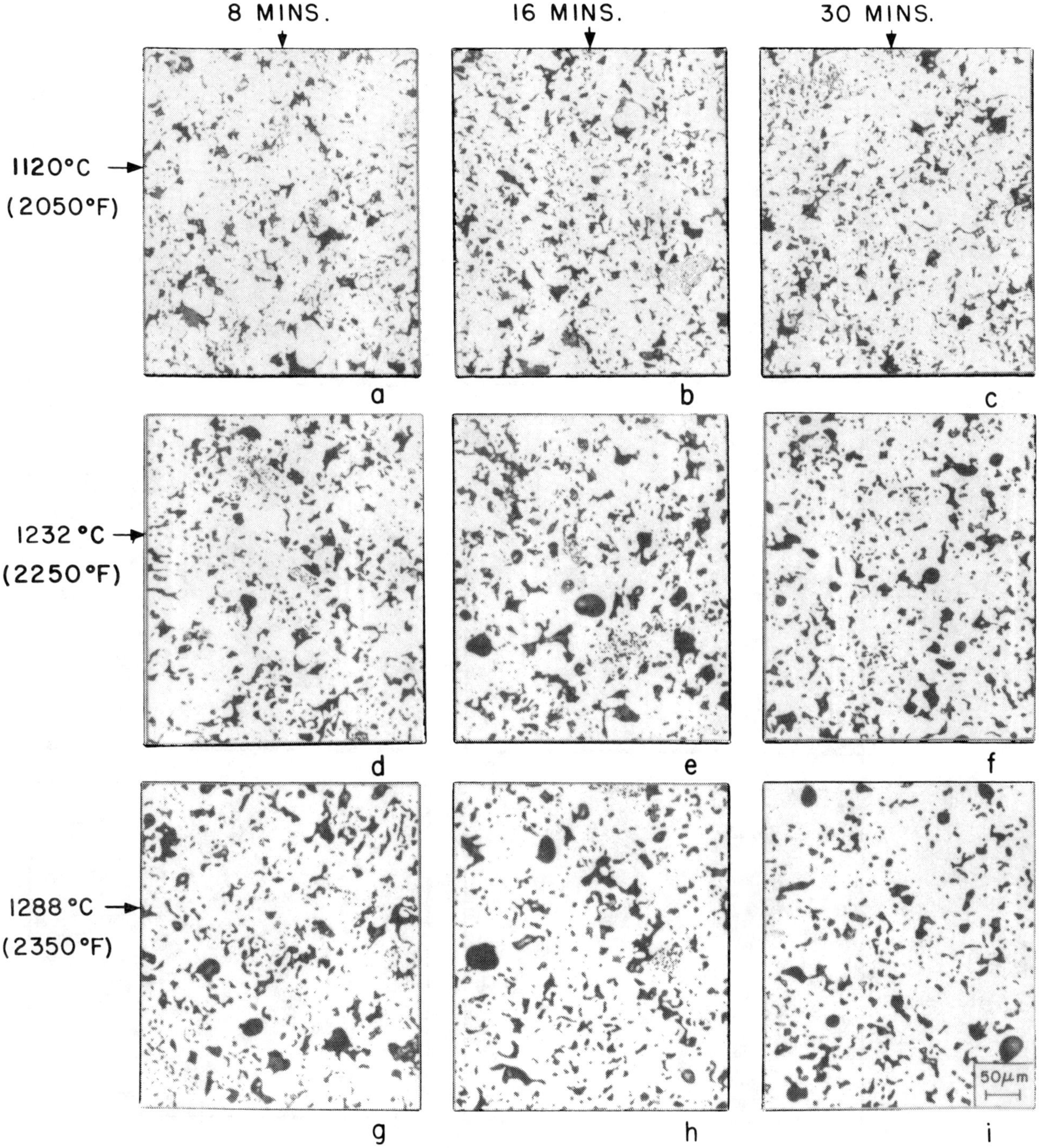

FIGURE 7. Pore Morphology of Sponge Iron, Density 6.4 g/cc Sintered at Times and Temperatures Shown. Original Magnification: 200x.

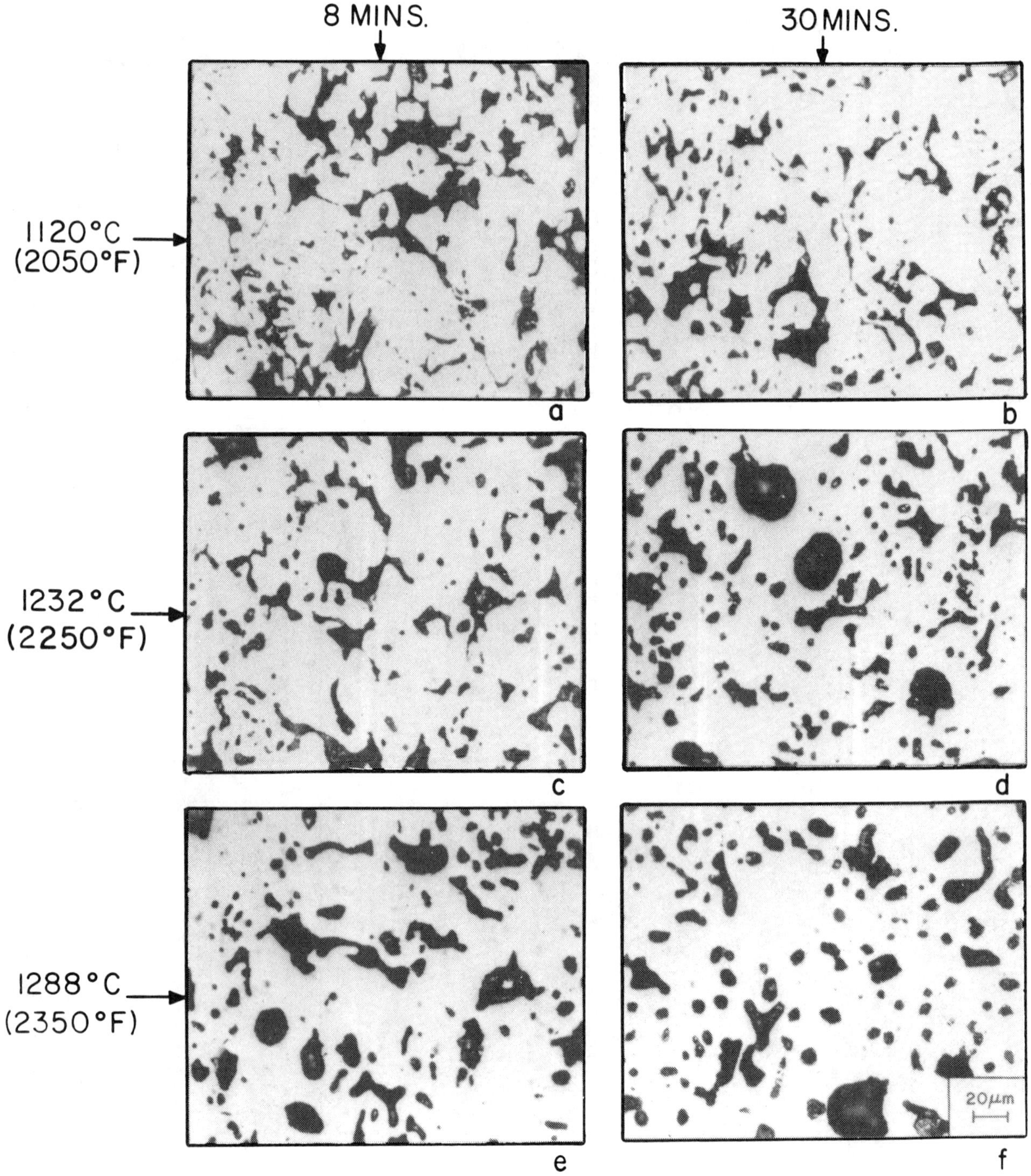

FIGURE 8. Pore Morphology of Sponge Iron, Same as in Fig.7 but at an Original Magnification of 500x.

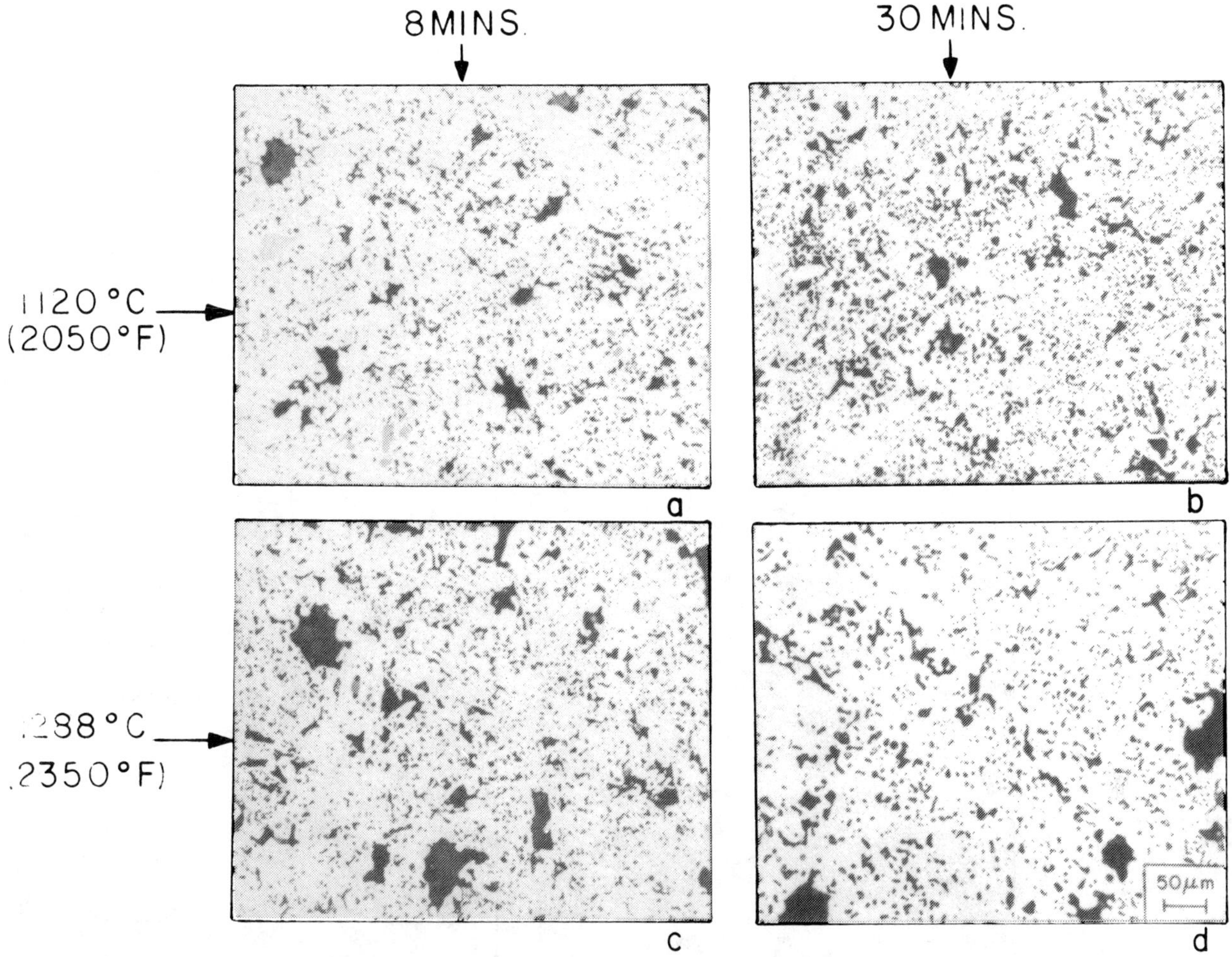

FIGURE 9. Pore Morphology of Reduced Mill Scale Iron Powder with 2.8% Cu, Density 6.15 g/cc, Sintered at Times and Temperatures Shown. Original Magnification: 200X.

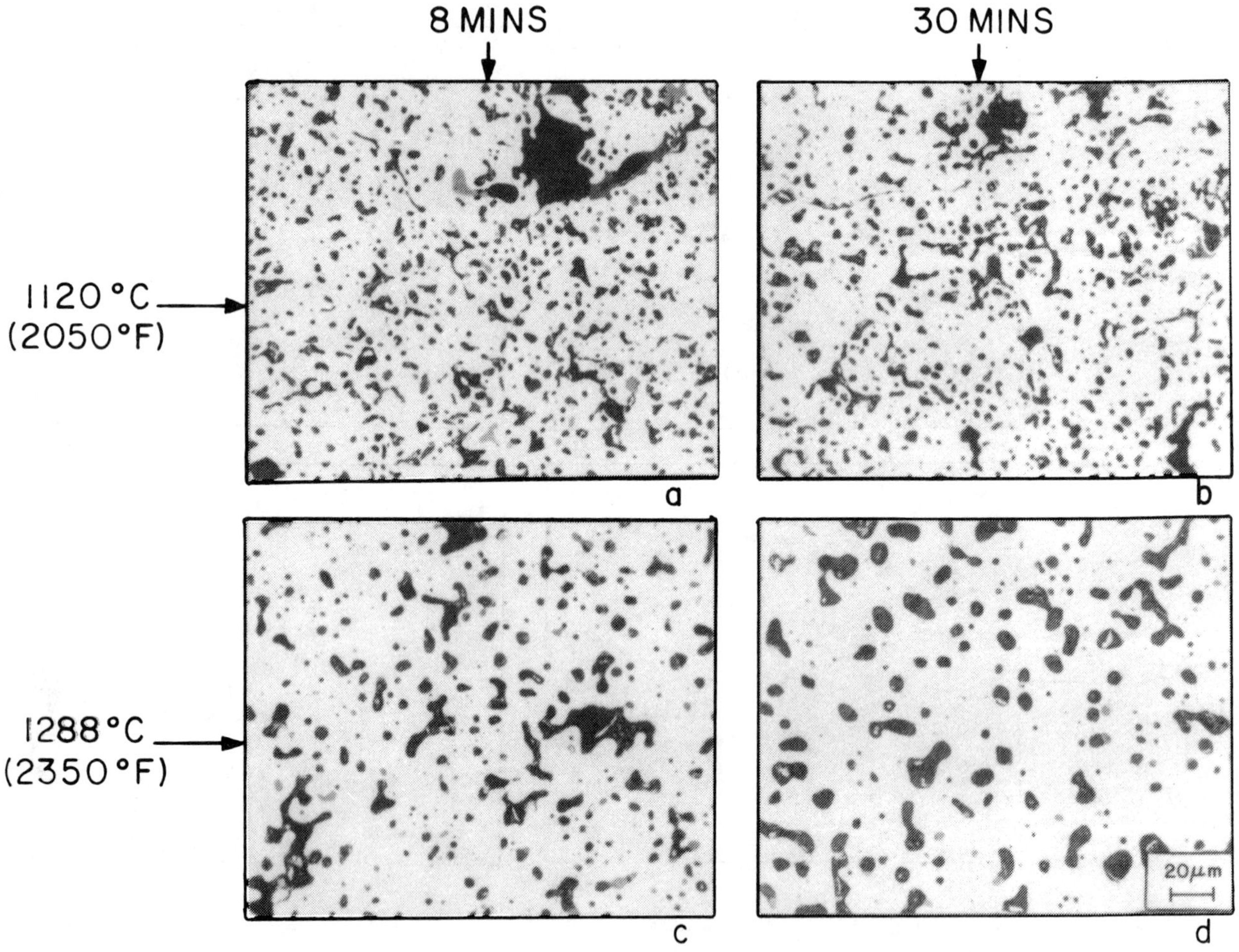

FIGURE 10. Pore Morphology of Reduced Mill Scale Powder, Same as in Figure 9 Except Original Magnification was 500x.

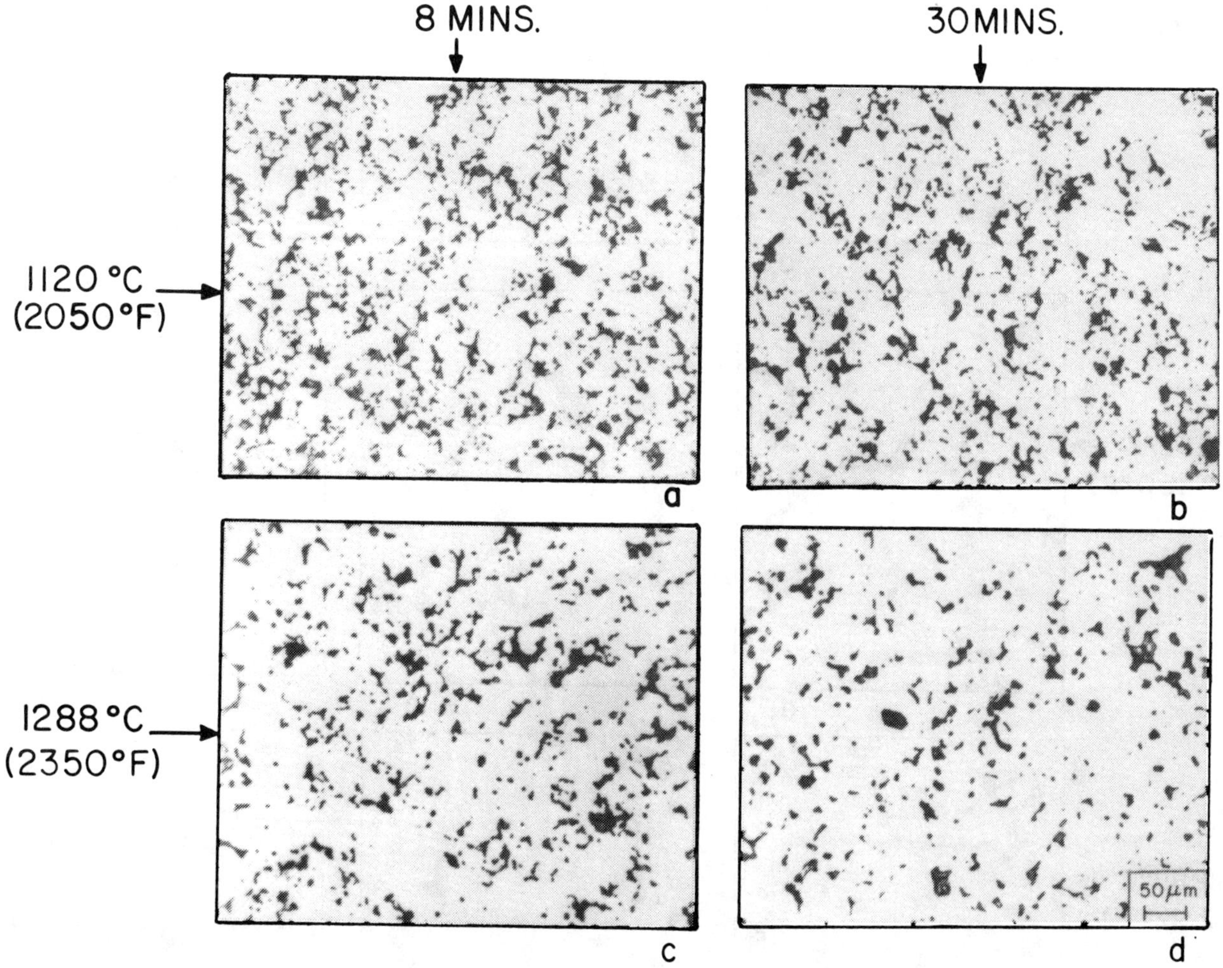

FIGURE 11. Pore Morphology of Water Atomized Low Alloy Mn-Ni-Mo Steel Powder, Density 6.8 g/cc, Sintered at Times and Temperatures Shown. Original Magnification: 200x.

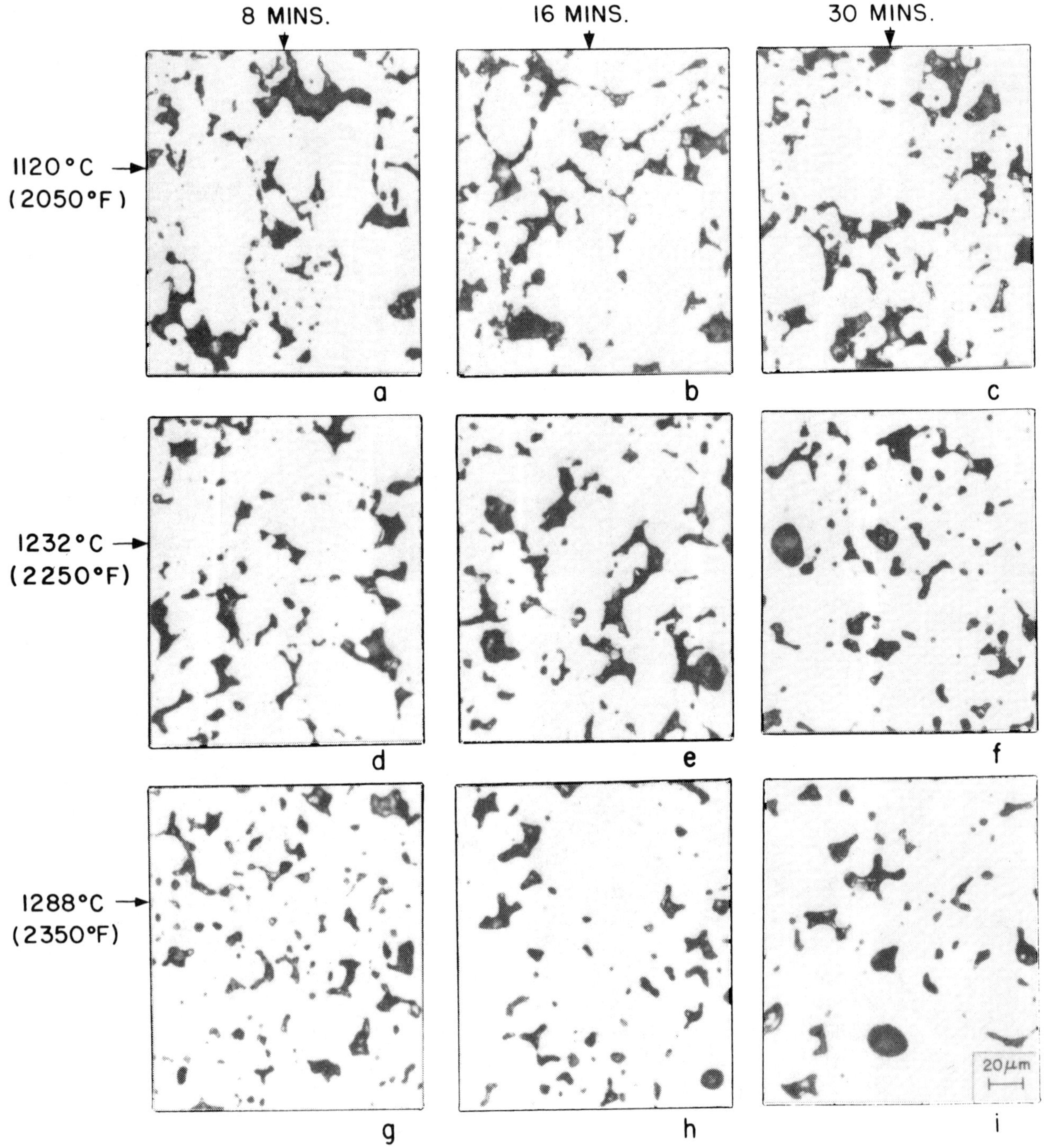

FIGURE 12. Pore Morphology of Water Atomized Low Alloy Steel Powder, Same as in Fig. 11, Except Original Magnification Was 500x.

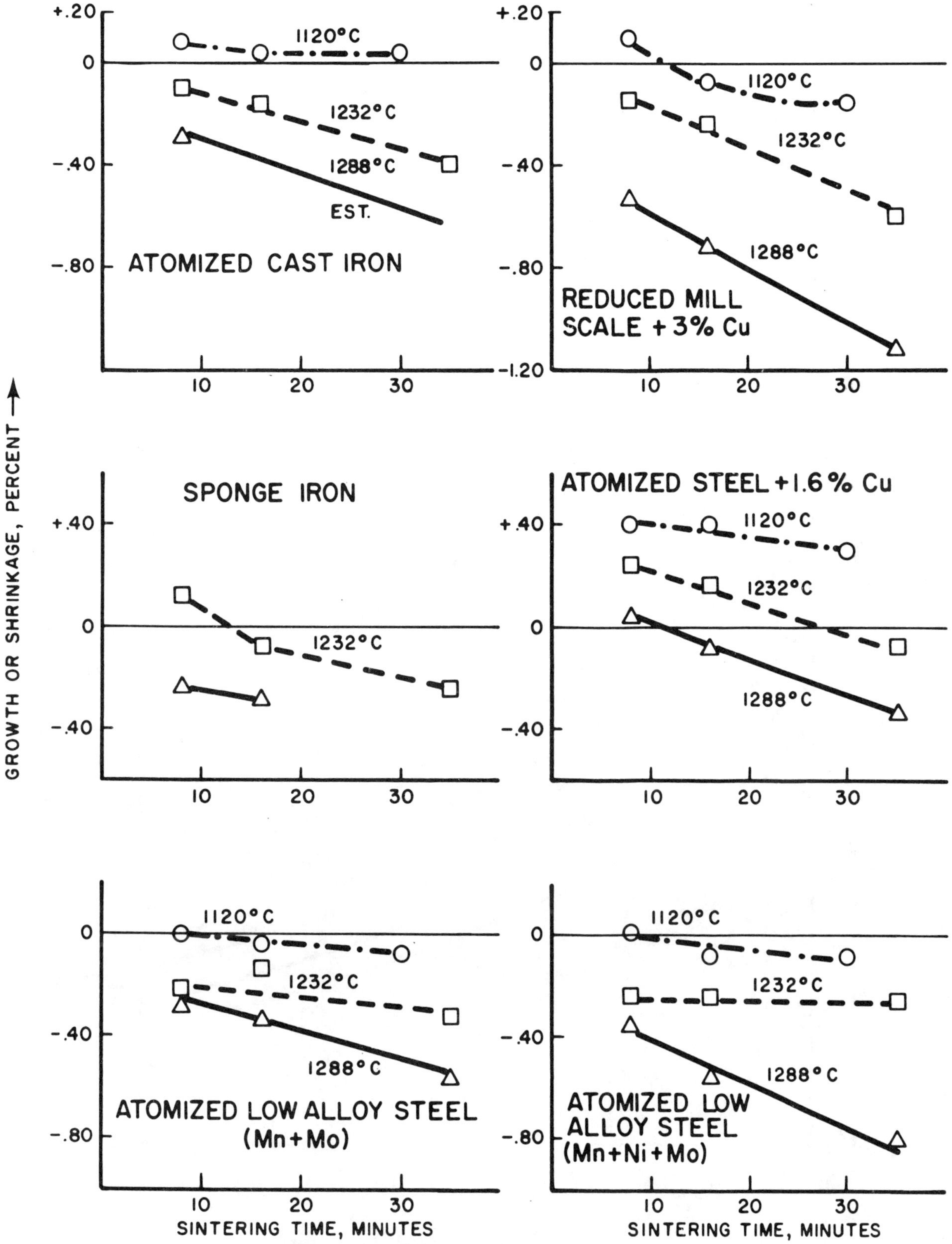

FIGURE 13. Dimensional Change (%) from Die Size for Listed Powders for Various Sintering Times and Temperatures.

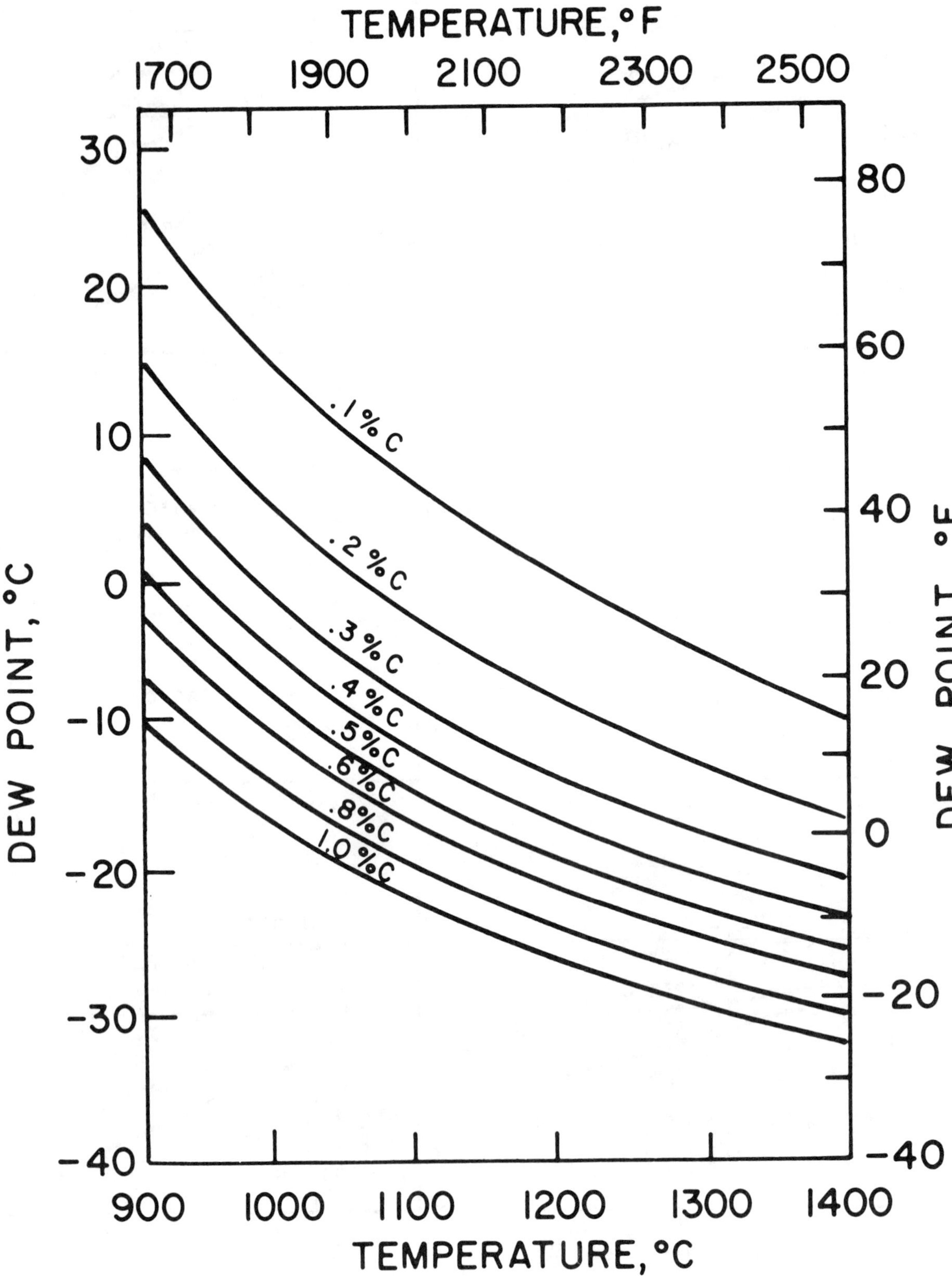

FIGURE 14. Carbon Potential of Endothermic Gas at Different Dew Points and Furnace Temperatures

SAFETY CONSIDERATIONS IN SINTERING ATMOSPHERES

W. McKinley and H. S. Nayar

Airco Inc. Central Research Laboratories
Murray Hill, New Jersey USA

INTRODUCTION

Sintering atmospheres normally consist of a mixture of gases, some of which are flammable, some toxic, and some both. Explosion, fire and poisoning are potential hazards in the use of these mixtures. This paper explores these intrinsic dangers in relation to sintering furnaces, then discusses National Fire Protection Association Standard 86C in relation to continuous belt sintering furnace design. The authors interpret the standard as applied to furnace atmosphere safety control systems, and extend the principles of NFPA 86C to nitrogen-based and nitrogen-backed atmospheres.

INTRINSIC DANGERS

Three high risk intrinsic dangers are associated with producing and using the common sintering atmosphere gases. First an explosive mixture of atmosphere gas with air might accumulate in a confined area and ignite. Second, relatively small quantities of atmosphere gas might unexpectedly burn or flash out of control; and third, personnel might be poisoned by carbon monoxide or ammonia. On the basis of personal experiences and experiences related by people directly involved, we can say with some confidence that all three of these hazards are a real part of life around the sintering furnace.

Figure 1 describes the composition of the common sintering atmospheres. Three of the five major constituents are flammable and two are toxic. The percentage by volume of the flammable parts ranges from a high of 75% hydrogen in dissociated ammonia to only a few percent in purified exothermic and nitrogen-based atmospheres. The dangers of toxicity and flammability are present even when sintering with atmospheres containing only relatively small percentages

FIGURE 1

ATMOSPHERE COMPOSITIONS [1,2,3]					
% By volume	Exothermic	Purified exothermic	Endothermic	Dissociated ammonia	Nitrogen based
Nitrogen	71 to 86	75 to 97	40 to 45	25	>90
Hydrogen	1 to 12	1 to 13	35 to 40	75	<10
Carbon monoxide	1 to 10	1 to 11	19 to 21	———	———
Carbon dioxide	5 to 10	———	.4	———	———
Methane	0 to .5	0 to .5	.3 to .8	———	0 to 1

FIGURE 2

Flammable range,
% by volume in air

H_2	———	4 – 74
CO	———	12.5 - 74
CH_4	———	5.3 – 14
NH_3	———	15 – 28

of hazardous ingredients because of the possibility that these gases will accumulate or concentrate. In these cases, however, the active ingredients (hydrogen, carbon monoxide, methane and carbon dioxide) are diluted below the volume percent which would constitute a flammable mixture even if all the nitrogen were replaced by air. The explosion, burning and poisoning hazards are reduced but not eliminated.

THE EXPLOSION HAZARD

Any flammable mixture of gas and air can explode. The upper and lower limits of flammability, shown in Figure 2, bracket what is called the explosive range. Any mixture between the high and low limits of flammability will burn when ignited and, under certain conditions, detonate or explode. The destructive power of the pressure wave from igniting a flammable mixture depends on the amount and heat of combustion of the fuel gas, the combustion mode (deflagration or detonation), and the geometry of the confinement space. The energy released is either absorbed by the surroundings or destroys them.

FIGURE 3

TNT EQUIVALENT FORMULA FOR ESTIMATING BLAST DAMAGE FROM CHEMICAL EXPLOSIONS[1]

TNT EQUIVALENT:

$$W_{TNT} = \frac{\alpha \Delta H \times W_F}{4.198 \times 10^6}$$

WHERE ΔH = HEAT OF COMBUSTION OF FUEL GAS, JOULES/KG

W_F = AMOUNT OF FUEL GAS, KG

α = EMPIRICAL FACTOR ($0 < \alpha < 1$)

4.198×10^6 = ENERGY OF TNT, JOULES/KG

FOR 50 CU. FT. OF H_2 AT STP:

$$W_{TNT} = \frac{\alpha \times 1.2 \times 10^8 \text{ J/KG} \times 0.11 \text{ KG}}{4.198 \times 10^6 \text{ J/KG}} = 3.14 \text{ KG} \times \alpha$$

ASSUME $\alpha = 0.25$:

$$W_{TNT} = 3.14 \text{ KG} \times 0.25 = 0.785 \text{ KG (1.73 LB)}$$

[1] J. H. LEE, C. M. GUIRAO, K. W. CHIU AND G. G. BACH, LOSS PREVENTION (CEP), VOL. 11, P. 59, SEPT. 1977

Figure 3 shows how to calculate the TNT equivalent of detonating 50 cubic feet of hydrogen at STP. The empirical factor alpha expresses the difference between a concentrated TNT explosion and the broader air-gas explosion. In the example, we have conservatively estimated alpha to be 0.25 from information in the article by J. H. Lee and associates. The calculation indicates 50 cubic feet of hydrogen equals almost 2 lb of TNT. By the same calculation, 50 cubic feet of gaseous methane would produce a blast equivalent of 6.34 lb of TNT.[4] The ordinary hand grenade contains only 19 oz of TNT.

The explosive potential of the large volumes of flammable gases used in sintering furnaces is obviously an important safety consideration. If sintering is to be safe, the equipment and systems that handle these gases must be designed, operated and maintained to exclude the possibility of explosive mixtures accumulating. Events leading to such dangerous accumulations can be subtle and elusive; therefore, it is not safe to rely on human operations to prevent them. Routine standard procedures, however, can be designed to check and isolate failure of the safety equipment. Properly designed safety systems maintained in this way and operated by well-trained and competent personnel, reduce the risks substantially.

THE BURNING HAZARD

Most sintering furnaces operate on the principle that equal volumes of atmosphere gas will leave the charge and discharge doors, mix with air and burn uniformly and completely. For a number of reasons, the atmosphere gas in a sintering furnace may not "burn off" this way. At such times, a combustible mixture of air and gas forms in the furnace throat, burns rapidly, and forcefully exhausts a volume of flame and hot gases. Unprotected hands or faces in or near the furnace doors risk painful burns. While flashes of this type are more common than explosions, fortunately they are potentially less destructive. Nonetheless, personnel who approach these areas without safety glasses, protective face shields, gloves and perhaps flameproof clothing, risk serious flash burns. The eyes are particularly sensitive to damage.

Operating procedures are more effective than furnace equipment in preventing such injuries. If safe methods are established and used for clearing jam-ups, inspecting the sintering, removing the product from the belt and so on, this hazard is substantially reduced.

Protective guards and shields, useful for routine operation, are no substitute for the necessary personal protection of the people who must occasionally work directly in the furnace doors. The possibility of flashback under nonflammable, nitrogen rich atmospheres is remote. The chance, however, still exists for the reasons mentioned earlier and cannot be ignored.

THE POISONING HAZARD

Ammonia vapor and carbon monoxide are highly toxic gases used routinely in P/M sintering. Ammonia is used to produce dissociated ammonia, and carbon monoxide is a constituent of exo and endothermic gases. Concentrations of less than one-half of one percent of either

FIGURE 4

PHYSIOLOGICAL EFFECTS OF VARIOUS CONCENTRATIONS OF AMMONIA

Courtesy: Matheson Gas Data Book, 1971, p. 18

Atmospheric Concentration	Physiological Effects
20 p.p.m.	First perceptible odor.
40 p.p.m.	A few individuals may suffer slight eye irritation.
100 p.p.m.	Noticeable irritation of eyes and nasal passages after few minutes exposure.
400 p.p.m.	Severe irritation of the throat, nasal passages, and upper respiratory tract.
700 p.p.m.	Severe eye irritation. No permanent effect if the exposure is limited to less than one-half hour.
1700 p.p.m.	Serious coughing, bronchial spasms, less than a half hour of exposure may be fatal.
5000 p.p.m.	Serious edema, strangulation, asphyxia; fatal almost immediately.

FIGURE 5

PHYSIOLOGICAL EFFECTS OF VARIOUS CONCENTRATIONS OF CARBON MONOXIDE

Courtesy: Matheson Gas Data Book, 1971, p. 100

Effect	CO Concentration
Allowable for an exposure of several hours	0.01%
Can be inhaled for one hour without appreciable effect	0.04-0.05%
Causing a just appreciable effect after one hour of exposure	0.06-0.07%
Causing unpleasant symptoms, but not dangerous after one hour of exposure	0.1-0.12%
Dangerous for exposure of one hour	0.15-0.2%
Fatal in exposure of less than one hour	0.4% and above

gas in air are considered fatal in less than one-half hour. Figures 4 and 5 emphasize this point, showing the physiological effects of various concentrations of these gases. The pamphlet, Threshold Limit Values, published by the American Conference of Government Industrial Hygienists, lists CO, CO_2 and NH_3 among commonly used toxic chemicals. Concentrations as small as 35 PPM for NH_3 and 400 PPM for CO are harmful, and only short exposures are allowed (Figure 6). CO_2 must also be mentioned here as one of the toxic constituents of exo and endothermic gas.

FIGURE 6

Threshold Limit Values

American Conference of Government Industrial Hygienists 1978

	TWA[1] ppm	STEL[2] ppm
CO	50	400
CO_2	5000	15000
NH_3	25	35

1 Time Weighted Average
2 Short Term Exposure Limit
Courtesy of ACGIH

Protection from poisoning by raw ammonia vapor or from contact with liquid ammonia is provided in part in the design of the storage and delivery system. The ammonia supplier is an excellent source of safety information and details concerning ammonia storage and supply systems and ammonia dissociators. Because it is to his best interest that ammonia be stored and used safely, he will generally be happy to provide information and guidance.

Because CO is almost as toxic as NH_2 but also odorless, it is the more dangerous of the two. Lethal concentrations can collect in isolated areas and be undiscovered until someone passes out. The general safety consideration here is straight forward. Piping between the generator and the furnace must be leak-tight, and all the gas that goes into the furnace must be either burned or properly vented.

To support this basic safety system, CO levels in and around the furnace and generator must be monitored periodically, except in problem areas where they should be monitored continuously. Many monitors designed for this purpose are available. In addition, personnel who must work in these areas should be familiar with the early syptoms of CO poisoning and trained in appropriate first aid. (The early symptoms may be nothing more than slight dizziness, weakness, or headache. In the later stages of poisoning, the victim's lips and skin will turn a characteristic cherry red.)

Finally, the danger of asphyxiation is not intrinsic in the natures of the gases used in sintering, but is just as real as if it were. While dangerous situations in the normal operations of sintering furnaces are hard to imagine, the hazards are present, and the people involved should be thoroughly familiar with them.

Reflecting for a moment on what has been said reveals some common points of consideration. Safety, it appears, depends heavily on well-trained operators who understand the intrinsic dangers of the gases they work with and the proper maintenance of the equipment.

ATMOSPHERE SAFETY SYSTEMS

This section deals with the atmosphere safety devices and systems outlined in the industrial standard NFPA 86C. This standard covers the design and operation of industrial furnaces which use special processing atmospheres. It represents a consensus of equipment manufacturers, users and insurers.

The atmosphere safety devices and systems found on most sintering furnaces are designed to prevent explosive mixtures from accumulating and thus to prevent fire or explosion. Most of the design and most of the equipment have been adapted from combustion safeguards. The basic design principle in both areas is that a positive and assured source of ignition at the point where the flammable gas mixes with air prevents explosive mixtures from accumulating. Absolutely no gas can flow unless this source is present.

This basic principle carries through the NFPA 86C guideline

and is applied to atmosphere gas in industrial furnaces. In discussing this standard we will refer to a typical continuous belt sintering furnace. This typical furnace is shown in Figure 7. It consists of an electric or gas fired preheat, usually operated between 1000 and 1600°F, which normally has a muffle. The hot zone of the furnace, also electric or gas fired, can either be muffle or brick lined and normally operates at 2050°F. These sections are followed by a slow cool and two water cooled sections. The atmosphere enters the furnace through the roof of the slow cool section. When exothermic, endothermic or dissociated ammonia are used, flame curtains are ignited at both doors. Vents over the doors exhaust the products of combustion generated at these points.

Standard 86C considers the furnace we have described as a class C furnace, "in which there is an explosion hazard due to the flammable special atmosphere being used for the treatment of material in process." Chapters 6, 7 and 8 of the standard refer specifically to the furnaces and their atmospheres. Chapter 7 does not apply in this case because it deals with furnaces without vestibules; the water cooled sections of our typical furnace would be considered vestibules.

In the furnace we are considering, the flammable atmosphere could be both introduced and removed by burning. That is, since no purge gas is used the incoming atmosphere gas must purge the air from the furnace by burning it out. Article 600 in Chapter 6, (See Figure 8) which governs such class C furnaces, classified this as a Type II furnace: "all zones within the furnace chambers above 1400°F (760°C) having vestibules without doors." Technically this descriptive classification would apply only if the preheat zone too was operated above 1400°F.

Two essential features of these Class C, Article 600, Type II furnaces permit atmosphere to be introduced and removed by burning. First, a substantial portion of the furnaces operate above 1400°F (760°C), and so provide the incoming atmosphere gas with a positive source of ignition. Second, the furnaces are designed to be relatively open and unrestrictive to the free flow of atmosphere gas. This design physically reduces the possibility of gas accumulating within the furnace.

Article 800 in Chapter 8 describes furnaces like our model furnace as a Type I furnace: "All furnace zones can be heated above 1400°F (760°C) before introduction and removal of flammable atmosphere gas, but all or part of the furnace zones normally operate below 1400°F (760°C)." (See Figure 9) An inert gas purge supply is required as an emergency back-up. The features which fix the classification are zones that operate below 1400°F (760°C), and a furnace design which tends to entrap atmosphere gas in compartments.

It must be obvious by now that our "typical" furnace doesn't exactly fit either category. The reason is that the NFPA standard, being a consensus standard, covers a broad range of furnace designs and applications, and should not be expected to precisely suit any one furnace. The actual classification and the related requirements for safety equipment and procedures derive from discussion between

FIGURE 7

Typical belt sintering furnace (6"- 36" belt)
NFPA Class C

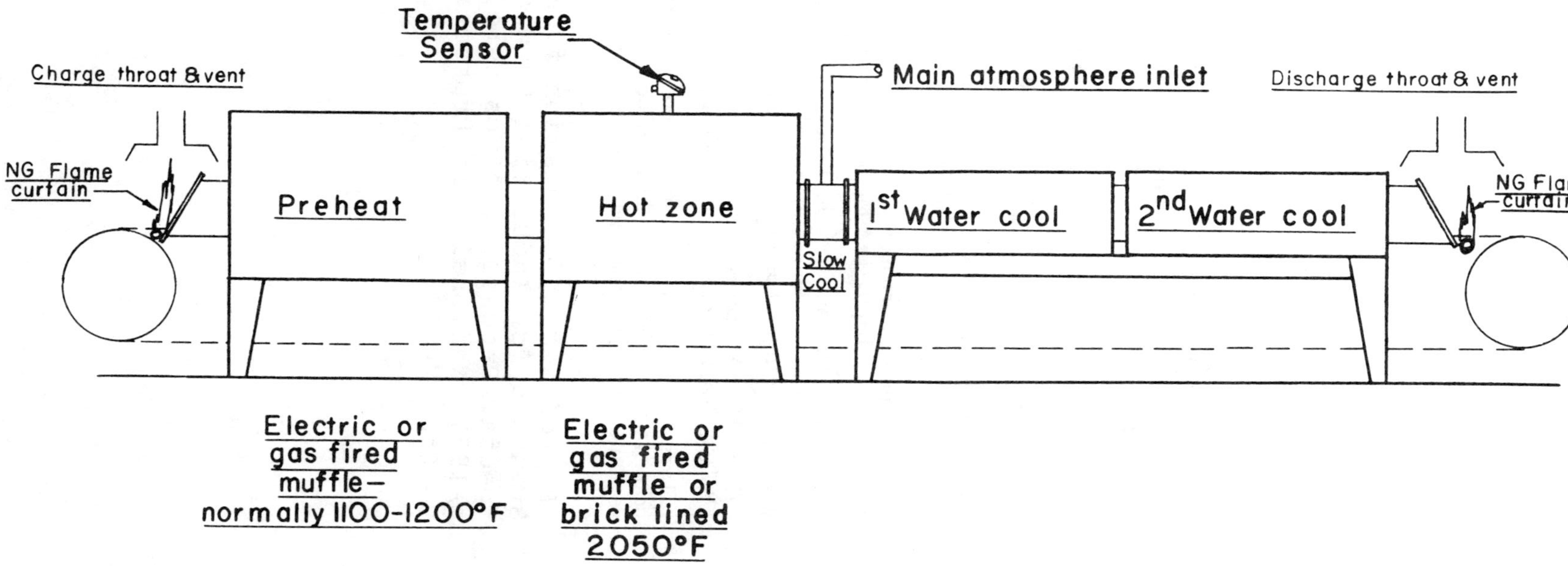

FIGURE 8

CHAPTER 6. CLASS C FURNACES WITH ZONES ABOVE AND BELOW 1400°F (760°C) ATMOSPHERE INTRODUCTION AND REMOVAL BY BURNING

ARTICLE 600. GENERAL INFORMATION

600-1. Scope. Procedures for "burn in" and "burn out" of indirectly heated atmosphere type furnaces having zones both above and below 1400°F (760°C), but where "burn in" and "burn out" procedures of flammable atmosphere gas may be safely used.

600-2. Type of Furnaces. The three general types of furnaces covered in this chapter (see Fig. 7) are as follows:

TYPE I. All zones within furnace chambers above 1400°F (760°C), having vestibules with doors. (Vestibule chamber utilized to prevent air infiltration into furnace chamber proper.)

TYPE II. All zones within furnace chambers above 1400°F (760°C), having vestibules without doors.

TYPE III. Batch-type furnace where temperature indication is below 1400°F (760°C), upon the introduction of a new load, having vestibules with doors and separate cooling section.

FIGURE 9

CHAPTER 8. FURNACES ABOVE AND BELOW 1400°F (760°C) — BURN-IN AND BURN-OUT NOT PERMITTED

ARTICLE 800. GENERAL

800-1. Scope. The two general types of furnaces covered are fuel fired or electrically heated continuous, semi-batch or batch furnaces having none, one or two vestibules attached to the furnace chamber(s) wherein:

TYPE I — All furnace zones can be heated above 1400°F (760°C) before introduction and removal of flammable atmosphere gas, but all or part of the furnace zones normally operate below 1400°F (760°C). Burn-in and burn-out procedures can normally be used; however, emergency conditions require inert gas purge.

TYPE II — All or part of the furnace zones operate below 1400°F (760°C), and cannot be heated to 1400°F (760°C), or the furnace is a batch-type and the flammable atmosphere gas must be introduced and removed below 1400°F (760°C), using the inert gas purge procedures.

FIGURE 10

610-9. The following safety equipment shall be provided for furnace Types I, II and III in conjunction with the flammable atmosphere gas system.

(a) Safety shutoff valve in the flammable atmosphere gas supply line to the furnace. This valve shall be energized when furnace temperature is above 1400°F (760°C). Operators attention shall be required.

(b) Low pressure switch to assure that flammable atmosphere gas supply is available. Loss of pressure shall shut off the safety shutoff valve and sound an alarm.

(c) Safety interlock with furnace temperature controller — extra contact to close at 1400°F (760°C).

the furnace manufacturer and someone known as the "local authority having jurisdiction." This authority is carefully defined in the standard.

Safety equipment required by NFPA for Class C, Article 600 furnaces is shown in Figure 10. Figure 11 diagrams our interpretation of the required atmosphere safety system. The left hand side of the diagram shows the location of the physical elements in the system, and the right hand side shows their electrical connections. The vertical lines at either end of the electrical "ladder diagram" represent common electrical potentials, usually 110v on one leg and ground on the onther. The key to reading the diagram is to imagine the series of events that results in a current path between the vertical legs. In this example, contacts in the temperature controller close when the furnace temperature is above 1400°F (760°C). Depressing the atmosphere start button then provides a current path through the atmosphere gas relay, which in turn will close its contact, AGR-1. This in turn energizes the magnetic coil of the Atmosphere Gas Safety Shut-Off Valve (AGSSOV). The valve can, as a result, be manually opened, simultaneously silencing the alarm and closing the City Gas Safety Shut-Off Valve (CGSSOV). The flow of atmosphere gas will then provide pressure to the Atmosphere Gas Low Pressure Switch (AGLPS). This switch, in conjunction with the limiting orifice valve, is adjusted so that an adequate minimum purging flow closes it contacts. With this contact closed the start button may be released, and atmosphere gas will continue to flow, provided there is sufficient purging flow to the furnace, the temperature remains about 1400°F (760°C), and the power does not fail.

Chapter 8 requires additional protective features for the Type I furnace. The system suggested here essentially augments the basic safety devices of Chapter 6. The features added are (1) an atmosphere gas flow monitoring device and flowmeter, (2) a bypass feature to allow some or all of the zones to operate below 1400PF (760°C), (3) a safety shutoff for flame curtain burners, (4) supervised pilots at outer doors and vents interlocked to door position, and (5) an emergency standby inert gas purging system. In our experience, we have yet to see a conventional sintering furnace designed with a 1400°F bypass for the preheat zone, safety shutoffs on the flame curtains, or supervised pilots at the outer doors. This particular equipment is more characteristic of heat treating or annealing furnaces than of continuous belt sintering furnaces used in the P/M industry.

The system shown in Figure 12 incorporates only the features suggested in Chapter 8 related to the atmosphere gas system. All that is required is adding a self-monitoring, automatic nitrogen (or other inert gas) purging system. The flowmeters are required. The key feature in the nitrogen back-up system is a normally open or, de-energized open, solenoid valve. Under normal operating conditions the solenoid is energized through the AGR and remains closed. With insufficient atmosphere gas flow, temperature below 1400°F (760°C), or electrical failure the system automatically shuts off the flow of all flammable gases (excluding those used for heating and flame curtains), introduces a purging flow of nitrogen, and

FIGURE 11

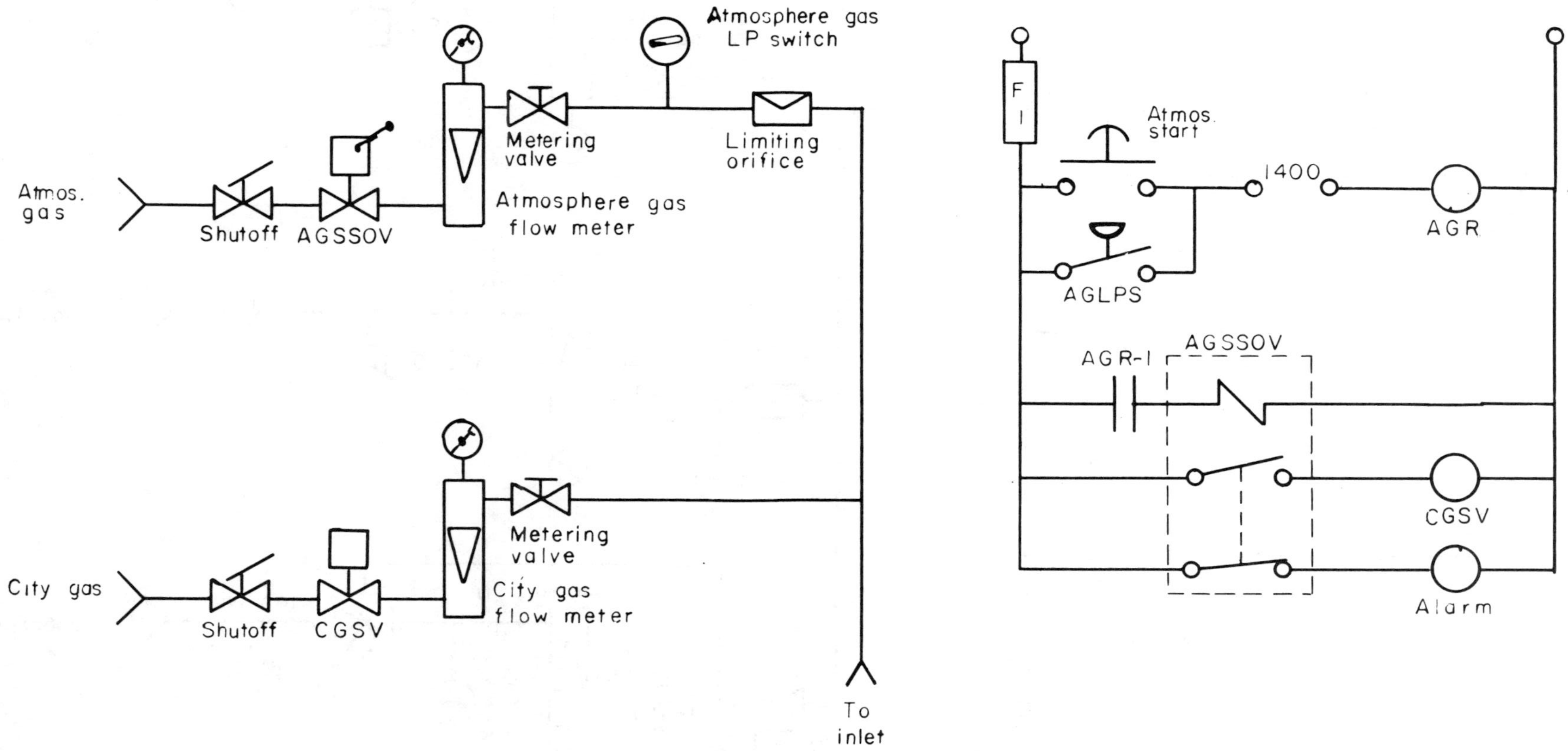

Atmos. gas
Shutoff
AGSSOV
Atmosphere gas flow meter
Metering valve
Atmosphere gas LP switch
Limiting orifice
City gas
Shutoff
CGSV
City gas flow meter
Metering valve
To inlet
F I
Atmos. start
1400
AGR
AGLPS
AGR-1
AGSSOV
CGSV
Alarm

FIGURE 12

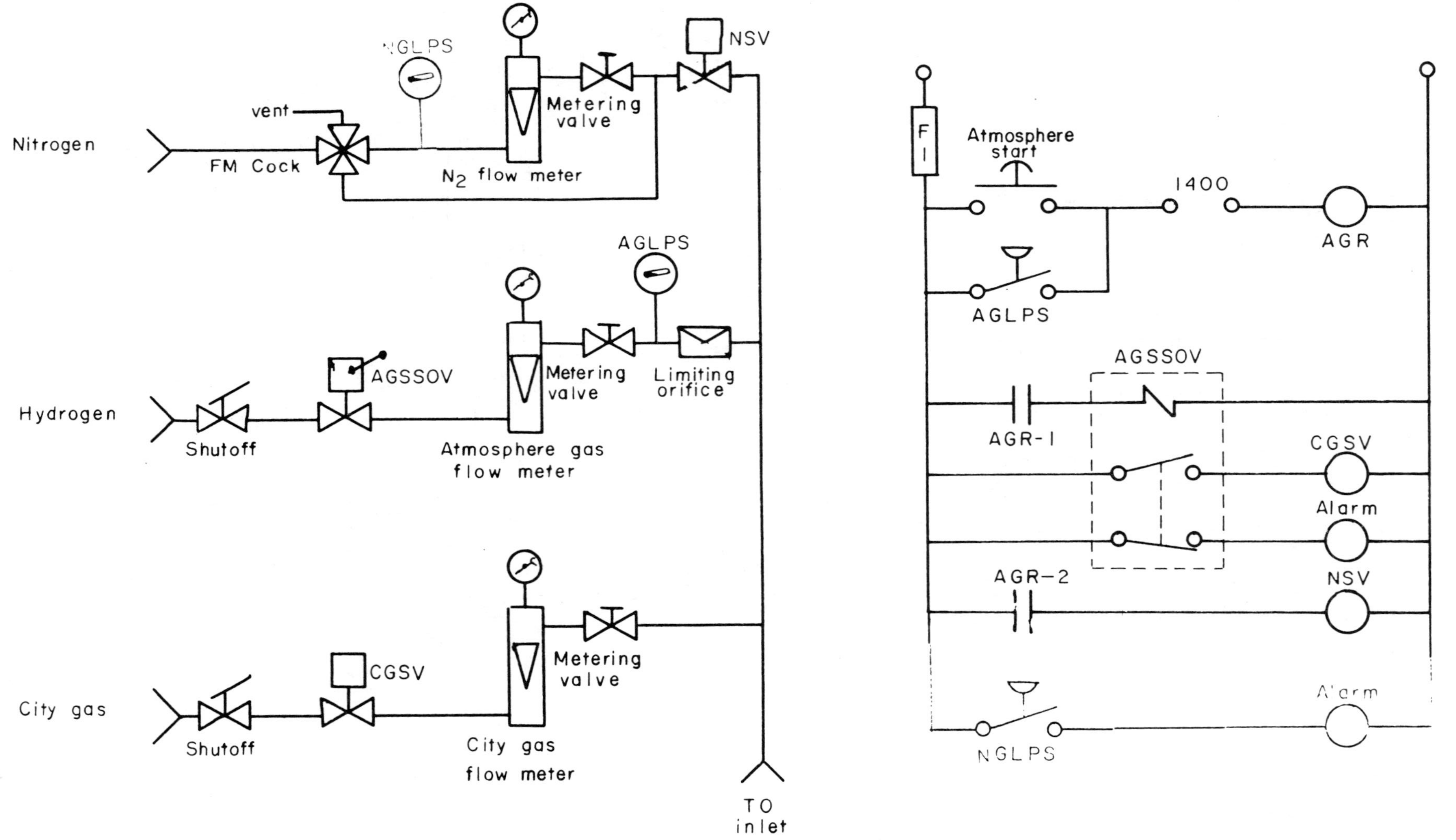
Nitrogen
vent
FM Cock
NGLPS
N_2 flow meter
Metering valve
NSV
Hydrogen
Shutoff
AGSSOV
Atmosphere gas flow meter
Metering valve
AGLPS
Limiting orifice
City gas
Shutoff
CGSV
City gas flow meter
Metering valve
TO inlet
F
Atmosphere start
1400
AGR
AGLPS
AGSSOV
AGR-1
CGSV
Alarm
AGR-2
NSV
NGLPS
Alarm

FIGURE 13

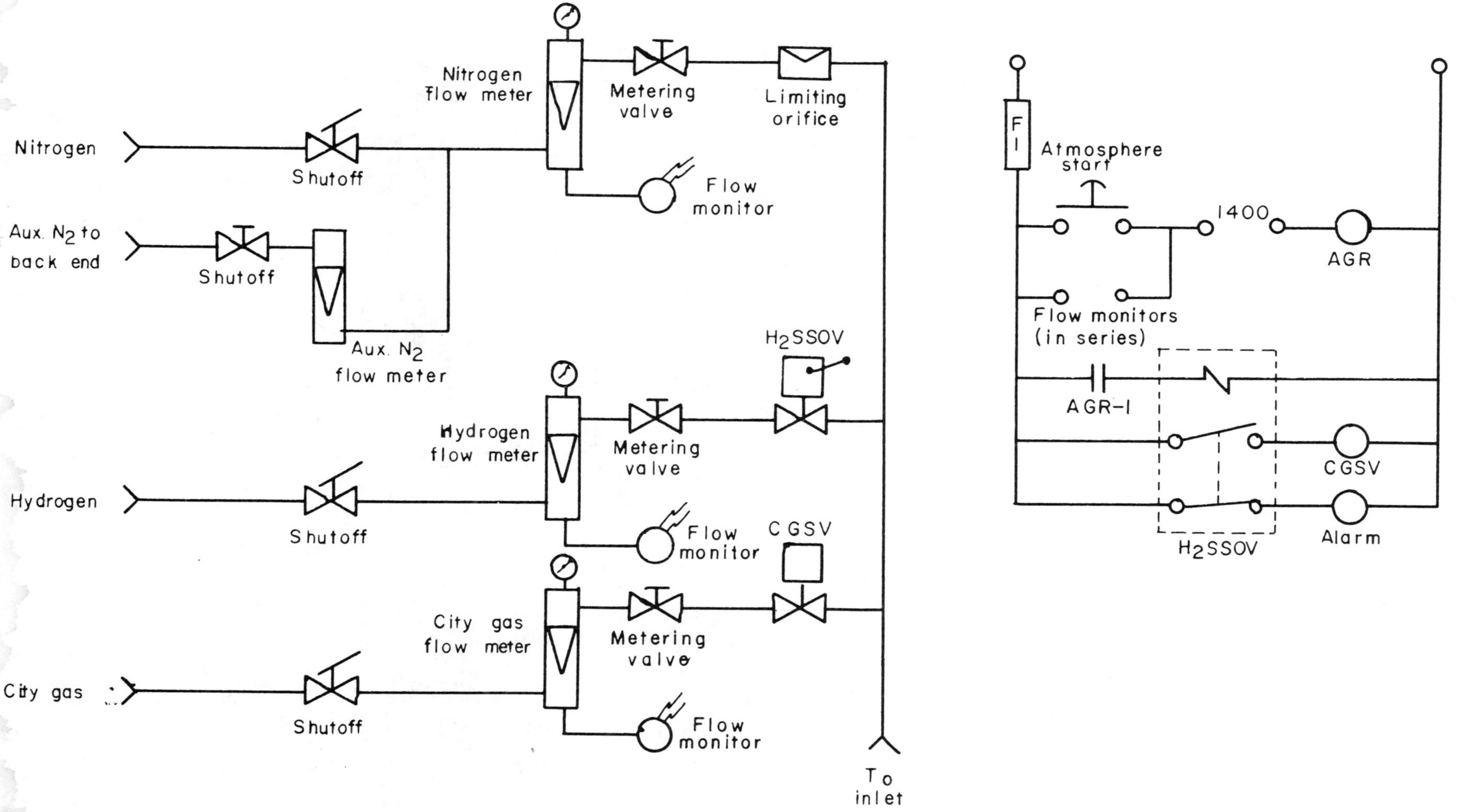
Nitrogen
Aux. N_2 to back end
Hydrogen
City gas
Shutoff
Shutoff
Shutoff
Shutoff
Nitrogen flow meter
Aux. N_2 flow meter
Hydrogen flow meter
City gas flow meter
Metering valve
Metering valve
Metering valve
Limiting orifice
Flow monitor
Flow monitor
Flow monitor
H_2SSOV
CGSV
To inlet
F I
Atmosphere start
1400
AGR
Flow monitors (in series)
AGR-1
CGSV
Alarm
H_2SSOV

sounds an alarm to warn the operator of the condition. If at any time the nitrogen supply pressure drops below a present level, the system sounds the alarm but does not shut off the atmosphere gas. The FM cock provides a bleed-off for gas trapped between it and the nitrogen solenoid valve (NSV) and so prevents accidental closure of the cock from producing a false pressure reading.

To extend these concepts to nitrogen-based atmospheres, which are not covered by NFPA, we should consider two possible cases. If, as is possible, the total volume percent of flammable constituents exceeds a level (approximately 7% by volume) which would be flammable when mixed with air, then the main nitrogen should be backed up by a reserve capable of purging the furnace. Such a reserve would be interlocked with the nitrogen flow monitoring system to be automatically activated if the main supply should fail. Its activation would close the safety shutoff valves controlling the flow of flammable gas into the furnace and sound an alarm.

In the second case, the level of flammables in the atmosphere gas is such that any mixture of this gas with air is nonflammable. But the nonflammable nature of this gas connot be assumed, it must be proven by analysis and continuously monitored. The continuous monitoring could be done by flowmeters with high and low limit alarms. The flow rate of the individual components would be limited so that a minimum flow of nitrogen with a maximum flow of flammables would combine to form a nonflammable mixture. Such a system is shown in Figure 13. The overall operation of the system would have to be verified by direct analysis of the gas, sampled from a representative area in the furnace. With this system, it would not be necessary to back the nitrogen supply with a secondary supply.

When modifying furnaces for use with nitrogen-based atmospheres we include an inlet for nitrogen near the discharge door, inboard of the curtain. Provision for this is included in Figure 13. Its use adds an extra margin of safety by continuously purging the water cooled sections with nitrogen.

The question might well be raised at this point whether we feel sintering with nitrogen-based atmospheres is safer than with conventional atmospheres containing high percentages of flammables. There are two answers. First, yes; nitrogen-based atmospheres are safer. Because they contain less flammable material the potential risk must be reduced to some extent. But the second answer is no. All atmosphere gases can be handled safely. Nitrogen can be mishandled as easily as any other gas. Nitrogen, either as a back-up or atmosphere base, will not eliminate the hazards of flammable and toxic gases, but when properly used it does substantially enhance the overall safety of the operation of sintering furnaces.

References

1. U.S. Patent 4,106,931, issued to Airco, Inc. Patent applied for in other countries.

2. Nayar, H.S., "Nitrogen-based Sintering Atmospheres", Modern Developments in Powder Metallurgy, 9 (213), Proceedings of the 1976 International Powder Metallurgy Conference (Chicago), MPIF, Princeton, New Jersey.

3. Nayar, H.S., "Sintering and Infiltration of Steel P/M Parts in Nitrogen-based Atmospheres", Progress in Powder Metallurgy, 33 (197), Proceedings of the 1977 National P/M Conference (Detroit), MPIF, Princeton, New Jersey.

4. Lee, J.H., Guirao, C.M., Chiu, K.W. and Bach, G.G., Loss Prevention (CEP), Vol. 11, P. 59, September 1977.

EFFECTS OF GAS COMPOSITION ON GAS-SOLID EQUILIBRIUM DURING SINTERING OF FERROUS POWDER

R. W. Heckel
Department of Metallurgical Engineering
Michigan Technological University
Houghton, MI 49931

Abstract

The gas-metal equilibrium associated with the use of five different gas compositions in the heat treatment and sintering of steels are analyzed. Both carburization/decarburizing and oxidation/reduction equilibria are considered. The analysis used provides a straightforward, graphical approach toward the understanding of the effects of gas composition changes in the temperature range of 1800 to 2000°F.

Introduction

The interactions between steels and furnace atmospheres are critical in most heat treatment operations. Undesirable alteration of carbon content (either carburization or decarburization) and/or oxidation can occur if the gas composition is not controlled properly. Difficulties with carbon control and/or oxidation can be especially serious in the sintering of steel compacts since the porosity which is present can allow gas-metal interactions at significant depths into the metal.

Heat treatment atmospheres are typically mixtures of CO, CO_2, H_2, H_2O, CH_4 and N_2. The molecular composition is determined not only by the source of the gas, but also by the temperature of the gas (1). Changes in the molecular composition will occur (shift in the gas phase equilibrium) as the temperature is increased due to the temperature dependence of the equilibrium constants for the reactions between various gas molecules. Thus, desired ratios of molecular species during heat treatment are a function of both the ratios fed to the furnace and the temperature of the heat treatment.

The response of a ferrous alloy to a gas atmosphere (carburization-decarburization and/or oxidation-reduction) is sensitive to the kinetic aspects of the gas-gas and gas-solid reactions as well as to the equilibrium aspects. For temperatures above about 1000°F, the achievement of equilibrium among the molecular species in a gas is usually rapid (2). Thermodynamic predictions are, therefore, often valid for normal heat treatment times and temperatures if the extent of the reaction necessary for equilibrium is not extensive.

Kinetics may play an important role, though, in the reaction between the gas and the metal. Certainly, in situations involving carburization or decarburization (intentional or unintentional), the diffusion of carbon is often the major factor in the determination of the overall kinetics of the process. In addition, gas-metal reactions all involve time-dependent processes associated with the alteration of the surface of the metal by the gas. Such processes include, for example, formation or reduction of surface oxides and/or the alteration of carbon concentration at the surface of the steel. If these surface reactions are slow, they could be rate-controlling relative to other reactions. Furthermore, if any of the reactions involved in the gas-metal system are slow relative to the time allowed for the process, the thermodynamic predictions of the (equilibrium) state of a system must be evaluated experimentally.

For many types of heat treatment, it is usually found that thermodynamic (equilibrium) predictions of the molecular composition of a gas and the surface state of the steel (carbon concentration and presence/absence of an oxide) are reasonable approximations of actual operating conditions. Kinetic effects*, which are far more complex and difficult to predict, are often evaluated as the "departure from equilibrium." In spite of the accuracy of a prediction based upon the assumption of equilibrium, thermodynamic analysis of the gas-metal equilibrium will be of value for indicating the change necessary in a gas composition necessary to achieve a desired change in the alloy system (e.g., higher carbon potential, greater protection from oxidation, etc.). Often the magnitude of the change in the alloy system is predictable from the change in the gas composition even if the absolute state in the system is not predictable due to kinetic factors.

Gurry has presented a most useful analysis** of the equilibrium between gas atmospheres and steels in the temperature range of 1000 to 1800°F (2). Several interesting features of the Gurry treatment serve to circumvent the inherent awkwardness which usually attends analyses of gas-metal equilibria:

i. The results are presented graphically, facilitating input of conditions and output of useful information. The gas compositions are presented in terms of atomic composition, rather than molecular composition, providing a composition basis which is independent of temperature.*** The active atoms in a typical gas are C, O, and H; the inactive atoms are N. The ternary gas composition can be expressed in terms of the independent variables C/S and O/S (along with the dependent variable H/S = 1-C/S-O/S) where:

$$C = \%CO + \%CO_2 + CH_4, \quad (1)$$
$$O = \%CO + 2 \cdot \%CO_2 + \%H_2O, \quad (2)$$
$$H = 2 \cdot \%H_2 + 4 \cdot \%CH_4 + 2 \cdot \%H_2O, \text{ and} \quad (3)$$
$$S = C + O + H. \quad (4)$$

*Aside from solid state diffusion of carbon into the metal which can usually be predicted rigorously from the surface and interior carbon concentrations and the diffusion coefficient.

** Assuming gas phase equilibrium and a total pressure of one atmosphere.

***Even though the molecular composition of a gas can vary with temperature, the atomic composition must remain constant.

The dilution effect of N_2 was accommodated by an additional dimension (nitrogen axis) defined by N/S where:

$$N = 2 \cdot \%N_2. \quad (5)$$

ii. Thermodynamic terms are not used in the Gurry analysis. The gas composition plots indicate the appropriate equilibria (carbon potential (equilibrium carbon concentration) and oxidation/reduction) over the temperature range from 1000 to 1800°F (in 200°F increments). At each temperature thermodynamic calculations have been summarized to indicate:

a. regions of gas composition which will be reducing or oxidizing to iron.*

b. the equilibrium carbon concentration (expressed as weight percent carbon) as a function of gas composition. Lines of specific carbon concentration in the austenite ranges were plotted as a function of gas composition for 1400, 1600, and 1800°F.

iii. The analysis considers all of the species present in a typical gas and their interactions with each other and with the steel.

The Gurry analysis, therefore, permits direct "reading" of the equilibrium carbon concentration of a steel heat treated in a specific atmosphere. Furthermore, the gas composition regions for oxidation of iron or reduction of Fe_2O_3 or FeO to iron are defined. Thus the diagrams provided (several at each temperature for various N/S values from zero to infinity) are fundamentally gas-metal phase diagrams and can be used as readily as the more conventional alloy phase diagrams. (An extension of the Gurry analysis has been presented recently by Harvey (3).)

The hydrogen rich corner of a Gurry gas composition diagram is shown in Fig. 1. This region of the overall C/S, O/S, H/S ternary diagram is of importance because it comprises the regime of compositions obtainable by mixing H_2, H_2O, CH_4, CO and CO_2. The molecular gas composition points indicated on the figure are determined by Eqs. (1-4). The circle in the center of the plot represents an atomic gas composition of C/S = 0.2 and O/S = 0.3. It is significant to reiterate that this composition point is temperature independent (since it is based upon atomic gas composition), even though the ratios of molecular species for the point do vary with temperature. Furthermore, a particular gas composition point does not represent a unique molecular gas composition. The arrows extending from the gas composition point in Fig. 1 indicate the directions of changes in atomic gas composition which would result from additions of specific types of gas molecules; the gas composition shifts in the direction of the composition of the molecule which is added.

The purpose of the present study was to apply the Gurry analysis to specific gas compositions of interest in the heat treatment and sintering of ferrous alloys at 1800 and 2000°F. Furthermore, the study focused upon the sensitivity of the gas-metal equilibria to variations in gas composition, including dew point.

Analysis

The specific gas compositions which were studied via the Gurry analysis were obtained from the Metals Handbook (4) and are presented in Table I. Dew

*Lowest oxides are considered (Fe_2O_3 at 1000°F and FeO at 1200°F and above).

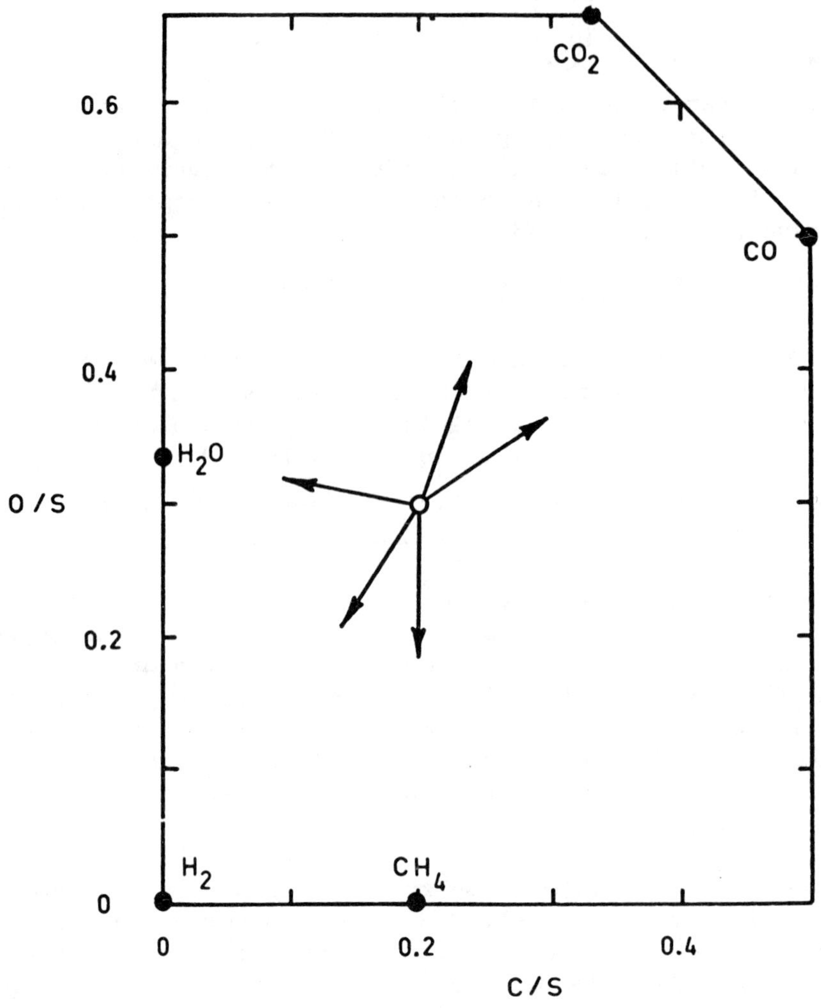

Figure 1. Ternary atomic gas composition plot indicating a specific gas composition and directions of composition shift due to additions of various types of molecules.

points were converted to water vapor content (4) and the contents of the other molecular species were adjusted accordingly prior to calculating the gas composition on an atomic basis via Eqs. (1-5). The sensitivity of the gas-metal equilibria to gas composition was obtained by consideration of the effects of increasing the amount of a molecular species by one percent (e.g., increasing the CO content of gas 102 (Table I) from 10.5 to 11.5%) with proportionate decreases in the other species. Dew point variations which were studied are listed in Table I.

Carbon potential curves at 2000°F were obtained by extrapolation of the lower temperature data given by Gurry and by thermodynamic calculations of the equilibrium:

$$C \text{ (in solution)} + CO_2 = 2CO$$

using the 2000°F equilibrium constant given by Gurry (2) for this reaction. More specifically, the 2000°F carbon potential data were extrapolated from the 1400, 1600 and 1800°F data and verified for CO/CO_2 gas mixtures (i.e., gases in the vicinity of C/S ≅ 0.4 and O/S ≅ 0.6 (see Fig. 1)). Interpolations of N/S values between those given by Gurry (N/S = 0, 1, 4, and ∞) was also necessary. The errors resulting from the extrapolation and interpolation techniques employed in the present work are judged to be insignificant relative to the overall errors associated with the graphical presentation of the results in the present paper.

TABLE I. Gas Compositions Investigated in the Present Study (4)

Class*	Description	Composition; Percent by Volume N_2	CO	CO_2	H_2	CH_4	Dew Point(s) (°F)
102	Rich exothermic	71.5	10.5	5.0	12.5	0.5	-50,40, and 80
201	Lean prepared nitrogen	97.1	1.7	-	1.2	-	-40
202	Rich prepared nitrogen	75.3	11.0	-	13.2	0.5	-40
301	Lean endothermic	45.1	19.6	0.4	34.6	0.3	20,35, and 50
302	Rich endothermic	39.8	20.7	-	38.7	0.8	-15,10, and 40

*American Gas Association classification

Results

The gas-metal equilibria for Fe-C alloys and the gas compositions given in Table I are presented in Figs. 2-6. Finer composition scales than those of the original Gurry treatment or that of Fig. 1 are used in order to provide more detail concerning the behavior of the specific gases. The figures show the Fe/FeO (OX/RED) equilibrium (at only 1800°F) for Figs. 3 and 4. Compositions below and to the right of these lines (richer in C, poorer in O) will reduce FeO to Fe. The Fe/FeO equilibrium lines are displaced to the upper left of the diagrams with increasing temperature; oxidation becomes a problem for the gases considered only at temperatures significantly less than 1800°F (2). The Fe/FeO equilibrium lines for the gases considered in Figs. 4, 5, and 6 are beyond the upper left corners of these graphs; the ranges of gas compositions considered in these figures are reducing at 1800 and 2000°F. The equilibrium carbon potentials at 1800 and 2000°F* are also shown in Figs. 2-6.

It is significant to point out that the Fe/FeO equilibrium lines and the lines of equal carbon potential all have slopes of about 45° to the C/S and O/S axes in the 1800-2000°F range for the five gases considered. Thus, increasing the reducing "power" of a gas or increasing its carbon potential depends almost equally on either decreasing O/S or increasing C/S.

The effects of variation in gas composition and, for some gases, dew point are also shown in Figs. 2-6. The open data points from which "radiate" the lines to the closed points represent the atomic compositions of the gases considered. The closed points represent the effect of a 1% increase in the molecular species noted. Additional open points on Figs. 2, 5 and 6 show the ef-

*Except for Fig. 3 (1800°F only) to eliminate crowding of this graph.

fects of dew point variations. Comparison of Figs. 2-6 to Fig. 1 (remembering that the 0.8% carbon line lies approximately between H_2 and CO points of Fig. 1) defines the strong effect of dew point (H_2O content) on the gas-metal equilibrium. Increasing the dew point (increasing H_2O content) moves the atomic gas composition in a direction almost perpendicular to the equilibrium lines for all of the gases considered. On the other hand, variations in H_2 and CO content can be seen (Figs. 2-6) to have little effect of the gas-metal equilibrium since changes in the compositions of these species move the atomic composition of the gas in a direction almost parallel to the equilibrium lines. The effects of variations in the CO_2 and CH_4 concentrations are intermediate between those of CO and H_2 and that of H_2O; the locations of the five gas compositions relative to the CO_2 and CH_4 points (Fig. 1) will determine the sensitivity of a specific gas to variations in CO_2 and CH_4 content.

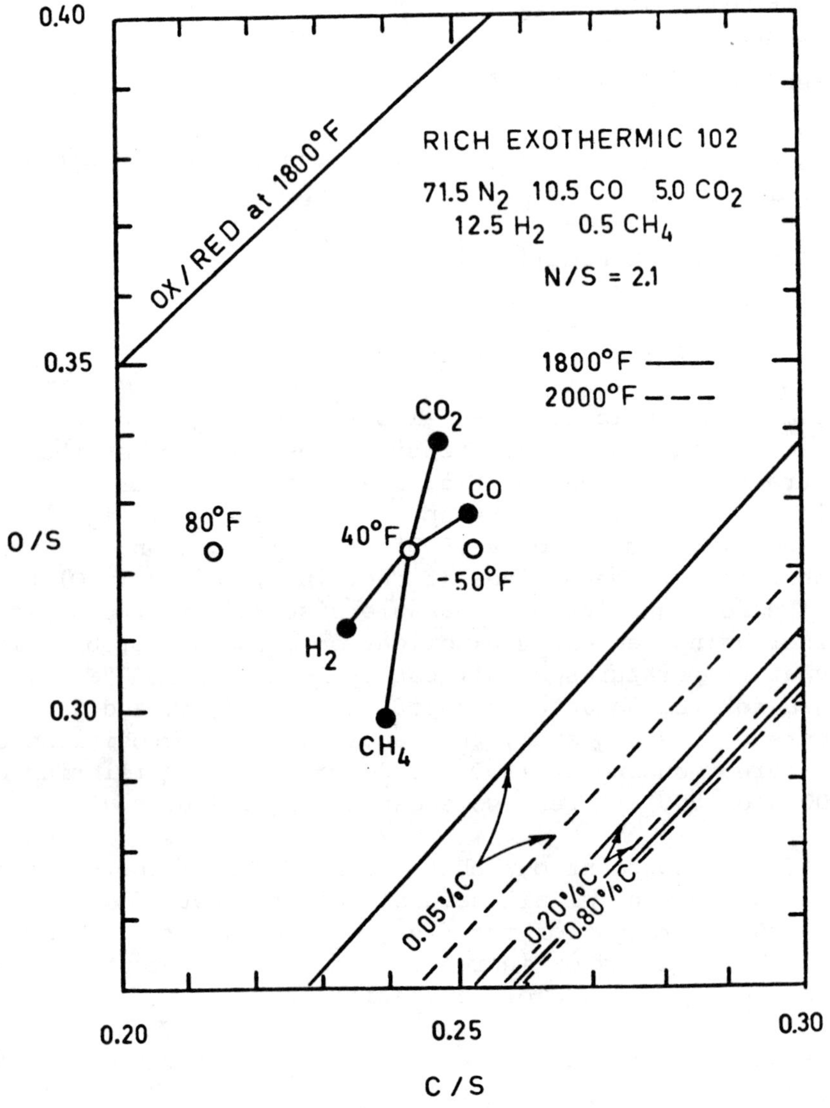

Figure 2. Gas-metal equilibria for a rich exothermic gas.

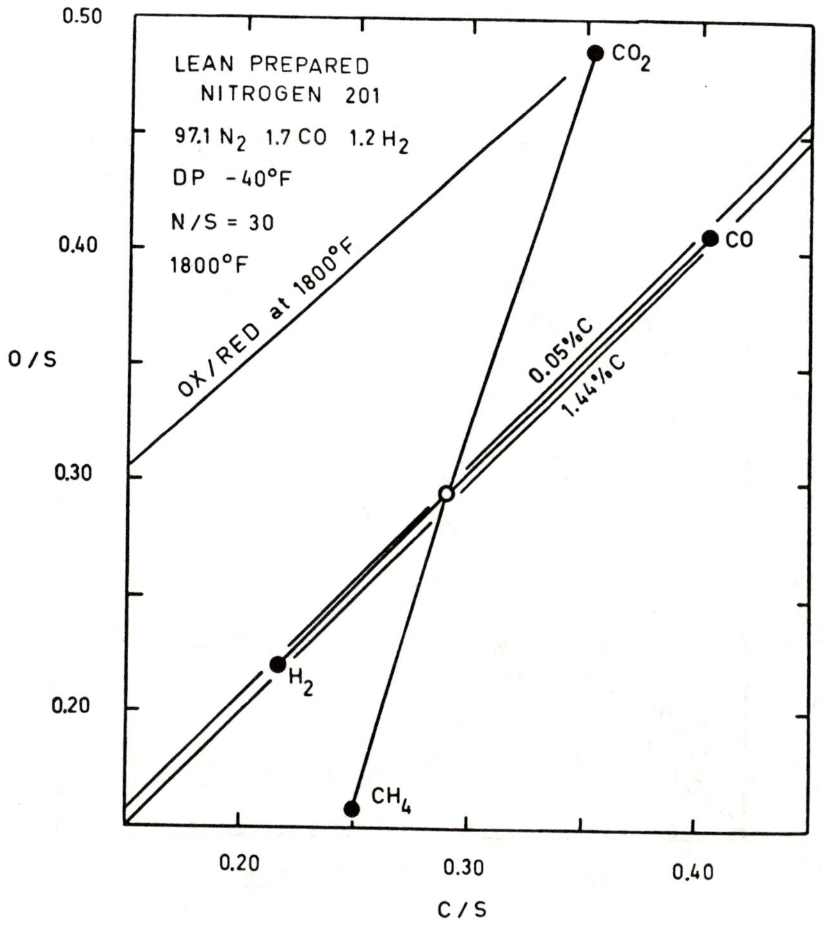

Figure 3. Gas-metal equilibria for a lean prepared nitrogen gas.

Figures 2-6 clearly show the effects of type of gas and heat treatment temperature on the equilibrium carbon concentration (Table II). The gases selected for study range from strongly decarburizing (rich endothermic (102); Fig. 2) to carbon potentials great enough to exceed the solubility of carbon in austenite and, therefore, possibly cause sooting in the furnace (rich prepared nitrogen (202); Fig. 4). The effect of temperature is indicated most clearly by the behavior of the lean endothermic gas (301) (Fig. 5 and Table II). Significantly less carbon potential is observed at 2000°F than at 1800°F. (This same temperature effect would be found with the other gases as well.)

The effects of a diluent gas like N_2 is also shown in Figs. 2-6. As the N_2 content increases (increasing N/S), the spacing between the equilibrium carbon concentration lines decreases. Figure 6 (N/S = 0.65) exhibits a wide spacing, whereas Fig. 3 (N/S = 30) exhibits a very narrow range of austenite carbon concentrations. The dilution of furnace gases with inert species such as N_2 necessitates more careful control of the active species in the gas in order to maintain control of the carbon concentration in the steel.

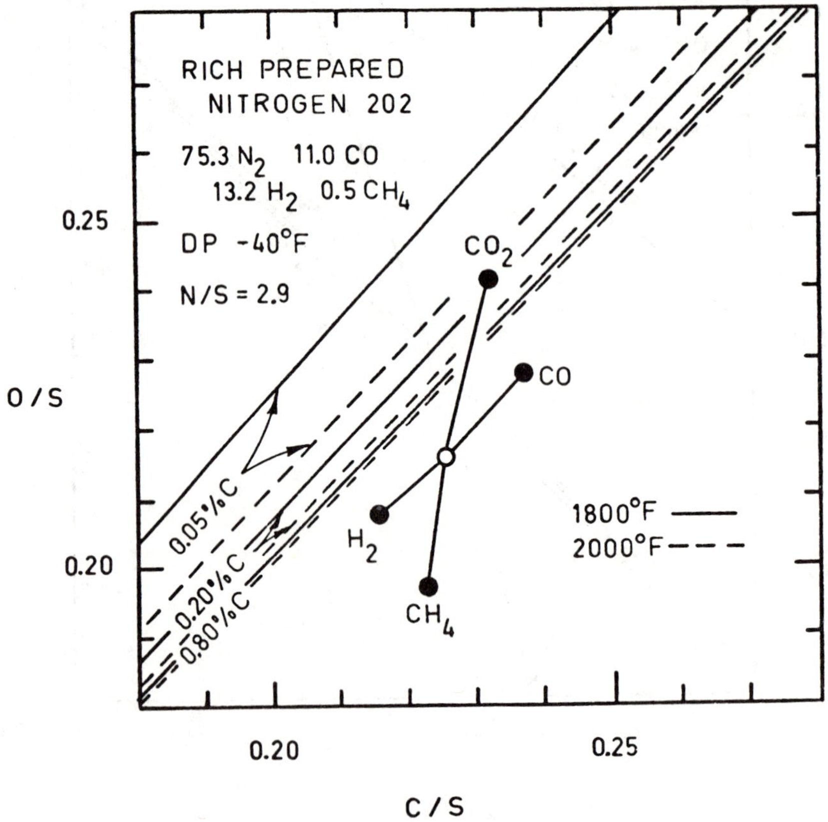

Figure 4. Gas-metal equilibria for a rich prepared nitrogen gas.

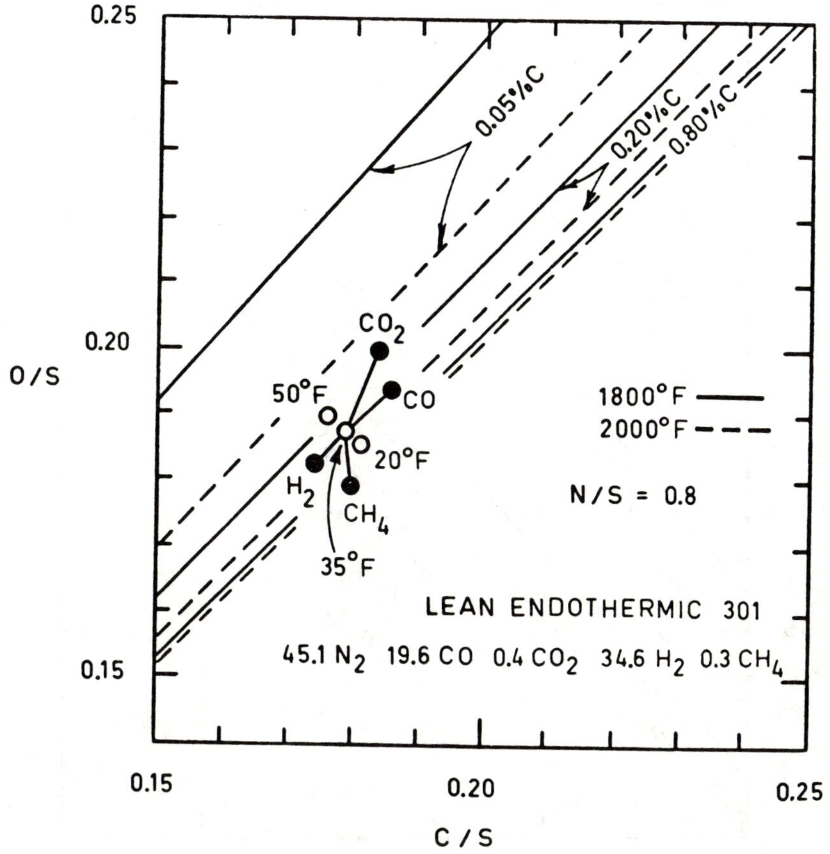

Figure 5. Gas-metal equilibria for a lean endothermic gas.

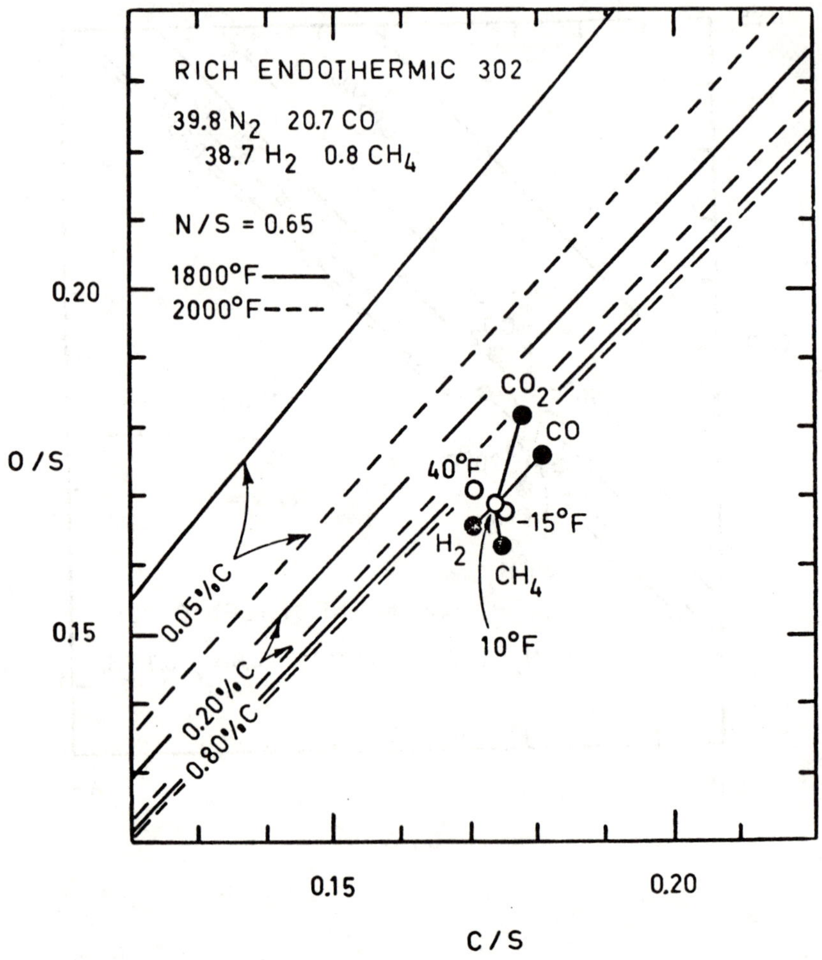

Figure 6. Gas-metal equilibria for a rich endothermic gas.

TABLE II. Equilibrium Carbon Concentrations for the Gases Defined in Table I.

Class	Description	Carbon Concentration (%) 1800°F	2000°F
102	Rich endothermic (40°F dew point)	< 0.05	< 0.05
201	Lean prepared nitrogen	(a)	
202	Rich prepared nitrogen	> 0.80(b)	> 0.80(b)
301	Lean endothermic (35°F dew point)	0.20-0.80	0.05-0.20
302	Rich endothermic (10°F dew point)	> 0.80(c)	> 0.80(c)

a. in the austenite range (0.05 to 1.44), but very sensitive to composition of the active species in the gas.

b. probably in excess of the maximum solubility of carbon in austenite (1.44 at 1800°F); furnace sooting may occur.

c. probably close to maximum solubility of carbon in austenite (1.44 at 1800°F).

Discussion

The Gurry analysis greatly facilitates the understanding of the effects of gas composition and temperature on the gas-metal equilibria involved in the heat treatment of steels. Under normal processing conditions where gas-gas and gas-metal reaction kinetics are rapid and/or heat treatment times are not too short, actual conditions should approach the equilibrium (thermodynamic) predictions. A most important feature of the Gurry analysis is that complex gas compositions (not just CO/CO_2 mixtures, for example) may be treated readily.

It is important to consider that the Gurry analysis is valid, strictly speaking, for binary Fe-C steel compositions and not for steels containing alloying elements in addition to carbon. However, alloying element effects may be treated quantitatively if the effect of the alloying element on the thermodynamic activity of carbon is known. The binary Fe-C percent carbon values on the Gurry charts can be converted to carbon activity values (from available data (2)) and then reconverted to the appropriate percent carbon values from the alloy data.

Treatment of alloy steels can also be accomplished by plotting measured carbon concentrations on the Gurry charts. The departures from the thermodynamic predictions can be noted and applied as a "correction factor." This correction factor would account for all departures (alloying element effects, kinetic effects, etc.) from the thermodynamic predictions and, therefore, should be used only after thorough comparisons between experimental and predicted information.

One of the most obvious uses of the Gurry analysis is in the alteration of gas composition to achieve a desired change in carbon potential. As mentioned previously, dew point and carbon potential are closely related because water content variations generally shift the atomic gas composition in a direction perpendicular to the equilibrium carbon concentration lines. Several possibilities should be considered if the other species of the gas are to be varied to change the carbon potential. If a minimum amount of a new gas is to be made to change the carbon potential, the gas composition should be moved perpendicular to the carbon potential lines. Thus, a small addition of a mixture* of H_2 and CO_2 will decrease the carbon potential significantly while a small addition of a mixture* of CH_4 and CO will increase the carbon potential significantly. If, on the other hand, better control of carbon potential is desired, the amount of gas added should be large enough so that metering, mixing, etc., are not critical. In such instances the alteration of the atomic gas composition should be at an acute angle to the carbon potential lines, such as by additions of pure CO_2 for decreasing carbon potential and additions of pure CH_4 (or other hydrocarbons) for increased carbon potential.

References

1. L. S. Darken and R. W. Gurry, Physical Chemistry of Metals, McGraw-Hill, 1953.

2. R. W. Gurry, Trans. AIME, 188 (1950) 671.

3. F. J. Harvey, Met. Trans., 9A (1978) 1507.

4. Metals Handbook, 8th edition, volume 2, Heat Treating, Cleaning, and Finishing, American Society for Metals, 1964.

*For most gases, approximately equal proportions of the gases indicated will be satisfactory, as can be seen from Figs. 1-6.

ENERGY ASPECTS OF IRON POWDER PRODUCTION

J.M. Capus

Quebec Metal Powders Limited, Sorel, Quebec, Canada

INTRODUCTION

In a recent paper, S.M. Kaufman (1) discussed the energy consumption in the manufacture of P/M parts from iron powder. In a detailed presentation he analysed all steps in the process, including the manufacture of the powder itself. His summation resulted in a total of 39.8 million Btu's for the various items of energy consumption required to make one ton of finished P/M parts. Of this total, 19.0 MBtu or 47.7% (see Table I) represented the energy used in making the iron powder. In his analysis, Kaufman considered the energy consumption for two processes of iron powder production that are commercially significant today in the United States.

TABLE I

Energy Consumption for One Ton of Finished Iron P/M Parts (S.M. Kaufman, Reference 1)

Item	MBtu	Percent
Powder manufacture	19.0	47.7
Blending & compacting	0.4	1.0
Sintering	17.3	43.5
Finishing	0.5	1.3
Miscellaneous	2.6	6.5
	39.8	100.0

These are: direct reduction of iron oxide and water atomization of remelted steel scrap. Neither of these processes is presently employed commercially in Canada, currently the source of a significant portion of the iron powder consumed in North America.

Iron powders manufactured in Canada are mostly produced by the process adopted by Quebec Metal Powders Limited for the conversion of liquid pig iron into high-purity iron powders.

The present paper is concerned chiefly with an analysis of the energy consumption in the manufacture of iron powder by the QMP process; some comparisons are made with Kaufman's data for alternative processes. Finally, some comments will be made on energy conservation projects in the manufacture of iron powder.

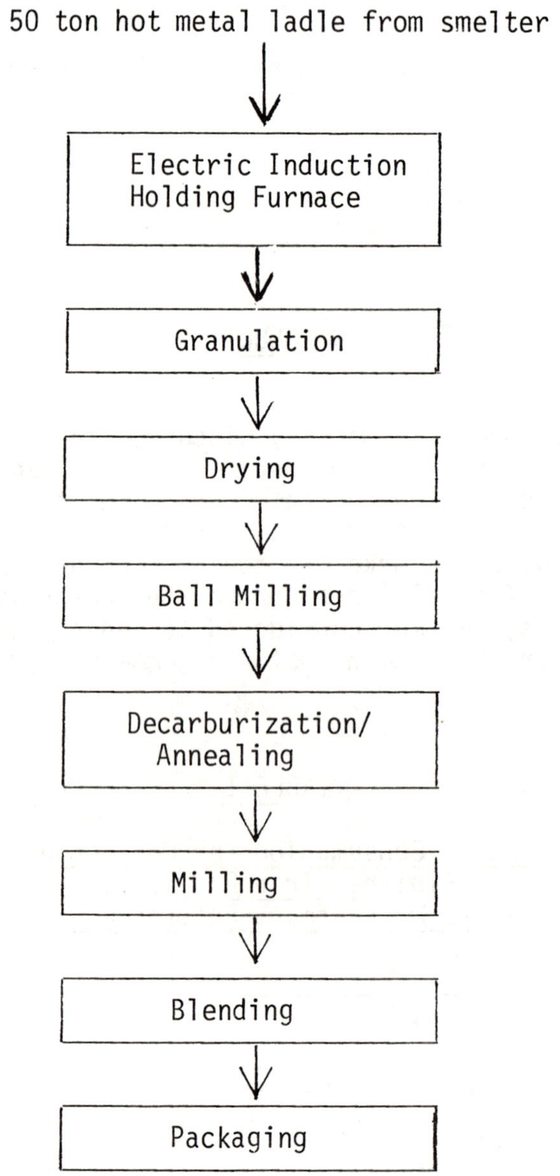

FIGURE 1 QMP Process Flow Chart (Schematic)

ENERGY CONSUMPTION IN THE QMP PROCESS

Although Quebec Metal Powders produces a variety of iron powder grades for the P/M industry, for welding electrodes, for cutting and scarfing, etc., all are made by the same basic process. These powders are produced from a unique starting material which provides a distinct advantage from the viewpoint of energy consumption. This raw material is the freshly-refined, high-purity pig iron obtained directly from the adjacent electric smelting plant of QMP's parent company. The smelter of QIT-Fer et Titane Inc. is connected by a rail spur along which are brought ladle cars of hot metal for the iron powder plant. As far as is known this is still the only iron powder plant using 100% virgin hot metal as raw material.

In 1978, the total energy consumed at QMP to convert molten pig iron into one ton of finished iron powder amounted to 13.9 MBtu. This energy figure was made up of 582 Kwh of electric power and 8030 cu. ft. of natural gas. The best way to explain where all this energy went is to follow the process through the plant.

As indicated in Figure 1, molten pig iron containing about 4% carbon is transferred from the smelter to the iron powder plant in 50-ton ladles. A certain amount of heat is lost in this short trip of about 2000 feet, partly due to the delay required to desulphurize the iron at an intermediate injection station. As a result it is usually necessary to raise the temperature of the iron in readiness for granulation, the first step in the QMP process. The molten iron is therefore first poured into one of the two 90-ton induction-heated holding furnaces shown in Figure 2.

FIGURE 2 90-Ton Capacity Electric Induction Furnaces for Holding Liquid Pig-Iron.

Hot metal is poured from the induction furnace into a large tundish fitted with four nozzles. The streams of iron falling through these nozzles are disintegrated with high pressure horizontal water jets (Figure 3). The resulting granulated iron is mostly in the form of coarse partially oxidised granules below 1/8" in dia. The wet slurry from the granulation chamber is pumped to a filter, then dried in a rotary dryer which is heated by natural gas.

FIGURE 3 View of Tundish from alongside water jets.

The energy consumed in these first two steps (granulation and drying) is chiefly that needed to energize the electric induction furnace, heat the tundish, pump water, and fire the dryer. These add up to 4.9 MBtu/ton: 4.1 MBtu for granulation and 0.8 MBtu for drying. These energy consumptions, like the others listed in the rest of this paper, are expressed in terms of finished and packaged powder with allowances made for chemical and physical losses.

Next the granulated iron is ball milled to powder. This is the step that determines the major differences between the various QMP iron powder grades; the ball mill circuit (Figure 4) may be set to grind down to specific particle sizes. This circuit consumes an average of 1.5 MBtu/ton.

The ball-milled pig iron is then passed to the reduction furnace section where carbon and oxygen are removed by heat-treatment. The powder is fed simultaneously to three decarburization or reduction furnaces. Powder is carried through the five heating zones of these furnaces on stainless steel belts. The furnaces are heated by natural gas radiant-burner tubes as seen in Figure 5. Inside the furnaces, a protective atmosphere of dissociated ammonia is provided to prevent the iron powder from re-oxidixing. This decarburization step is the most energy-consuming part of the QMP process. A total of 5.3 MBtu/ton is used in heating the furnaces, dissociating the ammonia, and transporting the powder from the ball mill circuit to the exit of the furnaces.

FIGURE 4 Partial view of ball-milling circuit showing discharge end of tri-cone ball mill.

After being heated to over 1800F and cooled again, the product emerges from the furnaces as a 'sintered' spongy cake of pure iron. The iron powder cake, in the form of one inch thick slabs, falls into crushers and is then passed to a series of mills for grinding back to powder. The dust from the freshly-reduced iron is highly combustible. As a result it is necessary to do the grinding in an inert atmosphere. This inert atmosphere is provided by the combustion of natural gas. Another 1.4 MBtu/ton has to be accounted for in the reduction of the cake back to powder, chiefly in the combustion of the natural gas.

Finally, after the powder has been passed through an 80,000 lbs. product blender to homogenize individual batches, a further 0.2 MBtu/ton is consumed in the packaging operations.

FIGURE 5 Heating zones of decarburizing furnace.

COMPARISON OF ENERGY CONSUMPTION

Energy consumption figures for the various steps in the QMP process are summarized in Table II.

TABLE II

Iron Powder Production:

Comparative Energy Requirements,

MBtu/ton of Product

Processing Steps	QMP Process	Direct Atomization*	Ore Reduction*
1. Melting and Refining	2.3	7.4	N/A
2. Granulating/Atomizing	1.8	5.0	N/A
3. Drying and Cleaning	0.8	3.4	N/A
4. Ball Milling	1.5	N/A	19.7 (steps 4–8)
5. Decarburizing/Annealing/Reduction	5.3	3.1 (steps 5–8)	
6. Finishing (Milling, Blending)	1.4		
7. Packaging	0.2		
8. Plant and Office Heating	0.6		
Total	13.9	18.9	19.7

N/A = Not applicable
* Data from S.M. Kaufman (1).

Alongside, in the same table, are data taken from Kaufman's paper for two alternative processes: remelting and atomizing of steel scrap, and production of iron powder by direct reduction of iron oxide.

Looking at the 'bottom line' of energy consumption in Table II it is evident that the QMP process consumes substantially less energy than the other two processes. The figure of 13.9 MBtu/ton for QMP powder is 26% less than the 18.9 MBtu/ton estimated for direct-atomized powder. The difference is mostly due to the energy of melting down the scrap for the direct-atomization process, estimated by Kaufman at 7.4 MBtu/ton. By comparison the electric induction holding furnace operation in the QMP process requires only 2.3 MBtu giving an immediate saving of about 5 MBtu. On the other hand, the production of powder by direct reduction of iron oxide, which also does not involve a melting step, appears to be a more energy-intensive process, requiring an estimated 19.7 MBtu/ton of powder.

For the remaining process steps there are some differences in Table II between Kaufman's figures and those for the QMP process that are more difficult to understand. In short, it would seem that he has over-estimated the energy requirements for atomizing, drying and cleaning, but under-estimated those for annealing and subsequent operations. This is not meant as a criticism of Kaufman's careful analysis since he made it clear that he did not have complete access to industry data, and so had to fall back on theoretical calculations. An attempt can be made to clarify the difference between the energy requirements for the QMP process and that for direct atomization. If Kaufman's figures for the individual steps in the process after melting and refining are adjusted in the light of QMP actual energy consumptions, his overall total would be reduced by 1.5 MBtu to 17.4 MBtu/ton (Table III).

TABLE III

Revised Estimate of Energy Consumption in the Production of Powder from Melting and Atomizing Steel Scrap

	MBtu/ton of Product
Melting and Refining	7.4
Atomizing	2.0
Drying and Cleaning	1.0
Annealing and Finishing; etc.	7.0
	17.4

Even after this adjustment the QMP process still uses 20% less energy than melting and atomizing of steel scrap.

It may be argued that the above analysis makes an unfair comparison by taking liquid pig iron as the starting material for the QMP process. While it is true that if the QMP plant were sited remotely from the smelter, then the energy for re-melting of pig iron would have to be added. However, the actual situation is that if QMP did not take its portion of hot metal from the smelter, then that liquid iron would have to be cast into pigs for sale and eventual remelting. The heat content of the hot metal would be lost entirely. Furthermore, scrap steel can also be considered to have an 'energy content' from its original smelting and processing. If that aspect was taken into account then the energy comparison would move even more in favor of QMP since its raw material is virgin iron on which only the energy associated with mining and smelting has been expended.

PROSPECTS FOR FURTHER ENERGY CONSERVATION

It is evident from the data presented in the previous section that the decarburization/annealing step is the most energy-consuming operation in the QMP process. Studies have been made from time to time on the thermal efficiency of the decarburization furnaces. A thermal balance analysis for one of these furnaces is shown in Table IV. As indicated in the upper part of this table, only 19% of the heat input is used in the process - for heating the powder, or for making the reaction. Most of the rest is lost up the stack in the flue gases from the radiant tube burners, or through the furnace walls.

TABLE IV

Thermal Balance for a Decarburization Furnace

	MBtu/Hr.	%
Heat input	12.71	100.0
Process Heat		
Iron Powder Heating	1.32	10.4
Reaction	1.10	8.6
Heat Losses		
Flue gas	5.74	45.2
Walls (by difference)	3.87	30.5
Atmosphere	0.25	1.9
Roll bearings	0.23	1.8
Belt	0.20	1.6

The high flue gas losses are, of course, due to the high operating temperatures of the burners. Heat-recuperative systems for these burners have been looked into to try to reduce the losses, but so far without finding an economic solution. Conversion to electric heating would do the trick, but would be extremely expensive for existing furnaces, as well as being rather impracticable from the viewpoint of the extended interruption of operations required. The construction of the furnaces allows little room for improving the furnace wall insulation either, although this item is still under investigation. Altogether, the best contribution made so far to energy saving in furnace heating has been in the scheduling of production to maximize on-stream time, and to concentrate down-time for maintenance into systematic and regular shutdown periods.

Now the thermal balance of Table IV does not include the energy for the generation of the dissociated ammonia atmosphere. This consists of the energy content of the liquid ammonia plus the energy spent in dissociation. Extensive tests have been made showing that energy can be saved by diluting the dissociated ammonia with high-purity nitrogen. So far, it was found that the atmosphere can be diluted until its hydrogen content drops from 75% to 45%, without adverse effects on the product. With further development a modest saving of energy could be achieved by this means.

A more interesting energy-saving project concerns the natural gas used for heating the runners and tundishes in the induction furnace section. At present traditional refractory linings are used for both runners and tundishes. As may be seen from Figure 3, these tundishes are rather large, and 1.1 MBtu/ton is currently consumed to heat them with natural gas. All this energy could be saved by changing to replaceable liners for both tundishes and runners that would not require heating. This project is still in the development stage, although the results of the tests made so far are very encouraging. At the present time, the additional cost of the replaceable liners would be just about balanced by the savings in natural gas. However, in future years the liner system could become more economical.

CONCLUSIONS

An analysis of energy consumption for iron powder production by the QMP process indicates that it is in a very favourable position compared with alternative processes. The comparison was based on actual 1978 energy consumption figures for QMP and recently-published estimates for the alternative processes of direct atomization of remelted scrap, and reduction of iron oxide. The advantage of QMP is chiefly due to the availability of high-purity molten iron on site.

Current energy-saving projects that are under development will, if succesful, permit reduction in unit energy consumption by up to 12%. Such savings would help to lessen the impact of future rises in energy costs.

ACKNOWLEDGEMENTS

The author wishes to acknowledge and thank Mr. T. Krantz and Mr. G. D'Amours for their assistance in compiling some of the data on which this paper is based. The author is also indebted to Dr. S.M. Kaufman for his valuable discussions and comments on the original draft.

REFERENCE

1. S.M. Kaufman, "Energy Consumption in the Manufacture of Precision Metal Parts from Iron Powder", International Journal of Powder Metallurgy and Powder Technology, 1979, 15, p.9.

— Notes —

— **Notes** —

- Notes —

— Notes —

— Notes —

— Notes —

- **Notes** —

- Notes —

— Notes —